SYSTEMS TROUBLESHOOTING HANDBOOK

INTEGRATED CIRCUITS APPLICATIONS HANDBOOK
Edited by A. Seidman
1984

R.C. ACTIVE FILTER DESIGN HANDBOOK
Edited by F.W. Stephenson
1985

MICROPROCESSOR HANDBOOK
Edited by J. Greenfield
1985

SYSTEMS TROUBLESHOOTING HANDBOOK
Edited by L. Faulkenberry
1985

FORTHCOMING VOLUMES:

ELECTRONIC COMMUNICATIONS APPLICATIONS HANDBOOK
Edited by S. Cheshier and R. Carter

OPERATIONAL AMPLIFIERS APPLICATIONS HANDBOOK
Edited by R. Thielking

ELECTRIC MACHINERY HANDBOOK
Edited by J. DeGuilmo

FIBEROPTICS SYSTEM DESIGN HANDBOOK
Edited by R. Hoss

HANDBOOK OF APPLICATIONS OF MEASURING INSTRUMENTS
Edited by R. Moore

MICROWAVE MEASUREMENTS HANDBOOK
Edited by J. Green

JOHN WILEY & SONS, Inc.
605 Third Avenue, New York, N.Y. 10158
New York • Chichester • Brisbane • Toronto • Singapore

SYSTEMS TROUBLESHOOTING HANDBOOK

Luces M. Faulkenberry, Editor
University of Houston

Contributors:

Theodore F. Bogart, Jr.
Lowell L. Winans
Leonard F. Garrett
Bernard McIntyre
Robert B. Thorne
Harold B. Killen
John Dynes
Dean Lance Smith
Farrokh Attarzadeh

JOHN WILEY & SONS
New York • Chichester • Brisbane • Toronto • Singapore

Copyright © 1986, by John Wiley & Sons, Inc.

All rights reserved. Published simultaneously in Canada.

Reproduction or translation of any part of
this work beyond that permitted by Sections
107 and 108 of the 1976 United States Copyright
Act without the permission of the copyright
owner is unlawful. Requests for permission
or further information should be addressed to
the Permissions Department, John Wiley & Sons.

Library of Congress Cataloging-in-Publication Data:

Main entry under title:
Systems troubleshooting handbook.

  Includes bibliographies and index.
  1. Electronic systems—Maintenance and repair.
2. Electronic systems—Testing.   I. Faulkenberry,
Luces M.

TK7870.2.S96 1986   621.381′1   85-22527
ISBN 0-471-86677-6

Printed in the United States of America

10 9 8 7 6 5 4 3 2 1

PREFACE

This book is the result of many requests to John Wiley & Sons for the inclusion of a chapter on systems troubleshooting in several books on varying topics in electronics. The topic seemed too large to be covered in a single chapter, so this book was planned. It has become clear that systems troubleshooting is a topic that is too large to be covered in a single volume of reasonable size. Thus, the purpose of this volume is to introduce the novice troubleshooter to the tools and techniques used to identify malfunctions in electronic systems. Additionally, the book will aid more experienced troubleshooters when they are forced to solve a problem in electronic equipment outside of their area of specialization.

The book is organized into three sections: test equipment and troubleshooting basics, analog system troubleshooting, and digital system troubleshooting. There is some inevitable overlap between the sections. The authors assumed the readers would be familiar with basic electronic theory and active devices. Two- or four-year electronic technology students or engineering students nearing the end of their curriculum should have sufficient background. Appendices on microprocessor and robotics basics are included for readers who need them.

Each chapter in this book is written by an author who has experience with the type of equipment or circuitry discussed in the chapter. The reader will note that many troubleshooting techniques are mentioned by several authors, even though the authors' backgrounds vary widely.

I would like to thank and acknowledge the people who helped make this volume possible: the authors who provided the material; Hank Stewart and the many people at John Wiley & Sons who provided advice, help, and support; my wife, Dixie, who typed two chapters and provided many secretarial services; and Art Seidman, who convinced me to do the project.

CONTENTS

Theodore F. Bogart, Jr.
University of Southern Mississippi

Lowell L. Winans
Heath Company

CHAPTER 3   The Spectrum Analyzers   63

Leonard F. Garrett
Tektronix, Inc.

CHAPTER 8   Stereo System Troubleshooting   202

Robert B. Thorne
Texscan Corporation

CHAPTER 9   Analog Communications Systems   230

Harold B. Killen
University of Houston

CHAPTER 12   **Robot Equipment**   **291**

Robert B. Thorne
Texscan Corporation

CHAPTER 13   **Troubleshooting a Robotic Control System**   **304**

John Dynes
Heathkit/Zenith Educational Systems

CHAPTER 14   **Personal Computers**   **318**

Dean Lance Smith
Engineering Consultant

APPENDIX A   Basic Microprocessor Concepts   379

Bernard McIntyre
University of Houston

APPENDIX B   Basic Robotic Concepts   387

Farrokh Attarzadeh
University of Houston

SYSTEMS TROUBLESHOOTING HANDBOOK

PART ONE

TEST EQUIPMENT AND TROUBLESHOOTING BASICS

CHAPTER 1

Basic Test Equipment

Theodore F. Bogart, Jr.
University of Southern Mississippi

1.1 MULTIMETERS

Multimeters are instruments that can be used to measure several different electrical quantities, usually including dc voltage and current, resistance, and ac voltage. Some are also designed to measure ac current, while others require a special adapter such as a clamp-on probe for ac current measurements.

Multimeters may be classified as being either of the *analog* or *digital* type, in reference to the method used for displaying the values of the quantity being measured; and as being either *active* or *passive,* depending on whether or not they contain amplifying circuitry. Those containing active circuitry are often referred to as *electronic* multimeters. Analog meters, in which a pointer moves in an arc across a continuous scale, are available in both the passive and active design, while digital meters always contain active circuitry.

A multimeter containing only passive circuitry (except for a small battery used for resistance measurements) is often referred to as a VOM (volt–ohm–milliammeter) and has the advantage of being more portable and rugged than its counterpart containing active circuitry. It is simpler in design and usually of more modest cost. On the other hand, a multimeter is usually less accurate and more limited in the ranges of values it can measure. In troubleshooting, VOMs are quite useful for "quick and dirty" field checks to detect the presence or absence of voltage, and opens and shorts, and for order-of-magnitude comparisons.

When selecting a multimeter for a given application, it is important to distinguish between the *accuracy* and the *precision* required in the measurements to be made. Accuracy refers to the closeness of the measured value to the actual value, while precision refers to the number of significant digits in which the measured value is expressed. Clearly, great precision is of little value in an instrument that has poor accuracy. Digital multimeters (DMMs) generally have greater precision than their analog counterparts but, depending on how they are used, they may or may not have greater accuracy. However, unlike digital instruments, the analog types require the user to interpret the location of a pointer on a scale, and therefore they are more susceptible to human, or operator, error. This kind of error arises from the operator's inability to interpolate accurately between scale divisions, to misinterpret or misapply multiplying factors, and to misread a pointer location because of *parallax*. The latter arises from improper alignment of the eye with the pointer when judging the position of the pointer against a background scale. Some analog instruments have mirrored scales that can be used to eliminate parallax error: the eye is positioned so that no pointer reflection is visible.

Accuracy specifications for voltage and current readings are usually quoted as being a certain percent of full-scale, meaning that any reading is accurate to plus or minus that percent of the maximum voltage or current in the measurement range selected. For example, a voltage reading made in the 0- to 10-V range with an instrument having 3% accuracy is accurate to ± 0.3 V. Thus, a measured value of 1 V may actually represent a voltage in the range from 0.7 to 1.3 V (*not* 1 V $\pm$ 3% = 0.97 to 1.03 V).

1.1.1 Portable, Passive-Type Multimeters (VOMs)

Most modern passive-type VOMs are constructed around a *d'Arsonval meter movement*, in which current flowing through a coil in the presence of a magnetic field (established by a permanent magnet) creates a torque that rotates the pointer in direct proportion to the magnitude of the current. Thus, the instrument is fundamentally a current-sensitive device; but through the proper switching of series and shunt resistors, it can be calibrated to indicate voltage and resistance as well.

The most important characteristic of a meter movement is its rated *full-scale current I_{fs}*, the current required to cause full-scale deflection of the pointer. The *sensitivity S* of the movement is the reciprocal of the full-scale current, and has the units of ohms/volt:

$$S = \frac{1}{I_{fs}} \ \Omega/V \qquad (1.1)$$

In general, the greater the rated sensitivity (the smaller the full-scale current) of a meter movement, the better the specifications of the instrument that employs it. Typical I_{fs} values range from 50 μA to 1 mA, corresponding to sensitivities from 20 to 1 kΩ/V.

Figure 1-1 shows how meter movements are incorporated into circuits for measuring resistance, dc current, and dc voltage. Note in each case that the current I_m in the meter depends upon the value of the unknown quantity R_x, V_x, or I_x. In Figure 1-1a, the series resistance R_s is chosen so that full-scale deflection occurs when $R_x = 0$:

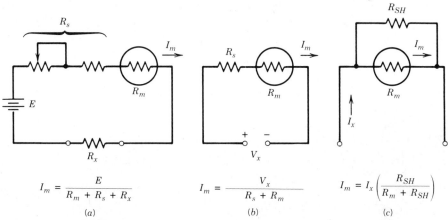

$$I_m = \frac{E}{R_m + R_s + R_x}$$

$$(a)$$

$$I_m = \frac{V_x}{R_s + R_m}$$

$$(b)$$

$$I_m = I_x \left(\frac{R_{SH}}{R_m + R_{SH}} \right)$$

$$(c)$$

Figure 1-1   VOM circuits that incorporate a meter movement for measuring resistance, dc voltage, and dc current. (*a*) Measuring an unknown resistance R_x. (*b*) Measuring an unknown voltage V_x. (*c*) Measuring an unknown current I_x. I_m = current in meter (that causes pointer deflection); R_m = resistance of meter movement (fixed); R_s = resistance connected in series with the meter movement; R_{sh} = resistance connected in shunt with the meter movement; E = fixed battery voltage, used for resistance measurements.

$$R_s = \frac{E - I_{fs}R_m}{I_{fs}} \tag{1.2}$$

As shown in the figure, a portion of R_s is made adjustable to permit the user to compensate for changes in the battery voltage E, that is, to permit "zeroing" of the instrument. In Figure 1-1*b*, R_s is chosen so that the maximum value of the unknown voltage V_x in a given range causes full-scale deflection:

$$R_s = \frac{V_{max}}{I_{fs}} - R_m \tag{1.3}$$

In Figure 1-1*c*, the shunt resistor R_{SH} is chosen so that I_{fs} flows in the movement when the maximum unknown current in a given range enters the circuit:

$$R_{SH} = \frac{I_{fs}R_m}{I_{max} - I_{fs}} \tag{1.4}$$

VOMs contain switch-selectable values of R_s and R_{SH} that permit the user to change the ranges (full-scale values) of voltages and currents measured. There is, however, only one resistance scale. For measuring large resistance values, some VOMs cause a higher voltage battery to be switched into the circuit.

   For ac voltage measurements, a half-wave or full-wave diode rectifier is switched into the basic voltage measuring circuit shown in Figure 1-1*b*, so unidirectional current flows in the movement in direct proportion to the magnitude of the ac voltage. It is an important fact that the meter movement responds to an *average* value. Since the average value of a half- or full-wave-rectified sinusoidal voltage is a fixed multiple of the peak or RMS (root mean square) value of the ac voltage, the ac scale of the meter can be calibrated in peak or RMS volts. Therefore, ac voltage readings are valid for sinusoidal voltages *only*.

1.1.2 Measurement Precautions

1. When performing resistance measurements:
   a. Always zero the instrument first, by touching the probes together and adjusting the "zero" control until the pointer indicates zero ohms. Best results are obtained when the battery is fresh, because reading error increases as battery voltage decreases, even after "compensation" using the zero adjust.
   b. Do not select a resistance range (multiplier) that results in a pointer deflection to the left of the center of the scale. Resistance scales are logarithmic and values are therefore crowded toward the high-resistance end, reducing the precision in those regions.
   c. Never touch or hold a resistor while making a resistance measurement, since skin resistance can affect readings.
   d. Never measure resistance in a circuit to which power is applied.
   e. For in-circuit resistance measurement, make certain that there are no other components connected in parallel with the resistor being measured. Transformers, transistors, diodes, and coils as well as other resistors connected in parallel with a resistor may affect the resistance measurement. When in doubt, open one terminal of the resistor being measured.
2. When performing dc voltage measurements:
   a. Always select the dc voltage function *before* connecting the meter in a circuit.
   b. Always connect the meter in *parallel* with the component whose voltage is to be measured.
   c. Always select the highest voltage range before connecting the meter. The range may then be reduced as necessary to obtain a readable pointer deflection.
   d. Be aware of the fact that the meter can *load* the circuit in which voltage measurements are made; that is, the resistance of the meter itself, being in parallel with the component across which voltage is measured, reduces the total resistance of the combination and may therefore reduce the voltage across it. This effect is illustrated in Figure 1-2. The actual voltage across R_1, without the meter connected, is $V_1 = VR_1/(R_1 + R_2)$. When the meter is connected, the meter resistance R_m is in parallel with R_1, so the measured voltage V_m is $V_m = V(R_1\|R_m)/(R_1\|R_m + R_2)$, which may be significantly smaller than V_1, unless $R_m \gg R_1$.

   When measuring voltage, the meter resistance R_m can be calculated using

   $$R_m = SV_{\max} \tag{1.5}$$

   where S is the rate sensitivity in Ω/V and $V_{\max}$ is the maximum voltage of the voltage range selected. For example, if a meter has a sensitivity

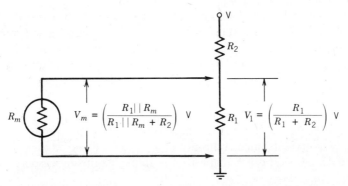

Figure 1-2 Loading caused by meter resistance. V_m = measured voltage; V_i = actual voltage.

of 20 kΩ/V and the 0 to 5-V range is selected, then the meter resistance is 100 kΩ. A voltage reading V_m can be corrected for loading by multiplying it by a correction factor, as follows:

$$V_1 = \left(1 + \frac{R_s \| R_1}{R_m}\right) V_m \tag{1.6}$$

where R_1 is the resistance across which the voltage V_1 is to be determined, R_s is the total resistance in series with R_1, and R_m is the meter resistance. If we let R_2 in Figure 1-2 be R_s, then Equation 1.6 is found by Thevenizing the series circuit of R_1, R_s, and V in Figure 1-2. If the Thevenin equivalent circuit is loaded with R_m, then

$$V_1 = V_m + I_m(R_1 \| R_s)$$

But

$$I_m = \frac{V_m}{R_m}$$

so

$$V_1 = V_m + \frac{V_m}{R_m}(R_1 \| R_s) = V_m\left(1 + \frac{R_1 \| R_s}{R_m}\right)$$

3. When performing dc current measurements:
   a. Always select the dc current function *before* connecting the meter in a circuit.
   b. Always connect the meter in *series* with the component in which current is to be measured.
   c. Always select the highest current range before connecting the meter. The range may then be reduced as necessary to obtain a readable pointer deflection.
   d. Be aware that the meter can load the circuit in which current is measured; that is, the total current that flows in a component may be reduced when

the meter resistance is inserted in series with it. In general, the meter resistance is small when large current ranges are selected, but may be as great as 1 kΩ in microampere ranges. A current reading I_m can be corrected for loading by multiplying it by a correction factor, as follows:

$$I = \left(1 + \frac{R_m}{R_s}\right)I_m \qquad (1.7)$$

where R_m is the meter resistance and R_s is the total series (or Thevenin equivalent) resistance of the circuit in which current I is to be determined. The development of Equation 1.7 is similar to the development of Equation 1.6.

4. When performing ac voltage measurements:
   a. Remember that the meter sensitivity is smaller for ac measurements than for dc measurements. Consequently, loading effects, as described in **1d** above, may be more serious. Consult the technical manual to determine the value for S in Equation 1.5, rather than using the relation $S = 1/I_{fs}$.
   b. Make certain that the frequency of the ac voltage being measured is within the manufacturer's specific range for the instrument. VOMs are usually not designed for high-frequency measurements, and the maximum permissible frequency in some instruments may be as low as 60 Hz.
   c. Remember that the meter responds to average value. Consequently, if the ac voltage has a dc component, the reading will be in error since it will not represent the RMS or peak value of the ac component alone. In meter terminology, an *output* voltage is the ac component of a waveform having a dc level, and some meters have an "output" terminal (actually used for input) that blocks the dc component of any waveform connected to it. An external capacitor can always be connected in series with the meter to block dc levels. Consult the technical manual for the proper-size capacitor.
   d. Remember that RMS or peak voltage calibrations are valid for sinusoidal inputs only. A "true RMS" instrument is required to measure nonsinusoidal waveforms.

1.1.3 Electronic Analog Multimeters

The principal feature of the analog-type electronic multimeter (EMM) is its use of active circuitry to provide buffering (isolation) between the circuit in which measurements are made and the measuring instrument itself. This characteristic is particularly valuable in voltage measurements because the isolation resulting from the use of a high-input impedance amplifier eliminates the loading problem that afflicts the passive-type VOM. The input amplifier in modern EMMs is either a field effect transistor (FET) connected as a source follower or a pair of FETs connected as a differential amplifier, both of which present a very high impedance to input signals. In practice, the input is a high-resistance voltage divider (typically 10 MΩ) that supplies different proportions of the input voltage to the amplifier, according to the range selected. See Figure 1-3. The

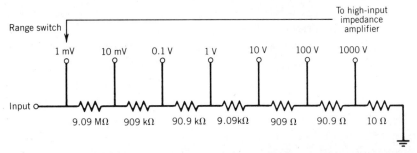

Figure 1-3 Voltage divider for voltage range selection in an electronic multimeter.

high impedance of the amplifier does not load the voltage divider, and the input resistance to the instrument is the same as the total resistance of the divider.

Methods used to perform resistance measurements in an EMM include the constant-current and the voltage divider techniques. In the former, the instrument supplies a fixed current from a constant-current source, so the voltage developed across an unknown resistor is proportional to its resistance. This voltage is measured as before and causes a meter deflection on a scale calibrated in ohms. In the voltage divider method, a fixed voltage is applied to a voltage divider consisting of an internal reference resistor and the unknown resistor. The voltage developed across the latter is proportional to its resistance.

Current is also measured by conversion to an equivalent proportional voltage. Current flowing into the instrument develops a voltage across a small internal resistance that is selected by the user according to the range desired. Small current ranges require larger resistors. Some inexpensive meters use the meter movement directly to cause a deflection proportional to current and have switch-selectable shunts to change ranges, as in the passive-type VOM.

For measuring an ac voltage, the electronic multimeter uses a rectifier to create a unidirectional current proportional to the magnitude of the ac voltage, as in the passive VOM. However, many modern instruments use an operational amplifier with diode feedback to provide better rectification and reduce the effects of diode nonlinearities. A low-pass filter is then used to create a dc voltage that is measured by the instrument. In some circuits, the voltage generated this way is proportional to the peak rather than the average value of the input; but in any case, the ac scale is calibrated for sinusoidal waveforms only (unless it is a true RMS instrument).

The principal disadvantages of the EMM in comparison to the VOM are:

1. A power source is required, so the instrument is not so portable as the VOM. Some EMMs have internal batteries to permit dc/ac operation, but these are by necessity much larger and more expensive than the small battery used in a VOM for resistance measurements.

2. In less expensive units, the input amplifiers may be subject to drift and will require frequent adjustment and recalibration. For resistance measurements, the pointer deflection must usually be adjusted for ∞ (open circuit) as well

as for the zero indication. More expensive units have chopper-stabilized dc amplifiers that correct drift and offset problems.

3. The more complex active circuitry makes the EMM somewhat less reliable and more difficult to repair and maintain than the VOM.

1.1.4 Digital Multimeters

The essence of a digital multimeter (DMM) is an *analog-to-digital convertor* (A/D converter, or ADC) that is used to generate a binary number proportional to the magnitude of an unknown voltage. The binary number is then converted to an equivalent decimal value whose digits are displayed in the readout. The typical inexpensive DMM employs analog signal conditioning circuitry, similar to that described for the EMM, to develop voltages that are proportional to resistance or current and can therefore be used for multifunction and multirange measurements. More elaborate and expensive laboratory-grade digital instruments use digital sampling techniques and extensive digital logic circuitry (including a microprocessor in some models) in lieu of analog signal conditioning. These instruments provide much greater accuracies than their analog counterparts, but they too must incorporate an A/D converter somewhere in the signal processing chain.

The most widely used method to perform A/D conversion in DMMs is called the *dual slope* technique. See Figure 1-4a. The electronic switch is initially in position 1, connecting the unknown voltage V_x to the integrator (typically an operational amplifier with capacitive feedback). The output of the integrator is a ramp voltage whose slope depends on the magnitude of V_x. As soon as the ramp is a few millivolts beyond zero, the voltage comparator switches and enables the AND gate, which gates the constant frequency clock signal into the binary counter. The counter counts until it reaches a predetermined value (such as overflow), at which time (t_1) the count detector logic causes the electronic switch to assume position 2. In position 2, a reference voltage having polarity opposite that of V_x is applied to the integrator, which then generates a ramp of opposite slope. The output of the integrator is therefore a ramp that will eventually reach zero (in t_2 seconds) and reset the comparator, which then stops the counter. The integrator voltage is reset to zero, and the process automatically repeats, a new count being developed if the unknown input voltage has in the meantime changed value. The principal advantage of this technique is that the integrator suppresses high-frequency noise, its output amplitude being inversely proportional to frequency.

Figure 1-4b shows the output of the integrator for two different unknown inputs, V_A and V_B, with $V_A < V_B$. Note that the time t_1, set up by the meter logic circuitry, is the same for both cases but that the ramp with slope $k_1 V_B$ reaches a greater voltage in time t_1 than the ramp with slope $k_1 V_A$. Since the negative ramps have equal slopes $-k_2 V_{ref}$, the time t_2 to reach zero is greater for the circuit having input V_B. It is easy to show that

$$t_2 = \frac{k_1 t_1}{k_2 V_{ref}} V_x \qquad (1.8)$$

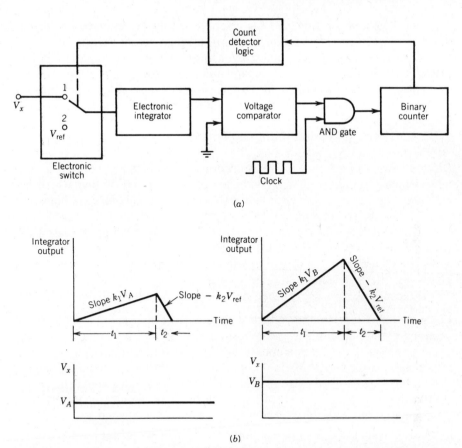

Figure 1-4 Dual-slope A/D conversion. (*a*) Block diagram of the converter.
(*b*) Ramp outputs from the integrator for inputs $V_x = V_a$ and $V_x = V_b$.

Since k_1, k_2, t_1, and V_{ref} are constants, Equation 1.8 shows that the time t_2 is proportional to the unknown voltage V_x. Therefore, the count stored by the binary counter is proportional to V_x.

Digital multimeter accuracy specifications are generally better than those of comparably priced analog instruments. Accuracy specifications may be given as a percentage of the reading plus a certain percentage of the full-scale value in the selected range, or they may be stated in combination with the phrase "plus or minus one digit". In the latter case, the "digit" refers to the *least significant digit* of the decimal readout. The greater the number of decimal digits displayed, the greater the resolution of the instrument, and also usually the greater the accuracy. Many DMMs are specified as providing a certain number of digits plus a half-digit, as, for example, $3\frac{1}{2}$ digits. The $\frac{1}{2}$ digit is the most significant digit in the readout and may have only the value 0 or 1. This $\frac{1}{2}$ digit permits readouts that are slightly greater than the nominal maximum value in the selected range, and this feature is called *overranging*. With overranging, a voltage of 1.052 V, for example, could be displayed in the 0 to 0.999-V range, instead of requiring that the range be changed to 0 to 9.99 V where it would be displayed at 1.05 V. Other features that may be found in a DMM include *autoranging* (automatic selection of the proper measurement range) and *autopolarity* (the ability to measure either positive or negative voltages and display the appropriate sign in the readout).

1.1.5 Component Testing Using Multimeters

Diodes A multimeter set for resistance measurements can be used to determine which terminal of a semiconductor diode is its anode and which is its cathode. The diode will exhibit low resistance when it is forward-biased (the positive lead of the ohmmeter connected to the anode and the negative lead connected to the cathode) and a high resistance when reverse-biased (lead connections reversed). If a very low or zero resistance is measured with the leads connected both ways, or if an infinite resistance is measured both ways, then the diode is defective (shorted or open).

Precautions to observe when making tests include:

1. Check the polarity of the voltage supplied by the instrument to its test leads. Some meters supply a negative voltage to the red, or "high," lead when they are set for resistance measurements. Also be certain that the voltage supplied to the test leads is not so great that the diode could be damaged by excessive forward current. Since the meter resistance limits current flow and the resistance of a VOM is low on the $R \times 1$ range, avoid using this range except for rectifier or power diodes that can sustain larger currents.

2. Make certain there is sufficient voltage available at the test leads to forward bias the diode (0.3 V for Ge diodes and 0.7 V for Si diodes). Instruments such as the DMM that supply a constant current when set for resistance measurements may not provide sufficient voltage. Some of these instruments have a special setting for diode resistance measurements.

3. Do not expect different instruments or different range settings on the same instrument to produce the same resistance measurements for a forward-biased diode. The nonlinear characteristic of the diode results in resistance values that depend on the amount of current flowing through it. To obtain an accurate measurement of the diode's resistance when a certain amount of current is flowing through it, it is best to connect it in a series circuit with a dc power source and a current-limiting resistor. Adjust the power source until the desired current I_D is measured in the diode, measure the voltage V_D across it, and calculate $R = V_D/I_D$ ohms. (R is its *static*, or dc, resistance.)

Transistors Since the base-to-emitter and collector-to-base junctions of a transistor are *pn* junctions, these can be checked and identified in the same way described for diodes. Observe the same precautions. Resistances measured between collector and emitter should be moderately high (but not necessarily equal) regardless of the direction (ohmmeter polarity) in which it is measured.

Figure 1-5 summarizes the kinds of resistance readings that should be observed when measuring *npn* and *pnp* transistors with both polarities of the ohmmeter voltage.

It should be noted that the "high" and "low" resistance measurements referred to above are strictly relative and that a large variation in values is to be expected among transistors of different types. In general, the *ratio* of reverse to forward resistance should be at least 30 to 1.

For in-circuit transistor testing of a transistor biased in its linear region, as in a class A amplifier, the dc collector-to-emitter and base-to-emitter voltages should be measured. The dc collector-to-emitter voltage should have some value *between* 0 and V_{CC}, the collector supply voltage (but not including those extremes). The base-to-emitter voltage should be in the vicinity of $+0.65$ V for *npn* transistors and -0.65 V for *pnp* transistors (± 0.3 V for Ge transistors). If this voltage is zero, the junction is shorted; and if it is larger than the specified values, the junction is open.

Figure 1-6 shows a circuit that can be used to determine if a transistor is operating properly and to measure its dc value of β (h_{FE}). For *pnp* transistors, reverse the polarities of the 10-V source and the milliammeters. If the transistor is operating properly, increasing and decreasing I_b by adjusting the 1-M potentiometer should cause I_c to increase and decrease correspondingly. To determine the dc β, adjust the potentiometer until I_c has some value between 2 and 8 mA and measure the value of I_b that results in that value of I_c. Then,

$$\beta = \frac{I_c}{I_b} \tag{1.9}$$

The value of α for the transistor can be calculated from

$$\alpha = \frac{\beta}{\beta + 1} \tag{1.10}$$

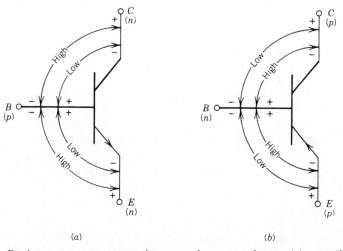

(a)   (b)

Figure 1-5 Resistance measurements in *npn* and *pnp* transistors. (a) *npn*. (b) *pnp*.

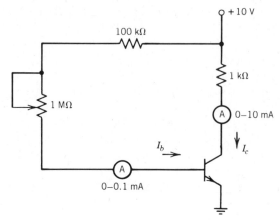

Figure 1-6   Circuit for testing the operation of a transistor.

Silicon-Controlled Rectifiers (SCRs)   When an SCR is not connected in a circuit (all terminals open), the resistance between all pairs of its terminals (anode, gate, and cathode) should be high regardless of the direction in which it is measured, except that the gate-to-cathode resistance should be low when the gate is positive. In other words, the only low resistance observed should occur when the positive ohmmeter probe is on the gate and the negative probe is on the cathode.

To test an SCR for proper operation, it is necessary to connect it in a circuit with dc power sources and current limiting resistors. See Figure 1-7. The resistor R should be chosen so that $I_h < E_A/R < I_{max}$, where E_A is a voltage less than the specified forward-breakover voltage of the SCR ($V_{F(BO)}$), I_h is the specified holding current at $E_{GC} = 0$ V, and I_{max} is the rated maximum continuous (dc) forward current. Set $E_G = 0$ V before connecting the source E_A. Then, connect E_A. The measured voltage V_{AC} should then be high (close to E_A). As E_G is increased gradually, a point should be reached where V_{AC} goes low (near 0 V), indicating the SCR has fired. Then, set E_G back to 0 V. The measured voltage V_{AC} should remain low. Decrease E_A to 0 V. Then, restore E_A to its original value. With E_G still set at 0 V, V_{AC} should once again be high.

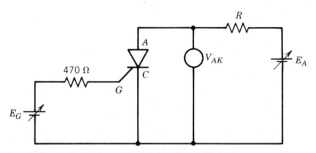

Figure 1-7   Circuit for testing the operation of an SCR.

If the results of the SCR test are as described above, then the SCR is operating properly. To check its value of $V_{F(BO)}$, connect the gate directly to the cathode and increase E_A until V_{AC} goes low. The value of E_A at which V_{AC} goes low is the value of $V_{F(BO)}$. To check the reverse breakover voltage $V_{R(BO)}$, reverse the anode and cathode connections and repeat this test, using a value of R about 10 times larger than before, to limit the reverse current.

Zener Diodes   The resistance of a forward-biased zener diode should be small and can be checked using an ohmmeter in the same way as a conventional diode. The reverse-biased resistance should be high, provided the reverse voltage is less than the rated zener (breakdown) voltage V_Z. The circuit shown in Figure 1-8 can be used to test the diode for proper operation and to determine the value of V_Z. Set R equal to approximately $V_Z^2/(0.5P_D)$ Ω, where V_Z is the rated zener voltage and P_D is the diode's rated power dissipation. (For example, for a 10-V $\frac{1}{2}$-W zener, $R = (10)^2/0.25 = 400\ \Omega$; use 390 Ω.) The power rating P_R of the resistor should satisfy $P_R > V_Z^2/R$. As the supply voltage E is increased gradually, the measured value V should increase correspondingly until the value V_Z is reached. As E is increased further, V should remain substantially constant at V_Z volts. Do not increase E beyond $2V_Z$ volts.

Capacitors   An ohmmeter can be used to determine if a capacitor is shorted. With the capacitor terminals open and the ohmmeter set for its highest resistance range, connect the ohmmeter probes across the capacitor. If the capacitor is electrolytic, be certain to connect the probe having positive voltage to the positive side of the capacitor. The resistance reading should rise as the capacitor charges (slowly, for large capacitance values) until a very large, essentially infinite resistance is indicated and maintained.

To test the leakage current in an electrolytic capacitor, use the circuit shown in Figure 1-9a. For best results, set the voltage E to a value near, but less than, the rated capacitor voltage. For electrolytic capacitors tested near their rated voltage, leakage currents should not appreciably exceed 0.1 mA at 100 V or less, 0.2 mA at 100 to 300 V, and 0.5 mA at 300 V or greater.

Nonelectrolytic capacitors should have much smaller leakage currents. In the leakage current test shown in Figure 1-9b, V should be essentially zero. If it is not, then compute the equivalent resistance R_{eq} of the capacitor using

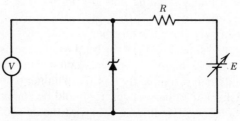

Figure 1-8   Circuit for testing a zener diode.

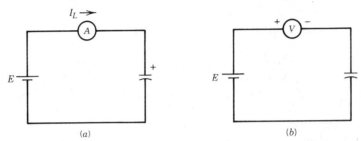

Figure 1-9 Capacitor leakage tests. (*a*) Electrolytic capacitor. (*b*) Nonelectrolytic capacitor.

$$R_{eq} = R_V\left(\frac{E}{V} - 1\right) \Omega \qquad (1.11)$$

where R_V is the resistance of the voltmeter. R_{eq} should not be substantially less than about 100 MΩ.

Capacitors in circuits where ac voltages are present, as, for example, coupling and bypass capacitors, can be checked in-circuit by measuring the voltage across them. In most of these applications, the ac voltage across the capacitor should be nearly zero, or very small in comparison to the ac voltage measured between the capacitor and ground. Remember that a dc component in an ac voltage may cause the ac meter reading to be in error, so dc components should be blocked (see Measurement Precautions, Section 1.1.2). If a large ac voltage is measured across a capacitor, it may be open.

Coils and Transformers Use an ohmmeter to verify that the resistance of a coil is approximately equal to the manufacturer's specified value, if known. (Polarity is of no consequence.) Coils have a wide range of resistance values, so no general rules can be given for acceptable values. Small coils having many turns of fine wire can be expected to have higher resistances than others. If the measured resistance is zero or infinity, the coil is shorted or open. If the resistance is appreciably less than the manufacturer's specified value, some windings may be shorted to others.

The primary and secondary windings of a transformer can be tested as separate coils in the manner described above. Except for autotransformers, which use the same coil for the primary and secondary, the resistance between any terminal of a primary winding and any terminal of a secondary winding should be infinite (open).

The resistance between a transformer tap and one end of a winding should be approximately the same percent of the total winding resistance as the percent of the tap. For example, the resistance between a center tap and either end of the winding should be approximately half the winding resistance. If resistance tests show a transformer to be suspect, it should be checked for conformance to specifications under ac conditions. Measure the ac secondary voltage with a resistive load connected across the secondary and an ac signal source applied across the primary. The RMS secondary voltage should equal the turns ratio (secondary turns divided by primary turns) times the RMS primary voltage.

Power Supplies and Voltage Regulators   Poor system performance or total system failure is often a problem that is traceable to the system's dc power supply, and this is one of the first components that should be checked. It is obvious that the first test that should be made is an in-circuit measurement of the supply's dc output voltage, but a low or zero voltage at that point does not necessarily mean the power supply has failed. Not infrequently there is a component, such as a filter capacitor, that has shorted the output to ground and overloaded the supply. Modern IC voltage regulators, such as the 7800 series fixed-voltage regulators and the 723 adjustable regulator, have built-in overload sensing circuitry that reduces the terminal voltage when the current reaches a prescribed value. If the power supply voltage is abnormally low, disconnect its output and measure its open-circuit terminal voltage. If the voltage is still low, then the supply can be reasonably suspected to be defective, though the failure may have been caused by the shorting of an external component. Do not replace a defective supply or regulator until the rest of the system has been checked for shorts.

Abnormal system behavior, such as excessive "hum" in an audio amplifier, or sporadic and intermittent shutdown, may be the fault of a defective filter in the power supply. "Leaky" or shorted electrolytic filter capacitors are often the source of such problems. Look for an excessive ac component (ripple) in the dc output. Use a multimeter to measure the ac voltage present in output, remembering to block the dc component (see Measurement Precautions, Section 1.1.2, for ac voltages). Generally speaking, the RMS ripple voltage should be less than 1% of the dc voltage at the output of the filter when a normal load is connected.

When testing a voltage supply for conformance to specifications, the terminal voltage should be measured under both no load (zero current) and full load (maximum rated load current). Some dc voltage sources, particularly some batteries, may have a higher terminal voltage under no load but virtually zero voltage under load.

To check a power supply's *percent voltage regulation* (% VR), measure both the no-load and full-load voltages, V_{NL} and V_{FL}, and calculate

$$\% \text{ VR} = \frac{V_{NL} - V_{FL}}{V_{FL}} \times 100\% \qquad (1.12)$$

Ideal voltage regulation is 0% (not 100%).

Many manufacturers provide an *output* impedance (or resistance) specification for power supplies, which is defined by

$$R_{\text{out}} = \frac{\Delta V}{\Delta I} \qquad (1.13)$$

where ΔV is the total change in output voltage over a specified range of loads (as, no load to full load) and ΔI is the total change in load current over that range. Note that no-load current is zero amperes. R_{out} should be quite small for well-regulated supplies, on the order of milliohms for laboratory-grade supplies. Given R_{out}, the voltage regulation can be determined from

$$\% \text{ VR} = R_{out} \frac{I_{FL}}{V_{FL}} \times 100\% \tag{1.14}$$

where I_{FL} and V_{FL} are the full-load current and full-load voltage, respectively.

Multimeter Probes   Many multimeters are available with accessory probes that expand the capabilities and versatility of those instruments. Some of these are described briefly below.

Clip-on current probes. Current probes equipped with an expandable, segmented ring (jaws) that clamp over a conductor are available for the measurement of ac current. Besides providing the ability to measure ac current in instruments not designed for that function, current probes are very useful for in-circuit testing because they eliminate the need for opening the circuit and inserting an ammeter in series with it. Furthermore, they do not load the circuit in which measurements are made since no meter resistance is inserted in the path of the measured current. The inductive loading of the circuit due to the transformer action of the conductor and the probe is usually negligible.

The ac current probe is essentially a step-up transformer in which the primary is the conductor carrying the current to be measured and the secondary is a winding on the clamp-on ring. The magnitude of the ac voltage induced in the secondary is proportional to the ac current in the primary, and the induced ac voltage can then be measured by an ac voltmeter. The appropriate conversion factor for interpreting a voltage reading as a certain amount of current must be known. The factor 1 mA/mV is frequently used.

At least one manufacturer supplies an analog-type electronic voltmeter equipped with a clip-on dc current probe. The Hewlett–Packard Model 428B is capable of measuring dc currents from 1 mA to 10 A full-scale in nine ranges, without requiring insertion of a meter in series with the test circuit.

High-frequency probes. Most multimeters are incapable of measuring ac voltages at frequencies in the rf range. Some are available with high-frequency probes, also called demodulator probes, that effectively extend the frequency range over which voltage measurements can be made. These probes contain the same kind of passive circuitry that is used for peak detection in AM demodulation, namely, a diode rectifier, resistor(s), and a filter capacitor. See Figure 1-10. Capacitor C_1 is used to block any dc component that may be present in the rf input. Diode D_1 rectifies the signal by passing the negative peaks to ground, and capacitor C_2 charges to the peak value of the rectified signal. The capacitor voltage is therefore proportional to the peak value of the rf input and will vary in accordance with any low-frequency amplitude modulation (the positive envelope) of the rf signal. The resistors are chosen to attenuate the peak voltage by the factor 0.707, thus providing a voltage proportional to the RMS value of the sinusoidal input.

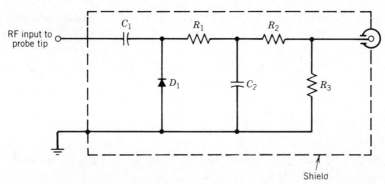

Figure 1-10   A high-frequency multimeter probe.

In some designs, capacitor C_2 and/or some of the resistors are omitted, their function being performed by impedance at the meter terminals. The simplest designs consist of no more than capacitor C_1 and diode D_1, but be aware that such probes (most probes) must be used with the particular meter for which they are designed if accurate readings are desired.

The limitations of the probe are determined primarily by the characteristics of diode D_1. The voltage required to forward bias the diode limits the minimum signal level that can be detected, and for that reason germanium diodes are often used. The maximum signal level is determined by the peak inverse voltage rating of the diode, usually 100 V or less. A typical rf probe (HP Model 11096B) will provide a response within ±1.2 dB from 100 kHz to 500 MHz (when used with a 10-MΩ meter input impedance) up to 30 V RMS.

High-voltage probes. For measuring very high voltages, such as those encountered in television picture tubes, a high-voltage probe can be used with a multimeter not designed to read high voltages directly. These probes contain a very large series resistance (typically 900 to 1000 MΩ) that divides down the voltage across the meter's input resistance. For a meter with a 10-MΩ input resistance, a 100-to-1 reduction in voltage is typical.

The plastic body of a high-voltage probe has protruding fins, similar in appearance to cooling fins, that increase the surface area between the handle and the tip and thereby provide more isolation between the high-voltage source and the user. These fins must be kept dry and free from grease, dirt, and other contaminants to maintain that isolation and protect the user from shock. High-voltage probes should not be used in a wet environment or at high altitudes where the thin air has low ionization potential. The probe manufacturer will specify the maximum altitude at which the probe can be used.

1.2 FUNCTION GENERATORS

1.2.1 Operation

Function generators are general-purpose signal sources that produce several kinds of waveforms, usually including sine, square, and triangular waves. The type of waveform obtained is switch-selected by a user, and the amplitude and frequency of the waveform are adjusted through the use of continuous or step-type controls, or combinations of these. In some models, the different wave-forms are available simultaneously at different sets of output terminals. The signal frequencies in modern function generators can be set over a very broad range, from as low as 1 mHz or 1 μHz to as high as several MHz. The range for a typical, general-purpose generator is 0.01 Hz to 10 MHz.

Audio-signal generators are designed to produce low-distortion sine wave signals over a more limited frequency range. Frequencies are in the audio range, 20 Hz to 20 kHz, but may also go as high as 100 kHz. Laboratory-grade audio generators are equipped with precision attenuators and/or output meters that allow the user to set the signal amplitude accurately. Many audio-signal generators also produce a square wave output over the same frequency range as sine waves. Unlike function generators, the source of oscillation is usually an audio oscillator such as an RC phase-shift or Wien bridge oscillator. The sine waveform is then converted to a square wave by driving a voltage comparator whose output switches state when the sine wave crosses through zero.

The source of oscillation in most function generators is a free-running (astable) multivibrator that generates square waves. The square wave is integrated by an operational amplifier having capacitive feedback to produce a triangular wave. This process is based on the fact that the integral of a constant is a straight line; that is, the integral of a fixed voltage level is a linearly increasing ramp voltage:

$$\int_0^t Edt = Et \tag{1.15}$$

If E is a positive constant, then Et is a ramp with positive slope E; if E is negative, then Et is a negative-going ramp. Since electronic integration usually causes a 180-degree phase inversion, the output ramp is actually positive-going when the input is negative, and vice versa. The alternating polarity of an input square wave causes the output to be a sequence of alternately positive- and negative-going ramps, that is, a triangular wave.

An inverting operational amplifier having a resistor input and capacitive feedback (i.e., a voltage integrator, as shown in Figure 1-11) can be regarded as a capacitor that is charged by a constant-current source. Since the capacitor voltage V equals Q/C, where Q is the charge stored on the capacitor, it follows that

$$V = \frac{Q}{C} = \frac{1}{C}\int i \, dt \tag{1.16}$$

Since the current i is constant, the voltage V increases linearly with time. In some function generators, the square wave alternately switches positive and negative current sources into a capacitor, thus alternately charging and discharging it to produce a triangular wave. As noted above, this is essentially the same principle as the integrating operational amplifier.

The multivibrator circuit is often composed of a voltage comparator whose output is switched by the level of the triangular wave, so square and triangular

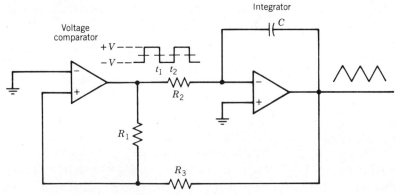

Figure 1-11   Square wave and triangular wave generation in a function generator.

waves are generated by the same circuit. A typical configuration for generating square and triangular waves in this manner is shown in Figure 1-11. The output of the voltage comparator switches to $+V$ when its noninverting input goes slightly positive and to $-V$ when its noninverting input goes slightly negative. Since $+V$ is fixed during the time it appears, t_1 to t_2, and the inverting input terminal of the integrator is approximately zero volts, the current charging the capacitor is constant. The same holds true when the voltage comparator output is at $-V$. Assuming the output is initially $+V$, the inverting integrator produces a negative-going ramp. Resistors R_1, R_2, and R_3 feed a combination of the integrator and comparator output voltages back to the noninverting comparator input. Because the integrator output is a negative-going ramp, this feedback voltage will eventually go negative and cause the comparator output to switch to $-V$. The integrator then generates a positive-going ramp until the comparator input once again becomes positive, causing it to switch back to $+V$, and the process repeats.

The sine wave output can be derived by filtering the square wave to recover its fundamental component (the sinusoidal component having the same frequency as the square wave). Of course, the cutoff frequency of this (low-pass) filter must be adjusted when the oscillation frequency is changed, and this is easily accomplished by ganged controls that simultaneously vary component values in the oscillator and filter circuits. This filtering process causes considerable attenuation of the sine waveform, so voltage amplification is necessary before the sine signal is delivered to the generator output.

In some models, the sine wave is produced by piecewise linear approximation. The triangular wave drives a resistor–diode network that shapes the wave into sinusoidal form. (An operational amplifier may be used in this process.) The instantaneous value of the triangular wave controls the bias on the diodes and determines which and how many of the diodes are conducting, which in turn determines the output voltage level. Resistor values are chosen so that the changing values of the triangular wave create an output that approximates a sinusoidal shape.

Figure 1-12 shows a generalized block diagram of a typical function generator.

1.2.2 Special Features

Many modern function generators are available with standard or optionally specified features that greatly expand their versatility. Following is a brief description of a number of these.

1. DC (OFFSET) LEVEL CONTROL. This control permits the user to add a dc component to the output waveforms. The dc level is adjustable and can usually be either positive or negative. A typical range is ±10 V dc.

2. GENERATION OF OTHER WAVEFORMS. Some generators can be set to produce sawtooth and pulse-type waveforms in addition to the traditional sine, square, and triangular waves. (In some models, they are obtained through symmetry control, described below.) Also available are square-waves that alternate between 0 V and a positive level, or between 0 V and a negative level (positive-only or negative-only pulses). Some models produce haversine outputs, haversine of $A = \frac{1}{2}(1 - \cos A)$, used for telephone line and vibration testing.

3. SYMMETRY CONTROL. This control permits the user to adjust the symmetry of the waveform with respect to a vertical axis. Triangular waves can thus be adjusted to produce sawtooth waveforms having prescribed rise and fall times. The duty cycle of squarewaves can be adjusted to produce pulse trains with prescribed duty cycles.

4. EXTERNAL FREQUENCY CONTROL. This feature allows the frequency to be controlled by an externally applied voltage; that is, it permits the unit to operate as a voltage-controlled oscillator (VCO). The frequency range that can be controlled may be as little as 10 to 1 (1 decade) or, in

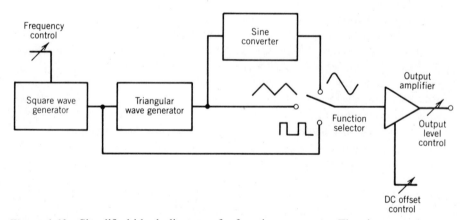

Figure 1-12 Simplified block diagram of a function generator. The sine wave is produced by a diode-shaping network.

some models, as great as 1000 to 1 (3 decades). Frequency control can be used to generate FM waveforms and to convert the unit into a swept frequency oscillator. Some models have automatically generated linear or logarithmic frequency sweeping.

5. EXTERNAL SYNCHRONIZATION. This feature allows the user to synchronize the waveform produced by the generator with another (external) signal. In some models, the generated waveform can be synchronized at a fraction or multiple of the external frequency. The phase angle between the external and the generated waves may also be adjustable.

6. EXTERNAL START–STOP CONTROL. This feature permits the user to start and/or stop the generator output using an externally applied signal. In the gate mode, the output signal is present when the external input is high and is absent when the input signal is low. In some models, a single cycle or a half-cycle of the output waveform can be triggered by an externally applied pulse. By selecting a pulse-type output and adjusting its duty cycle, the triggered operation allows the unit to be used as a monostable multivibrator (one-shot). In advanced models, a burst consisting of a preset number of cycles can be triggered by an external pulse.

7. AMPLITUDE MODULATION. Some function generators can be set to produce amplitude-modulated (AM) waveforms. The modulating signal can be internally or externally generated, or both. Some models having internally generated modulating signals permit the modulating frequency to be adjusted over a wide range, while others have a fixed modulating frequency, such as 400 Hz or 1 kHz. The modulation index, or depth of modulation, can usually be adjusted between 0 and 100%.

1.2.3 Specifications

Following are descriptions and typical values of some important manufacturers' specifications for function generators.

Output impedance. The (Thevenin) equivalent output impedance of the generator. Usually 50Ω or 600Ω.

Output level. The maximum output voltage level. This specification may be given as a peak, peak-to-peak, or RMS (sinewave) value, and usually refers to no-load (open terminal) conditions. The maximum output level when a load equal to the rated output impedance is connected across the terminals is one-half the no-load value. A typical maximum, no-load output is 20 V p–p.

Generators may also have outputs rated in dBm (decibels referenced to 1 mW in a 50-Ω load), VU ("volume units": decibels referenced to 1 mW in a 600-Ω load), or dBmV (decibels referenced to 1 mV RMS across 75 Ω). By these definitions,

$$dBm = 10 \log_{10} \frac{P_o}{P_1} \tag{1.17}$$

where $P_1 = 10^{-3}$ W and P_o = output power in watts across 50 Ω;

$$VU = 10 \log_{10} \frac{P_o}{P_1} \qquad (1.18)$$

where $P_1 = 10^{-3}$ W and P_o = output power in watts across 75 Ω;

$$dBmV = 20 \log_{10} \frac{V_o}{V_1} \qquad (1.19)$$

where $V_1 = 10^{-3}$ V and V_o = output voltage in volts across 75 Ω. Note that $P_1 = 1$ mW corresponds to 0 dBm and 0 VU, and that $V_1 = 1$ mV RMS corresponds to 0 dBmV.

Given D_1 dBm, the RMS voltage across 50 Ω may be calculated from

$$V_{RMS} = 0.2236 \times 10^{D_1/20} \qquad (1.20)$$

Given D_2 VU, the RMS voltage across 600 Ω may be calculated from

$$V_{RMS} = 0.7746 \times 10^{D_2/20} \qquad (1.21)$$

Given D_3 dBmV, the RMS voltage across 75 Ω may be calculated from

$$V_{RMS} = 10^{-3} \times 10^{D_3/20} \qquad (1.22)$$

Dial accuracy. Most general-purpose function generators have a range switch that is used to select a multiplying factor ($\times 10$, $\times 100$, etc.) which in turn multiplies a value set on a rotary dial to determine the frequency. The specified dial accuracy can be used to determine the maximum possible difference between the actual output frequency and the frequency set by the dial and range switch. This accuracy is generally specified as a percent of the maximum frequency in the selected range. Typical values range from 3 to 5%. For example, a setting of 15 kHz in the 10-kHz to 100-kHz range on an instrument having 5% dial accuracy could have an actual output frequency of 15 kHz $\pm$ 0.05 (100 kHz); that is, it could be in the range from 10 to 20 kHz. Dial accuracy is sometimes specified as a percent of setting plus a percent of full range.

Distortion (sinewave). This specification refers to percent total harmonic distortion (THD). Many manufacturers supply different distortion figures for differing frequency ranges, the larger distortions applying to higher frequencies. Inexpensive units may have a distortion specification given as <5% at all frequencies, while a typical specification for a laboratory-grade instrument is <0.25% up to 20 kHz and <0.5% up to 100 kHz.

Linearity. This specification refers to the linearity of triangular and saw-tooth waveforms. It may be given as a percent deviation from linearity over a specified portion of the waveform (a typical value is 5% deviation between $0.1 \times$ peak value and $0.9 \times$ peak value), or in terms of percent linearity. Manufacturers rarely supply precise definitions of terms such as these. Unless otherwise defined, assume percent deviation refers to the maximum difference between actual voltage and ideal (linear) voltage at any point, expressed as a percent of the *maximum* (peak) voltage of the wave. Note that this interpretation permits the percent difference between actual and ideal values to be quite large at small voltages.

Stability. This specification refers to the maximum *drift* (change) in frequency that may occur after a particular frequency has been set. Although it is usually expressed as "percent stability," it means percent change in frequency and, unless otherwise noted, it refers to percentage of the maximum frequency in a selected range. Different specifications are given for different periods of time. A typical specification is <0.05% for 10 minutes, <0.1% for 1 hour, and <0.5% for 24 hours. Constant temperature conditions are assumed.

Amplitude flatness. The specification called "amplitude flatness" actually refers to the maximum variation in output level that can be expected when frequency (only) is changed. It is given under the assumption of a constant load and is usually specified for different ranges of frequency variation. A typical specification is ±3% to 100 kHz and ±6% to 1 MHz. The percentages are of a particular level (such as the maximum rated output) at a particular reference frequency, such as 1 kHz. The flatness specifications may be given in dB (±0.5 dB is typical) instead of percent and may differ according to the waveform selected.

Attenuator calibration. Laboratory-grade function generators have a calibrated attenuation control that allows the user to change the output level in steps. The steps are usually calibrated in 20-dB increments, where a 20-dB step corresponds to a 10:1 change in voltage level. These are usually specified in dB below the maximum output, which can be considered 0 dB. A typical specification is 80 dB of attenuation in 20-dB steps, with error <5%. All function generators have an uncalibrated variable amplitude control that permits the user to adjust the output amplitude continuously over some specified range.

Squarewave characteristics. Specifications related to the squarewave output include rise and fall times: the times required for its transitions between the 10% and 90% of peak value points and "total aberrations": the percent deviation from an ideal squarewave, due to overshoot, ringing, droop, and so on. Specifications for rise and fall times, also called transition times, vary from a few nanoseconds to 0.1 μsec, depending on the quality of the generator. Typical specifications for aberrations, also called perturbations, range from 1 to 10% of the peak-to-peak value.

1.2.4 Function Generator Applications

Signal tracing.   Signal tracing is a troubleshooting procedure in which signal levels are measured (or observed on an oscilloscope) at successive points in the path through which a signal flows in a circuit. For example, a signal can be traced through a multistage amplifier by observing the input and output of each stage, beginning with the first. A defective stage is located by determining the first point where the input is normal and the output is absent or otherwise abnormal. Figure 1-13 illustrates these ideas. In the three-stage system shown, stage 2 is found to be defective.

A function generator can be used as the source of the signal that is traced through a circuit. The advantage of using a function generator in this application is that the frequency and level of the signal can be set to specific, fixed values,

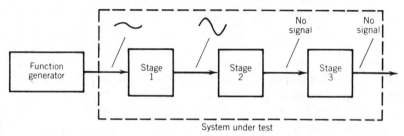

Figure 1-13   Signal testing in a three-stage system. The second stage is defective.

making it easy for the troubleshooter to identify the signal at various points in the circuit and to recognize distortion or abnormal magnitudes. This is particularly valuable when the normal signal is a complex waveform, such as music or speech, that continually changes shape and level.

The function generator can be used to replace the normal signal origin (a microphone or tape head, for example), or it can be used to *inject* a signal at some intermediate point in the circuit under test, as for example at the input to the power amplifier. An external capacitor should always be connected between the function generator output and the point in the test circuit where the signal is injected. The generator output is a short circuit to dc, and failure to isolate it with a capacitor can damage or seriously affect the operation of the circuit being tested. Adjust the function generator signal level to obtain the desired magnitude, as measured at the point in the test circuit where the signal is applied, rather than at the output terminals of the generator. Setting the level at the generator may not result in a sufficient level in the circuit, due to the ac drop across the capacitor.

Measuring frequency response.   The frequency response of a network or system is the manner in which its gain changes as the frequency of the signal applied to it is changed. In some applications, the change in the phase shift between input and output as frequency is changed is also an important aspect of its frequency response. Frequency response characteristics are often displayed by a graph showing changes in gain or phase as a function of frequency, and the axes of the graph often have logarithmic rather than linear scales. So-called log–log paper is used because the vertical axis is easily interpreted in decibels and because the horizontal axis permits a very wide frequency range to be shown on a practical size of paper.

A function generator is the ideal test instrument for obtaining frequency response data because its frequency can be changed over a wide range while maintaining a substantially constant signal amplitude. The voltage gain G of the circuit under test is defined as the ratio of the amplitude of its output e_{out} to the amplitude of its input e_{in}, $G = e_{out}/e_{in}$, where e_{out} and e_{in} are both measured in RMS volts, peak volts, or peak-to-peak volts. In terms of decibels,

$$G = 20 \log_{10}\left(\frac{e_{out}}{e_{in}}\right) \text{ dB} \qquad (1.23)$$

By changing the frequency of the signal e_{in} produced by the function generator, while maintaining its amplitude constant, the measured values of e_{out} can be used to compute gain at the different frequencies. To obtain very accurate data, the amplitude of e_{in} should be checked at each new frequency setting, since the generator output level may not remain perfectly constant at all frequencies (see the "amplitude flatness" specification).

When making frequency response measurements, the output signal from the circuit being tested should be monitored on an oscilloscope as the signal frequency is changed. If the signal appears distorted at some frequency (for example, clipped), then the level of the function generator should be reduced as necessary. If the signal becomes very small, to the extent that it is obscured by noise, then the level of the function generator should be increased. In either case, the new input level should be used in the gain calculation.

Of particular importance in frequency response data are the lower and upper cutoff frequencies f_1 and f_2 of the circuit being tested. Figure 1-14 shows typical frequency response plots in which these frequencies are identified. The cutoff frequencies are the frequencies at which the gain G has value $0.707G_m$, where G_m is the midband gain, that is, the gain in the region between the cutoff frequencies. At the cutoff frequencies, the gain is 3 dB below its midband value. The bandwidth is defined to be the difference between the cutoff frequencies:

$$BW = f_2 - f_1 \qquad (1.24)$$

In the case of a tuned amplifier or bandpass filter (Figure 1.14b), the Q of the circuit is an important characteristic of its frequency response. Q is defined by

$$Q = \frac{f_0}{BW} \qquad (1.25)$$

where f_0 is the "center" frequency, actually the frequency at which the gain is maximum. In high-Q circuits ($Q > 10$), the center frequency is located approximately midway between the cutoff frequencies.

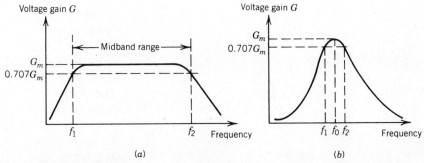

Figure 1-14 Typical frequency response plots. (a) Broadband amplified response. (b) Bandpass filter response.

Squarewave testing.   Squarewave testing is a technique used to test a circuit's ability to pass a broad band of frequencies. A squarewave consists of an infinite number of sinewave components (harmonics) that include the fundamental (the sinewave having the same frequency as the squarewave) and all odd multiples of the fundamental. If a circuit attenuates low- or high-frequency components, or both, then its output when driven by a square wave will be a distorted version of the squarewave. The shape and degree of the distortion reveal whether low- or high-frequency signals are attenuated, and the degree to which this signal rejection occurs. Figure 1-15 shows some representative output waveforms that result when either low or high frequencies are attenuated. In each case, the square wave frequency is f Hz. It is difficult to show exact waveforms because the shapes depend on the *rate* of attenuation caused by the circuit. The waveforms shown in the figure are based on a rate of 20 dB/decade. Sharply tuned circuits may cause an oscillatory (ringing) output when driven by a squarewave.

1.3 FREQUENCY COUNTERS

1.3.1 Theory of Operation

A frequency counter is a digital instrument used to measure and display the frequency of a signal applied to its input. Many frequency counters are also capable of measuring time intervals, as for example the duration of one cycle of an input signal (i.e., its period), and these are called *universal* counters.

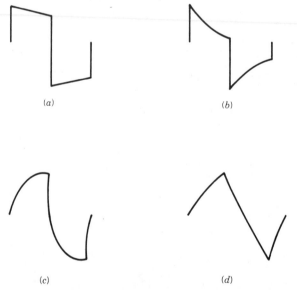

(a)   *(b)*

(c)   *(d)*

Figure 1-15   Square wave distortion resulting from attenuation of low or high frequencies in a test circuit. f = frequency of input square wave. (*a*) Frequencies below $0.1f$ are attenuated. (*b*) Frequencies below $0.5f$ are attenuated. (*c*) Frequencies above $10f$ are attenuated. (*d*) Frequencies above $5f$ are attenuated.

Every frequency counter has an interval oscillator, usually crystal controlled, that serves as a reference, or *time base,* for both frequency and time measurements. Figure 1-16 shows a simplified block diagram of a typical universal counter. The input signal is conditioned by an input amplifier and an adjustable attenuator control. It is then supplied to a Schmitt trigger that provides sharp pulse outputs synchronized with the rising and falling portions of the input waveform. An adjustable trigger level control (and possibly trigger slope control) is often provided so the user can set the points on the input waveform where the edges of the Schmitt trigger pulse occur. The input frequency dividers divide the frequency of the input signal by a user-selected decade multiple: 1, 10, 100, and so on. For frequency measurements, the dividers are not normally used, so when the "frequency-period" selector is in the "frequency" position, the pulses are applied directly to the main gate from the divide-by-one position. The time base oscillator output is divided by a selected decade, thus creating a pulse train whose period is a decade multiple of the original time base period. For frequency measurements, these longer-duration pulses set and reset the main gate flip-flop. The latter, operating in conjunction with the gate control logic, enables (or "opens") the main gate for corresponding time intervals. During each of these time intervals, the pulses derived from the input signal are gated into and counted by the counting register. The count is then displayed in a decimal readout.

To illustrate this method for determining frequency, suppose the input signal has frequency 1 MHz and the time base oscillator has frequency 10 MHz. If the time base frequency is divided by 1000, then the period of the result is 1000 times the period of the 10-MHz frequency, namely,

$$(10^3) \; 1/(10 \times 10^{-6}) = 0.01 \text{ seconds}$$

During this 0.01-second time interval, the main gate passes $(0.01 \text{ sec}) \, (10^6$ pulses/sec$) = 10^4$ pulses of the input, so the counter registers 10^4 (decimal). The frequency is the number of pulses counted divided by the total counting time, or $10^4/0.01 = 10^6$ Hz. In most counters, changing the time base frequency-dividing factor automatically positions the decimal point in the readout or changes the units to which it applied (Hz, kHz, or MHz).

For measuring periods, the "frequency-period" selector is placed in the "period" position, thus reversing the roles of the input signal and time base. In this case, the period of the input determines how long the main gate is enabled, and the time base pulses are counted by the counting register. Since the frequency of the counted pulses is known, the total number of pulses counted corresponds to a known period of time. For example, suppose the input has frequency 1 kHz and the time base frequency is 10 MHz. Each time base pulse then represents $1/10^7 = 10^{-7}$ second. The gate is open for one period of the input signal, that is, for $1/10^3 = 10^{-3}$ second. Therefore, the total number of pulses counted is $(10^{-3} \text{ second}) \, (10^7 \text{ pulses/second}) = 10^4$ pulses, corresponding to a time of $(10^4 \text{ pulses}) \, (10^{-7} \text{ second/pulse}) = 10^{-3}$ second, which is the period of the 1-kHz input. Some counters have a "period averaging" option, in which the main gate is enabled for multiple periods of the input rather than for just one period. The total time registered is then divided by the total

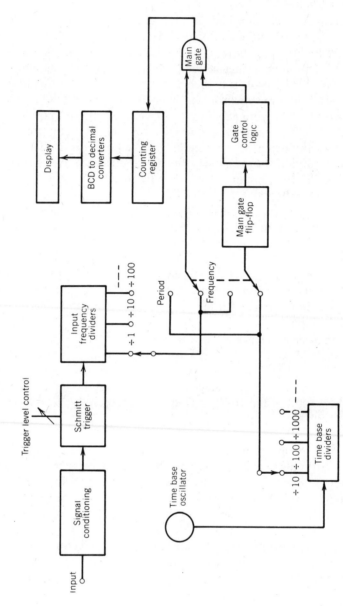

Figure 1-16 Block diagram of a universal frequency counter.

number of periods to obtain the average period. The input frequency dividers shown in Figure 1-16 are used for this purpose.

Another feature found in many counters is the capability of measuring the time between events. In this "stopwatch" mode, the counting of time base pulses is started by an externally applied start pulse and ceases on application of an externally applied stop pulse. Not shown in Figure 1-16, the start and stop inputs directly set and reset the main gate flip-flop, thus enabling the main gate and admitting pulses for the duration of time between those events.

An instrument having a *totalizing* capability can be used to count externally occurring events for a period of time selected by the user. This function is similar to frequency measurement, but the user enables and disables the main gate for whatever time is desired. Totalizing can be used to count randomly occurring pulses, mechanical shaft revolutions, and similar events.

Still another feature available in some counters is *ratio measurement,* in which two external signals are connected to the counter. The lower-frequency signal is used to enable the main gate while the pulses of the higher-frequency signal are counted. Thus, the higher-frequency signal replaces the function of the internal time base oscillator, and the latter is not used. In effect, the counter is used to measure frequency using an external time base.

1.3.2 Error Sources and Measurement Precautions

±1 Count Error   In most counters, the signal that causes the main gate to be enabled and disabled is not synchronized with the signal whose pulses are passed through the gate and counted. This lack of synchronization means that one pulse may or may not be counted, depending on the arbitrary timing of the two signals. Thus, the final count in both frequency and period measurements may be in error by ±1. Figure 1-17 illustrates how two different counts can result from the same pair of signals, depending on the time relationship of one with respect to the other.

In frequency measurements, the measurement error resulting from ±1

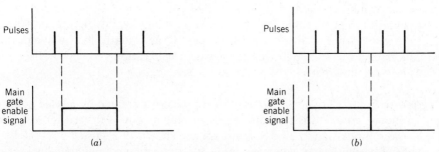

Figure 1-17   An illustration of ±1 count error. The main gate is enabled for the same time interval in each case, but two different counts result. (*a*) Two pulses counted. (*b*) Three pulses counted.

count error is inversely proportional to the frequency being measured. It can therefore be a serious source of error at low frequencies. For example, in Figure 1-17, suppose the main gate is enabled for 1 second. Then, the count obtained in Figure 1-17*a* would indicate a frequency of 2 Hz, while that obtained in 1-17*b* would indicate a frequency of 3 Hz. If the frequency is actually 2 Hz, and a count of 3 is registered, the resulting measurement error is 50%.

On the other hand, the measurement error resulting from the ±1 count problem in period measurements is inversely proportional to the interval time base frequency. Therefore, it is more accurate to determine the frequency of a low-frequency signal by measuring its period and calculating the reciprocal (frequency = 1/period) than it is to measure frequency directly. Depending on the design of the instrument and on the amount of frequency division used, there will be some frequency below which period measurements are more accurate and above which frequency measurements are more accurate. This "crossover" frequency equals the square root of the frequency of the time base if no frequency division is used on either the input signal or the time base. Some sophisticated modern counters that use the *reciprocal* frequency measuring technique actually measure the period and perform the reciprocal computation necessary to display frequency. Some instruments automatically switch to the reciprocal or direct method of frequency measurement, depending on which is the more accurate for the frequency being measured. Advanced, high-precision models eliminate the ±1 count error entirely, using a variety of techniques, including the "double-vernier" method. In the latter, the ambiguous time interval resulting from lack of synchronization is computed, and the final count is adjusted accordingly (see Reference 3).

Time Base Error   It is obvious that an accurate, highly stable time base oscillator is required for accurate frequency or period measurements. The most accurate models use an oven-controlled crystal oscillator to minimize the effects of temperature changes on oscillation frequency. Other models use a room temperature crystal oscillator (RTXO) or a temperature-compensated crystal oscillator (TCXO). The latter incorporates temperature-sensitive devices whose parameters vary with temperature in such a way that they cancel other parameter variations in the oscillator. Crystal oscillators tend to exhibit a cumulative frequency drift over a longer period of time due to aging. For a good RTXO, a typical drift is 3 parts per 10^7 per month, while a high-quality, oven-controlled crystal oscillator may drift only 1.5 parts per 10^8 per month. In any event, the manufacturer's recommendations for periodic calibration should be observed to ensure specified instrument accuracy.

Triggering Error   Triggering error refers to the error created by missed counts or extraneous counts arising from improper adjustment of the trigger level or noise in the input signal. The Schmitt trigger shown in Figure 1-16 switches state at one voltage level when the input is increasing and switches back at a different voltage level when the input is decreasing. The difference between these levels is called the *hysteresis* and determines the instrument's trigger sensitivity. Changing the trigger level effectively moves the hysteresis "window" up and down with respect to the input signal. Many counters have calibrated attenuation controls that effectively increase the sensitivity (enlarge the window), and many also have a variable trigger level control for moving the window.

Because of the fast switching times required in the input channel of a counter, the input must necessarily have broadband frequency characteristics. This in turn makes the input susceptible to noise and means that it is undesirable to have a very low trigger sensitivity or hysteresis. Consequently, the user must be aware of the limitations imposed by nonzero hysteresis and must set trigger levels in a manner that is optimum for the nature of the signal being measured.

If the input has a dc level, it is possible that one or both of the trigger points (the limits of the window) will be greater or less than any level reached by the input signal. This fact is illustrated in Figure 1-18a. In the example shown, the trigger level should be increased to move the hysteresis window into the ac region of the signal. Note, however, that the dc level may be so great that no trigger level adjustment will cure the problem, and increasing the input signal level (overdriving) only makes it worse. If this is the case, then the input should be ac coupled, a capability that is switch-selectable on many counters. However, ac coupling should not be used with signals that have a time-varying duty cycle because, in those cases, the trigger levels vary with the duty cycle.

Figure 1-18b shows how noise (in this case, ringing) can cause false counts to be generated due to improper setting of the hysteresis window. The remedy

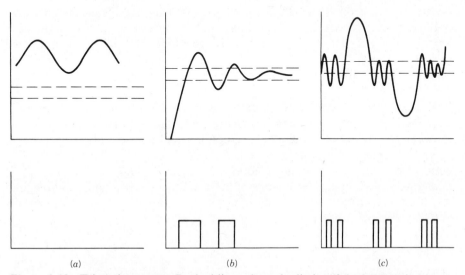

(a) (b) (c)

Figure 1-18 Triggering errors. Dashed lines show the limits of the hysteresis trigger window. (a) No pulses generated because of dc level in signal. (b) Extra pulse generated because of ringing. (c) Extra pulses generated by noise spikes.

in this case is to reduce the trigger level so the window falls below the ringing.

Figure 1-18c shows how noise "spikes" occurring at the input signal crossovers can create false counts. This problem can be remedied by setting the trigger level more positive or more negative, or by increasing the sensitivity so that the noise spikes do not cross the limits of the window.

1.3.3 Frequency Counter Specifications

Frequency range. The range of frequencies that can be measured. Different ranges may be specified for dc and ac coupling. Depending on cost and sophistication, the maximum frequency may be several MHz or several hundred MHz; 100 MHz is typical. Microwave counters may register up to several GHz.

Input impedance. The input impedance is typically 1 MΩ or 50 Ω (selectable on some units). The broadband nature of the input makes a very large input impedance undesirable, because of the consequent susceptibility to noise.

Sensitivity. The minimum signal level required to cause triggering. Typical values range from 10 mV RMS to 100 mV RMS. The specification may vary with frequency range and may be given as a peak-to-peak value for nonsinusoidal waves. Attenuator controls increase the sensitivity by a multiplying factor. Typical attenuator settings are labeled $\times 1$, $\times 10$, $\times 100$. If the sensitivity is 20 mV RMS in the $\times 1$ position, then it is 200 mV RMS in the $\times 10$ position.

Trigger level. The dc level that can be added to the sensitivity to raise or lower the hysteresis "window" through which the signal must pass (the sensitivity) in order to cause triggering. A typical range through which the trigger level can be set is ± 3 V.

Resolution. The smallest change in the measured quantity that will cause a change in the display (readout). The resolution is usually quoted as $\pm$ LSD (least significant digit), which means ± 1 in the least significant decimal digit of the display. The corresponding change in the value of the measured quantity must then be calculated, based on the decade divider setting and the units that this setting infers for the display. For example, in period measurements, a typical value is 100 nsec/N, where N is the decade divider step (1, 10, 100, etc.).

Time base characteristics. Specifications for the internal time base oscillator include its frequency and the change in frequency that can be expected to occur as a result of aging and temperature, and possibly also line voltage. A typical frequency is 10 MHz. Specifications for frequency change vary widely with instrument quality, manufacturer, and cost. Typical specifications for a moderately priced unit are: change of <3 parts in 10^7 per month due to aging; $< \pm 10$ parts in 10^6 over a temperature range of 0 to 50°C; $< \pm 1$ part in 10^7 for $\pm 10\%$ variation in line voltage.

1.4 REFERENCES

Carr, J. J., *Elements of Electronic Instrumentation and Measurement*. Reston, VA: Reston Publishing, 1979.

Cooper, W. D., *Electronic Instrumentation and Measurement Techniques*. Englewood Cliffs, NJ: Prentice-Hall, 1970.

Hewlett-Packard, *Fundamentals of the Electronic Counters*. Application Note 200, Hewlett-Packard, Palo Alto, CA: 1978.

Jones, L., and A. Foster Chin, *Electronic Instruments and Measurements*. New York: Wiley, 1983.

Kantrowitz, P., Kousourou, G., and Zucker, L., *Electronic Measurements*, Englewood Cliffs, NJ: Prentice-Hall, 1979.

Lenk, J. D., *Handbook of Electronic Meters, Theory and Applications*. Englewood Cliffs, NJ: Prentice-Hall, 1981.

Lenk, J. D., *Handbook of Practical Electronic Tests and Measurements*. Englewood Cliffs, NJ: Prentice-Hall, 1969.

CHAPTER 2

Oscilloscopes

Lowell L. Winans
Heath Company

2.1 INTRODUCTION

The oscilloscope is perhaps the most unique troubleshooting instrument. It visually displays both voltage and time. This allows the operator to see the amplitudes of voltages and shapes of signals so that their average, peak, RMS, or peak-to-peak values can be easily determined. Also, frequency and phase relationships can be measured; and with a dual-trace oscilloscope, critical timing relationships between two signals can be compared.

2.2 HOW IT WORKS

2.2.1 Basic Oscilloscopes

Figure 2-1 is a block diagram of a basic oscilloscope. It contains a vertical amplifier, horizontal amplifier, sweep generator, cathode ray tube (CRT), and high- and low-voltage power supplies.

Vertical Amplifier   This amplifier increases the input signal from as low as 1 mV to the desired value to be displayed on the CRT.

Horizontal Amplifier   This amplifier has two inputs. One, from the sweep generator, produces a horizontal trace across the CRT. The other input allows

Note: Portions of this Chapter are reprinted with permission of Heath Company, Benton Harbor, MI, from the following documents: IO-4235 (595-2237-02) 35 MHz Dual Trace Oscilloscope Copyright 1979. EE-3105-02 (595-2230-02) Test Equipment, Oscilloscopes Copyright 1979.

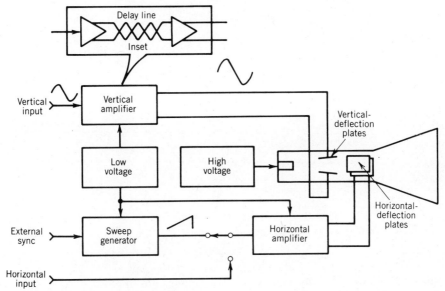

Figure 2-1   Oscilloscope block diagram.

an external signal to be amplified and displayed on the CRT only on the horizontal axis.

Sweep Generator   The output from the sweep generator is a linear ramp. This, when applied to the horizontal deflection plates of the CRT, causes the CRT electron beam to be driven from left to right across the CRT. As the beam travels, the signal on the vertical deflection plates causes the beam to move up and down, and thus displays the shape of the input signal.

When the trace reaches the right side of the CRT, the ramp voltage drops to its starting point, which causes the electron beam to return to the left side of the CRT. The beam returning across the screen (called *retrace*) is not seen because internal circuitry turns off the CRT (called *blanking*) during this time.

Triggering   In today's oscilloscopes, the sweep is started by a trigger signal as shown in Figure 2-2. When the incoming signal exceeds the trigger level, a trigger pulse is generated which starts the sweep generator. The sweep time is selected by the operator. In this case, it is equal to one cycle of the input signal.

The triggering can be selected to start on either the positive or negative (+/−) portion of the input signal. Also, a *LEVEL* control allows the sweep to be started at any point on the selected portion of the waveform.

Some inexpensive oscilloscopes are not triggered, but have a recurring sweep as shown in Figure 2-3. Here the sweep continues to operate as long as the circuit has power. This type of display has the major disadvantage that it can be difficult to make the display stationary on the screen.

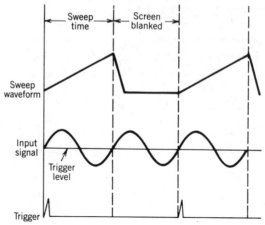

Figure 2-2 Triggered sweep signals.

Power Supplies Well-regulated power supplies are important. Improper reg-ulation can cause errors in time or frequency measurements, brightness and focus can be affected, and both vertical and horizontal sensitivities can be significantly changed.

2.2.2 Advanced Oscilloscopes

Advanced oscilloscopes have more sophisticated features. Delay lines allow the entire leading edge of a signal to be displayed, while the magnifier and delayed time base features allow a selected portion of the trace to be expanded and viewed. The dual-beam, or dual-trace, oscilloscope permits two signals to be viewed at once. The storage oscilloscope displays signals with low re-occurrence rates, and the sampling oscilloscope displays extremely high-fre-quency signals.

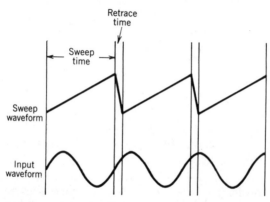

Figure 2-3 Free-running sweep signals.

Delay Lines   The incoming signal first triggers the sweep generator. Then, the signal is coupled through a delay line (see the inset drawing in Figure 2-1), amplified, and displayed on the CRT. In an oscilloscope without delay lines, a portion of the first leading edge of the signal is not displayed because it has already happened by the time the sweep circuits start the display. However, with delay lines in the vertical-amplifier circuits, the leading edge of the signal (the part that triggered the sweep in the first place) arrives later and is therefore displayed.

Magnifier   The magnifier, or expanded sweep feature, enlarges a portion of the waveform for closer observation. This enlargement is usually by a factor of 5 or 10.

Figure 2-4a shows a normal display. The sweep is 10 divisions long as shown. To see more detail of the wave shape, the length of the sweep in Figure 2-4b has been increased to 50 divisions. The sweep time remains the same; only the length is changed. Then, by using the horizontal-position control, the waveform can be moved until any portion of the waveform is visible on the CRT. This effectively increases the sweep speed of the selected portion of the waveform.

The same result could not be obtained by changing the time-per-division control. This would allow the first portion of the waveform to be expanded, but the second and third portions of the waveform would be pushed off the screen. Also, a portion of the leading edge of the first pulse would be lost, depending on the setting of the trigger level.

Magnifiers do have drawbacks. Since the magnification is obtained by increasing the gain of the horizontal amplifier, it is limited to about 10. Even at this magnification, it is sometimes difficult to locate the exact portion of the waveform to be displayed. Also, any errors in the magnification are added to the normal timing errors, which may approach 10%.

(a)

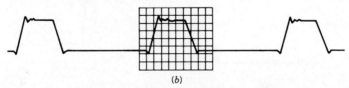

(b)

Figure 2-4   Magnifier feature. (a) Normal display. (b) Length of sweep increased to 50 divisions.

Delayed Sweep   Delayed sweep uses a separate time base to generate a faster sweep, which displays a selected portion of the normal waveform. Since both the start and the length of the delayed sweep can be controlled by the operator, any given portion of the display may be examined in detail. Figure 2-5 shows a delayed-sweep system. Waveform (a) is applied to the vertical input of the oscilloscope, and the normal-sweep waveform is as shown at (b). Under these conditions, all of the waveform that occurs from T_0 to T_3 will appear as shown in the "normal display." If we want to take a closer look at the glitch that appears after the third pulse, the delayed-trigger-start point would be set to a point just before the glitch occurs. When the normal-sweep voltage passes through the delayed-trigger level, a trigger is generated which starts the delayed sweep. A delayed-sweep time is selected that will allow just the desired portion of waveform to be displayed. In this case, the delayed-sweep time is between T_1 and T_2. When delayed sweep is selected, the normal sweep is disconnected from the CRT and the delayed-sweep circuits are used for the display, which will appear as shown in "delayed display." Selecting the portion of the display to be studied is usually done by intensifying the desired portion of the display. As shown in (d), intensifying is done by increasing the voltage on the control grid of the CRT. The intensified portion is then used to produce the delayed

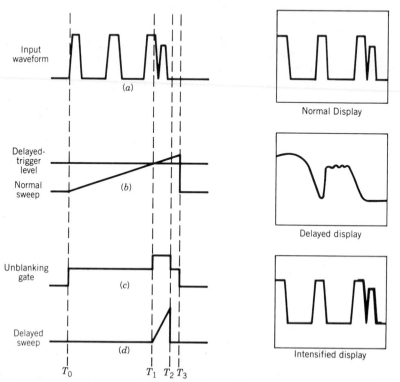

Figure 2-5   Delayed-sweep signals. (a) Input waveform. (b) Delayed-trigger level, normal sweep. (c) Unblanking gate. (d) Delayed sweep.

sweep (*d*). By varying the delay time and the delayed time per division, any portion of the input waveform can be expanded. This feature allows you to select a portion of a waveform and then accurately measure or observe it.

Dual-Trace Oscilloscopes   There are two ways of producing two traces on the CRT. One is with a *dual-beam oscilloscope,* which effectively has two electron guns in it, and the other method is by electronically controlling a single beam to make it produce two displays. This second is more common and is called *dual-trace oscilloscope.* Dual-trace displays from a single electron beam can be obtained by two methods. One is *chop,* and the other is *alternate.* The chop method "time shares" between two vertical amplifiers at a very rapid rate. As shown in Figure 2-6, the multivibrator causes the electronic switch to switch back and forth between channel A and channel B. As shown in Figure 2-7, input A is displayed for a time and then input B is displayed. It would appear that the chopped display might cause viewing problems. However, this is rarely the case and only occurs when the chop rate is an exact multiple of the input frequency. The alternate method uses the same circuits as shown in Figure 2-7. However, this time the switch connects each channel for the duration of one entire sweep. This way, there is no chopping between channels during a sweep, and the channel signal can be cleanly displayed. If the alternate mode is used at low sweep rates, the operator will see first one trace and then the other. To eliminate this annoyance, the chop mode can be used at low frequencies and the alternate mode, at higher frequencies. Some oscilloscopes will automatically switch at a predetermined point.

Storage Oscilloscope   The most common type is the *variable persistence/ storage cathode ray tube* storage oscilloscope. The CRT of this type of oscilloscope, in addition to normal CRT parts, contains a flood gun, collimator, collector mesh, and storage mesh as shown in Figure 2-8.

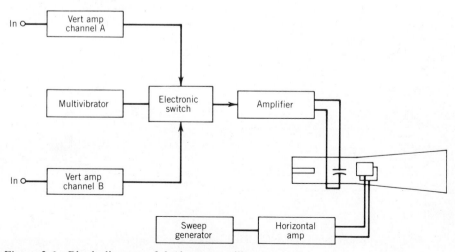

Figure 2-6   Block diagram of dual-trace oscilloscope.

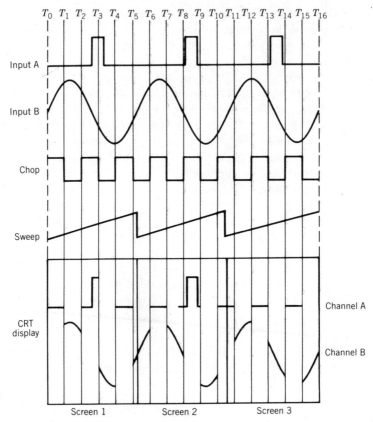

Figure 2-7   Dual-trace signals (chop method).

The normal electron gun is the *write* gun that emits a high-velocity stream of electrons. The flood gun emits a low-velocity flood of electrons that covers the entire screen. The collector mesh is a fine wire screen mounted parallel to the face of the CRT. The storage mesh is similar to the collector mesh but is mounted closer to the CRT face and is covered with insulative material. The material is such that it gives off electrons and becomes positively charged in

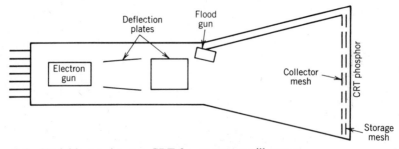

Figure 2-8   Variable-persistence CRT for storage oscilloscope.

that area when it is struck by high-velocity electrons. The result is that a high-velocity electron will penetrate the storage mesh, cause the CRT phosphor to glow, and the storage charge on the mesh to remain for some time causing the low-velocity flood of electrons to strike the CRT and maintain the display.

Sampling Oscilloscope   The sampling oscilloscope starts at a reference point and samples the amplitude of the incoming signal at differently timed intervals each time the signal passes through the reference point. After the sampling is complete, the sample values are recalled from memory and plotted on the CRT. The result is a reasonable reproduction of an input waveform that would otherwise be too fast to be displayed. The major drawback of this system is that the input signal must be repetitive. Figure 2-9 shows a repetitive signal being sampled and its resultant display.

Digital Storage Oscilloscopes   Digital oscilloscopes, which use digital memory for storage, are an increasingly popular evolution of the sampling oscilloscope. The samples of the input signal are digitized with an analog-to-digital converter and stored in memory. The storage information is then displayed on command by the user. The stored information is converted to a CRT driving signal by a digital-to-analog converter.

Many digital oscilloscopes can simultaneously store and display input signals in real time, up to the limit of their analog bandwidth. Most digital oscilloscopes can capture and display transient signals. Digital oscilloscopes are available with analog bandwidths as high as 50 MHz, with the ability to

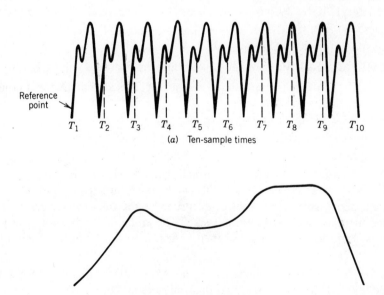

Figure 2-9   Sampling oscilloscope waveforms. (*a*) Ten-sample times. (*b*) Resultant display.

display transient events as short as 10 ns in duration, and sampling rates as high as 200 Megasamples/sec. Most digital oscilloscopes have more modest specifications, but their unique ability to store information in digital form has made them popular for many measurement applications.

2.3 CONTROL FUNCTIONS

The following paragraphs explain the major control and switch functions found on most oscilloscopes. As an example, Figure 2-10 shows the Heath IO-4235 35MHz dual-trace oscilloscope. This oscilloscope is not only dual trace but has the vertical-channel delay lines and magnifier features. Oscilloscopes with lesser features operate basically the same but do not have as many controls and switches. Other oscilloscopes however may have interchangeable vertical amplifiers and time bases for particular measurement needs. The numbered paragraphs below correspond to the circled numbers in Figure 2-10.

1. INTENSITY. Clockwise rotation increases the brightness of the display.
2. FOCUS. Varies the size of the beam striking the face of the CRT. Adjust for the sharpest display.
3. ON–OFF. Turns the oscilloscope on and off.
4. POWER INDICATOR. Glows when ac power is connected and turned on.

CHANNEL Y1 CONTROLS

5. Y_1 POSITION. Positions the channel Y_1 trace vertically on the screen.
6. DC BAL. This is not a normal operating control. Use it when necessary to touch up the calibration.
7. PULL TO INVERT. When pulled out, the Y_1 trace will be inverted.
8. Y_1 INPUT. This is the Y_1 input connector. It is also the X input connector during X–Y operation.
9. VOLTS/CM. Each position of this attenuator switch is marked for the number of volts (peak to peak) required to produce a pattern 1 cm high on the graticule.
10. VARIABLE. This control is normally operated in its fully clockwise (CAL) position where the VOLTS/CM switch positions are calibrated. Vertical gain decreases as the control is turned counterclockwise, permitting adjustment of the vertical-trace size. The trace will not show the correct volts/cm, however, because the control is no longer in the calibrated position.
11. AC-GND-DC (Input Switch). In the AC position, this switch blocks the dc level of the input signal so that only the ac component is displayed. In the GND position, the input is disconnected and the vertical amplifier input is grounded. Use this position when you wish to set the baseline (trace) at

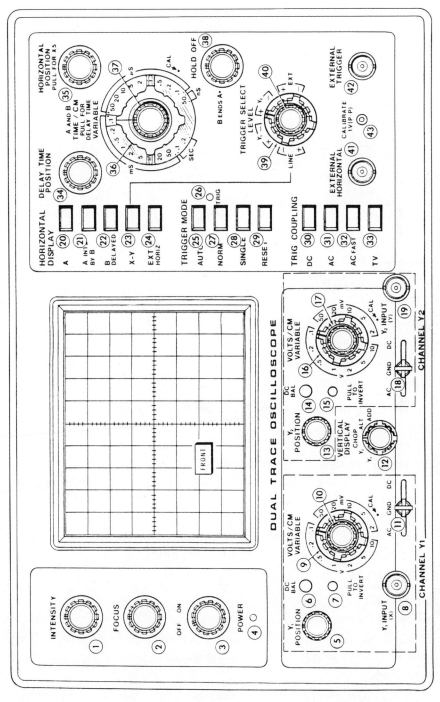

Figure 2-10 Oscilloscope control locations.

a desired position without disconnecting the input signal. In the DC position, both the dc and ac components of the input signal are displayed.

BOTH CHANNELS

12. VERTICAL DISPLAY. In the Y_1 position, only channel Y_1 is displayed on the screen. In the Y_2 position, only channel Y_2 is displayed. In the CHOP position, both Y_1 and Y_2 are displayed. In the CHOP position, both Y_1 and Y_2 are displayed in the CHOP mode. In the ALT (alternate) position, Y_1 and Y_2 are displayed alternately at the end of each trace. In the ADD position, the algebraic summation of channels Y_1 and Y_2 is displayed.

CHANNEL Y_2 CONTROLS

The following channel Y_2 controls perform the same functions as the corresponding channel Y_1 controls, only for channel Y_2:

13. Y_2 POSITION.

14. DC BAL.

15. PULL TO INVERT.

16. VOLTS/CM.

17. VARIABLE.

18. AC-GND-DC (Input Switch).

19. Y_2 INPUT.

HORIZONTAL DISPLAY

20. A. Selects normal, single-time-base operation.

21. A INTEN BY B. Intensifies the trace during the time that the B time base is running.

22. B DELAYED. Selects the B time base.

23. X–Y. Selects the X–Y function when the TRIGGER SELECT switch is in one of the LINE positions.

24. EXT HORIZ. When this function is selected, the voltage at the EXTERNAL HORIZONTAL input connector will move the trace horizontally. The voltage of the Y_1 or Y_2 input (or both) will move the trace vertically.

TRIGGER MODE

25. AUTO. When selected, a baseline will be automatically displayed in the absence of a trigger signal.

26. TRIG. Indicates when the oscilloscope is being triggered.

27. NORMAL. When selected, a trace will appear only when the oscilloscope is triggered.

28. SINGLE. The oscilloscope will sweep only once when triggered and then wait until after it is RESET. Then, on the next trigger signal, it will sweep again.

29. RESET. Resets the single-sweep function.

TRIG COUPLING

30. DC. Allows triggering on a dc or very slow ac signal.

31. AC. Triggers on most normal signals, but blocks dc and very low-frequency ac signals.

32. AC FAST. Rejects low-frequency signals and triggers only on high-frequency signals.

33. TV. Rejects high-frequency trigger signals and triggers on alternate field trigger signals. This provides a stable display when viewing vertical TV signals.

TIME BASE

34. DELAY TIME POSITION. Determines the starting point of the B time base. When in the A INTEN BY B mode, the control will position the intensified portion of the trace. When in the B DELAYED mode, the control determines the portion of the A sweep to be expanded.

35. HORIZONTAL POSITION, PULL FOR X5. Positions the trace horizontally on the CRT. It is also the X position control in the X–Y mode. When pulled out, it expands the trace by a factor of 5.

36. A AND B TIME/CM, PULL FOR DELAY TIME. Determines the time required for the beam to sweep. The clear knob (closest to the panel) sets the speed of the A time base; the middle knob selects the time for the B time base. The two knobs are normally locked together. However, if the center knob is pulled out and turned clockwise, then the A and B time bases will be at different settings. Then, when the two knobs are aligned again, they will automatically lock together.

37. VARIABLE. Provides a continuous adjustment of the A sweep time between time base ranges. When fully clockwise, the time base is calibrated.

38. HOLD OFF. Adjusts for a stable display if the input signal is an exact multiple of the time base repetition rate. Normally, it is in the full counterclockwise position. In the full clockwise position (which is the B ENDS A position), the A sweep will be stopped when the B sweep stops (in the A INTEN BY B or B DELAYED modes). This increases the writing speed and produces a brighter display, especially at low settings of the DELAY TIME POSITION control.

TRIGGER SELECT

39. TRIGGER SELECT. This control selects the source and polarity of the triggering signal:

Line $(+/-)$	Trigger signal is a portion of the 60-Hz line frequency.
Y_1 $(+/-)$	Triggers on a signal from channel Y_1 only.
Y_2 $(+/-)$	Triggers on a signal from channel Y_2 only.
EXT $(+/-)$	Triggers on a signal applied from an external source.

40. LEVEL. Adjusts the trigger circuits so the sweep can be started at any position on the input signal waveform. The sweep can be started on either a positive or negative slope, depending on the position of the TRIGGER SELECT switch.

EXTERNAL

41. EXTERNAL HORIZONTAL. With the EXT HORIZ switch pushed in, a positive voltage at this connector will cause a left-to-right deflection on the CRT.

42. EXTERNAL TRIGGER. An external signal can be applied through this connector to trigger the sweep circuits when the TRIGGER SELECT switch is in either EXT position.

CALIBRATE

43. CALIBRATE. This 1-V (peak-to-peak) squarewave signal (approximately 1000 Hz) can be used to periodically check vertical calibration. The rise time of this signal allows it to also be used for oscilloscope probe compensation.

2.4 MAKING MEASUREMENTS

2.4.1 Probes

The oscilloscope probe can be crucial to making accurate measurements. The typical oscilloscope has a high input resistance shunted by a low value of capacitance. When the oscilloscope is connected to a circuit, it will load the circuit to some degree. The loading effect depends on the ratio of oscilloscope input resistance (R_{in}) to signal source resistance (R_g) as follows:

$$\text{Percent error} = \frac{R_g}{R_g + R_{in}} \times 100$$

Therefore, if R_{in} is 10 times R_g, there is less than a 10% loading error. For a 1% loading error, R_{in} has to be approximately 100 times R_g.

However, input frequency greatly affects the loading error. For example, if the input capacitance (C_{in}) is 20 pF, the loading effect for the following frequencies f is as follows:

f	$X_c = 1/2\pi f_C$
500 Hz	16 MΩ
1 kHz	8 MΩ
10 kHz	800 kΩ
100 kHz	80 kΩ
1 MHz	8 kΩ
10 MHz	800 Ω
100 MHz	80 Ω

Thus, the effect of C_{in} is often more important than the effect of R_{in}. Consequently, the percentage error is

$$\% \text{ Error} = \frac{Z_{in}}{Z_{in} + Z_g} \times 100$$

where Z_{in} is the oscilloscope input impedance and Z_g is the signal source impedance.

Cable Capacitance   Usually, a coaxial cable is used to connect the circuit under test to the oscilloscope. This keeps 60-Hz line frequency and stray rf fields from being picked up. However, this type of cable has 28.5 pF of capacitance per foot. If a 3-foot cable is used, at 1 MHz, the total capacitive reactance of both the cable and oscilloscope is a low 1509 ohms. Then, if this scope lead is connected to a circuit that has an R_g of 1000 ohms, the measured voltage will only be 60% of the actual voltage. Should the input frequency be 10 MHz, then X_c drops to 151 ohms and the measured voltage is only 13% of the actual value. This is obviously unacceptable. For more accurate measurements at high frequencies, a $\times 10$ probe should be used.

$\times 10$ Probe   Because the cable capacitance cannot be eliminated, it must be compensated for. To do this, a capacitor (C_p) is placed in series with the scope input as shown in Figure 2-11. If C_p is one-ninth the value of C_c and C_{in} combined, the result will be a 10:1 capacitive voltage divider. Now, at 1 MHz, X_c is 15,157 ohms, which is 10 times the previous X_c of 1508 ohms, and results in a loading error of only 6%. Because there is a 10:1 voltage division between C_p and $C_{in} + C_c$, the volts/cm setting must be multiplied by ten to get the actual volts per division. As the input frequency decreases, the reactance increases until, at some frequency below 1 MHz, R_{in} becomes the determining factor for Z_{in}. Therefore, while X_{cp} increases, Z_{in} will tend to remain constant, and the voltage division ratio will change. To overcome this ratio change, a resistor is added in parallel with C_p as shown in Figure 2-12. If this resistor is nine times the value of R_{in}, the voltage ratio will be essentially constant at all frequencies. For the division ratio to be constant, the probe must be properly compensated

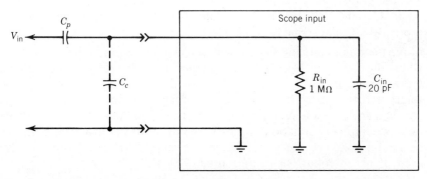

Figure 2-11   Equivalent circuit of oscilloscope and probe.

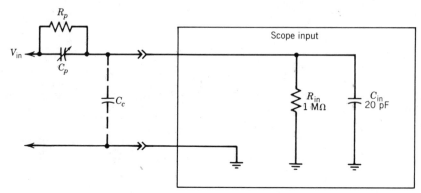

Figure 2-12 Equivalent circuit with probe resistor added.

so that the RC time of the probe equals the RC time of the scope input and cable capacitance combined. This compensation is usually done by adjusting C_p (a variable capacitor) until the corners of a displayed incoming squarewave are square, and no longer spiked or rounded, that is, C_p too high or too low, respectively.

2.4.2 DC Voltage

To measure dc voltage, the oscilloscope must have a dc input:

1. Set the VOLTS/CM switch to produce approximately five full divisions of deflection on the screen's graticule. (If you have no idea how much voltage will be measured, then set the switch at 1 volt/cm and readjust the switch later.) Be sure the VARIABLE control is in its CAL position.

2. Establish the ground reference. Either connect the probe leads together, or set the input switch to GND if your oscilloscope has an input switch with a GND position. Then, use the VERT POS control and position the trace toward the top of the screen if the voltage will be negative, toward the bottom of the screen if the voltage will be positive, or in the center of the screen if you don't know if the voltage will be positive or negative.

3. Use the proper probe and connect the voltage to be measured to the vertical input. Use the ×10 probe if you don't know which one to use. This will protect the oscilloscope from high voltage.

4. Move the input coupling switch from the GND position to the DC position, adjust the VOLTS/CM switch as necessary, and adjust the VERT/POS control as necessary.

5. Interpret the display by counting the number of major divisions on the screen's graticule and multiplying this number times the VOLTS/CM switch times the probe attenuation, normally ×1 or ×10. For instance, if the total deflection is 4.5 cm, the VOLTS/CM is set at 0.5, and a ×10 probe is being used, then the voltage being measured (V_{in}) is

$$V_{in} = 4.5 \times 0.5 \times 10 = 22.5 \text{ V dc}$$

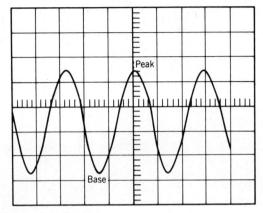

Figure 2-13   Peak-to-peak voltage.

2.4.3 Peak-to-Peak Voltage

To measure the peak-to-peak voltage of a signal, first make sure the VARI-ABLE control is turned to the CAL position. Then, multiply the vertical deflection (number of lines on the screen's graticule) by the setting of the VOLTS/CM switch.

EXAMPLE: As shown in Figure 2-13, the displayed amplitude is 4.2 cm. If the VOLTS/CM switch is at 0.2 V, then

$$\text{peak-to-peak voltage} = 4.2 \text{ cm} \times 0.2 \text{ V/cm} = 0.84 \text{ V}$$

If the measured waveform is a sinewave, then the RMS or average value can be calculated by the following formula:

$$\text{RMS} = 0.354 \times \text{peak-to-peak value}$$

and

$$\text{Average} = 0.318 \times \text{peak-to-peak value}$$

If the waveform is a squarewave, the RMS and average values are the same. The average value may be found by multiplying the peak value times the duty cycle. Duty cycle (DC) is a ratio of pulse width (PW) to pulse repetition time (PRT). Figure 2-14 shows an example squarewave. If the peak voltage is 10 V, the PW is 10 μsec, and the PRT is 30 μsec, then the voltage is present for 10 μsec and absent for 20 μsec. Therefore, the duty cycle is

$$\text{DC} = \frac{\text{PW}}{\text{PRT}} = \frac{10}{30}$$
$$= 0.333$$

From this, the average or RMS value can be found by

$$\text{Average} = \text{PP} \times \text{DC} = 10 \times 0.333$$
$$= 3.33 \text{ V}$$

If a waveform is other than a sinewave or squarewave, it is very difficult to find the average or RMS value. It is seldom necessary, and most troubleshooting charts show a picture of the waveform with peak values given.

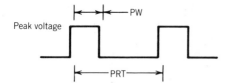

Figure 2-14   Square wave peak voltage.

2.4.4 Instantaneous Voltage

Instantaneous voltage is the voltage at any point on the waveform with respect to ground. To measure it,

1. Set the AC-GND-DC switch to GND and adjust the trace to some reference line. See Figure 2-15.

2. Set the AC-GND-DC switch to DC. If the waveform is above the reference line, the voltage is positive. If the waveform is below the reference line, the voltage is negative.

3. Count the number of divisions from the reference line to the point that you want to measure on the waveform. Then, multiply this value by the setting of the VOLTS/CM switch.

EXAMPLE: The VOLTS/CM switch is at 1 V/cm, and a ×10 probe is being used:

   Instantaneous voltage = vertical deflection × V/cm × probe attenuation

Thus,

$$3.2 \times 1 \times 10 = 32 \text{ V}$$

2.4.5 Time Duration and Frequency

The time duration between any two points on a waveform is found by multiplying the horizontal distance between the two points by the setting of the

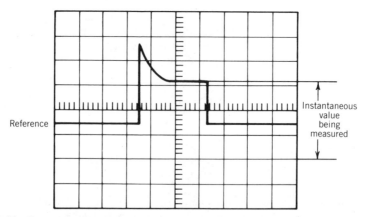

Figure 2-15   Instantaneous voltage.

TIME/CM switch. Frequency is the reciprocal of the time duration of one cycle.

EXAMPLE: Refer to Figure 2-16. The horizontal distance measured is 8.3 cm. The TIME/CM switch is set to 2 msec/cm. Therefore,

$$\text{time duration} = \text{number of divisions} \times \text{sweep setting}$$
$$= 8.3 \text{ cm} \times 2 \text{ msec/cm}$$
$$= 16.6 \text{ msec}$$

and

$$\text{frequency} = \frac{1}{\text{time duration}} = \frac{1}{16.6 \text{ msec}}$$
$$= 60 \text{ Hz}$$

2.4.6 PULSE MEASUREMENTS

Figure 2-17 shows the various times of a pulse waveform:

- T_r = rise time. The time required for the voltage to change from 10% of V_{max} to 90% of V_{max}.

- T_f = fall time. The time required for the voltage to decrease from 90% of V_{max} to 10% of V_{max}.

- *PW* = pulse width. How long the voltage remains above the 50% point.

- *PRT* = pulse repetition time. The time from the 50% point on the leading edge of one pulse to the 50% point on the leading edge of the next pulse. The *PRT* of a pulse waveform is analogous to the period of a sinewave. Therefore, the number of pulses-per-second (known as the pulse repetition frequency, *PRF*), can be calculated by

$$PRF = \frac{1}{\text{PRT}}$$

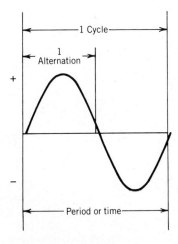

Figure 2-16   Measuring time.

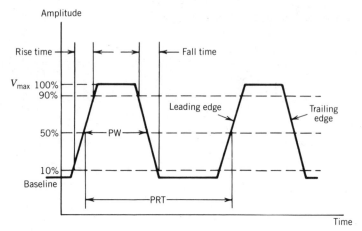

Figure 2-17   Pulse measurements.

See Figure 2-18 for some common types of distortion. This distortion must be compensated for when making pulse measurements by measuring from the baseline and 100% points.

When measuring very fast rise times, the rise time of the oscilloscope becomes a determining factor in the accuracy of the measurement. This is shown by

$$T_{rd} = T_{rs} + T_{rw}$$

where: T_{rd} equals the displayed rise time, T_{rs} equals the rise time of the oscilloscope, and T_{rw} is the rise time of the waveform being measured. Thus, to find the rise time of the waveform being measured, use

$$T_{rw} = T_{rd} - T_{rs}$$

EXAMPLE: If $T_{rd} = 120$ nsec and $T_{rs} = 70$ nsec, then

$$T_{rw} = T_{rd} - T_{rs} = 120 \text{ nsec} - 70 \text{ nsec}$$
$$= 50 \text{ nsec}$$

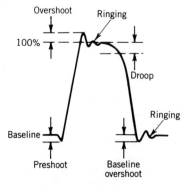

Figure 2-18   Square wave distortion.

If the displayed rise time is at least three times the oscilloscope rise time, then the rise time of the oscilloscope may be ignored and the measurement will be within 5%. If the displayed rise time is five times the oscilloscope rise time, then the accuracy will be within 2%. However, if the displayed rise time approaches the rise time of the oscilloscope, then an accurate measurement is impossible to obtain. Fast rise time may not be necessary to measure other waveform parameters such as *PW* and *PRT*.

2.4.7 X–Y Measurements

X–Y Measurements are made with a normal signal at the vertical input; but, instead of a linear ramp, a signal from an external source is applied to the horizontal input. If both input signals are sinewaves of the same frequency, amplitude, and phase, they produce a display as shown in Figure 2-19. However, if the signals are 90 degrees out of phase, then the display will be a circle as shown in Figure 2-20.

When the two inputs are of different frequencies, a more complicated display is produced, as shown in Figure 2-21. This display results when the horizontal signal is twice the frequency of the vertical signal. These displays are called *Lissajous figures* and can be used to compare the phase and frequency differences between two waveforms. Usually, the standard signal is applied to the horizontal input and the unknown signal is applied to the vertical input.

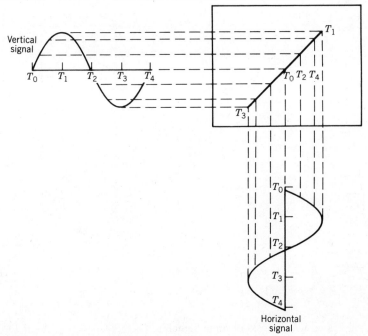

Figure 2-19   X–Y measurements.

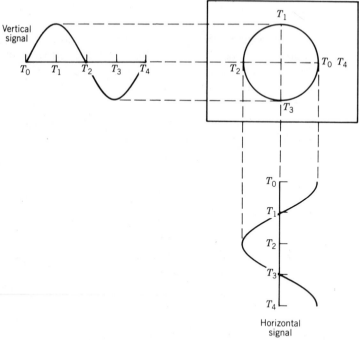

Figure 2-20   Inputs 90° out of phase.

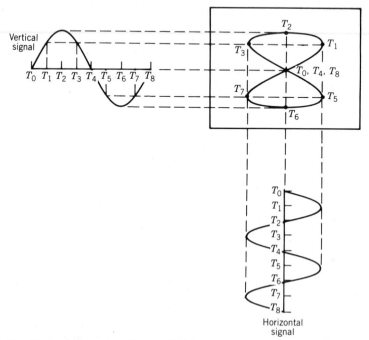

Figure 2-21   Two-to-one ratio of input signals.

Phase Measurements   You have already seen how to determine if signals are in phase or 90 degrees out of phase. Other angles can be calculated from two simple measurements. See Figure 2-22. If ϕ is the difference in phase between two signals,

$$\sin \phi = \frac{AB}{X} = \frac{CD}{Y}$$

In Figure 2-22, *AB* equals about 10 divisions and *X* equals about 14 divisions. Therefore,

$$\sin \phi = \frac{10}{14}$$
$$= 0.714$$

A table of trigonometric functions will show that the angle that corresponds to a sine value of 0.714 is nearest to 46 degrees. Thus, the apparent difference between the two input signals is 46 degrees. Actually, it may be 46, 136, 226, or 316 degrees, as this procedure does not tell us which quadrant the angle is in.

Part of the problem can be resolved by looking at Figure 2-23. Here, (*a*) is a 0- or 360-degree difference, (*b*) is 45 or 315 degrees, (*c*) is 90 or 270 degrees, (*d*) is 135 or 225 degrees, and (*e*) is 180 degrees.

Frequency Measurements   The Lissajous pattern can be used for frequency measurements, though it can be difficult and require considerable skill.

Figure 2-24 shows some typical displays when the vertical signal is less than the horizontal signal. If the reference frequency is applied to the horizontal input, then the unknown frequency can be found by dividing the reference frequency (f_r) by the number of horizontal loops (N_h) and then multiplying by the number of vertical (N_v) loops. For instance, *C* in Figure 2-24 has four horizontal loops and one vertical loop. If the reference frequency is 1000 Hz, then the unknown frequency will be

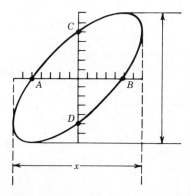

Figure 2-22   Phase measurements.

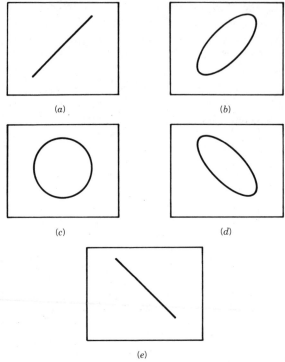

Figure 2-23   Phase relationships.

$$f_u = \frac{f_r}{N_h} \times N_v = \frac{1000}{4} \times 1$$
$$= 250 \text{ Hz}$$

2.4.8 Dual-Trace Measurements

The major application of the dual-trace oscilloscope is to compare two signals for phase, shape, amplitude, timing, and so forth.

Phase Measurements   Set the oscilloscope for dual-trace operation. Then, adjust the TIME/CM switch and HORIZONTAL GAIN control so that either one alternation or one complete cycle of the reference waveform requires exactly nine major horizontal divisions. Either channel can be used for a ref-

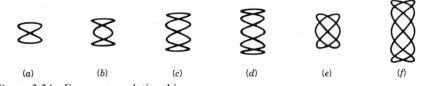

(a)   (b)   (c)   (d)   (e)   (f)

Figure 2-24   Frequency relationships.

erence; however, the oscilloscope should be triggered off the reference waveform. Also, it is normal to use the lower-frequency signal when the inputs are of a different frequency. The horizontal sweep does not need to be calibrated for this measurement, so the horizontal gain control can be adjusted as necessary for a proper display.

Figure 2-25 shows a typical display with one alternation equal to nine divisions. Since one alternation is equal to 180 electrical degrees, each division will equal 180/9, or 20 degrees. Thus, in the waveform shown, waveform A leads waveform B by three divisions, for 60 degrees. However, you must be sure that both waveforms are vertically centered.

Time Difference   To make time difference measurements, the time base control must be in the CALIBRATE position. See Figure 2-26. The time difference between the two pulses is the number of divisions from the leading edge of the first pulse to the leading edge of the next pulse multiplied by the TIME/CM setting. For example, in Figure 2-26, the difference between leading edges of the waveforms is three divisions. If the TIME/CM switch is at 2 μsec, then three divisions times 2 μsec per division equals a time difference of 6 μsec.

2.5 PRACTICAL APPLICATIONS

2.5.1 Isolation Transformers and ac Power Circuits

WARNING: Be very careful when you connect an oscilloscope or other line-operated test instrument to an ac operated product (such as a television) that does not use a power transformer. A severe electrical shock can result. *When you troubleshoot such items, always use an isolation transformer.*

Figure 2-27 shows what can happen if an isolation transformer is not used. If the item being tested is plugged into an ac outlet such that the chassis is connected to the "hot" side of the ac line, then the moment the ground lead

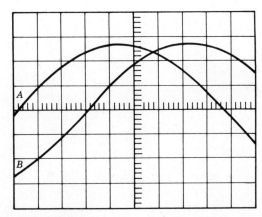

Figure 2-25   Phase measurements.

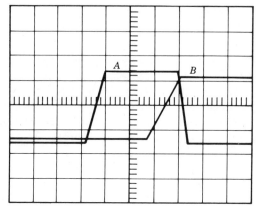

Figure 2-26 Time difference.

of the oscilloscope contacts the chassis of the test item, the line voltage can be shorted to ground through the oscilloscope's chassis and line cord. Of course, the grounding problem is eliminated if an isolation transformer is used as shown.

2.5.2 Digital Circuits and Probe Ground Lead Placement

Modern digital circuits, with their very fast operating speeds and rise times, can produce erratic displays on the oscilloscope screen even when the circuits are operating properly. To ensure the best possible displays, connect the ground lead of the oscilloscope probe to a ground point as electrically close to the point being tested as possible. Ground loops and ground noise may otherwise produce an unusable display.

2.5.3 Using External Trigger Signals

Figure 2-28 shows a simple flash circuit and the Heath IO-4235 connected in the single-sweep configuration. When switch SW1 is closed, the oscilloscope

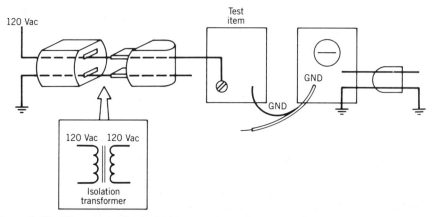

Figure 2-27 The reason for isolation transformer use.

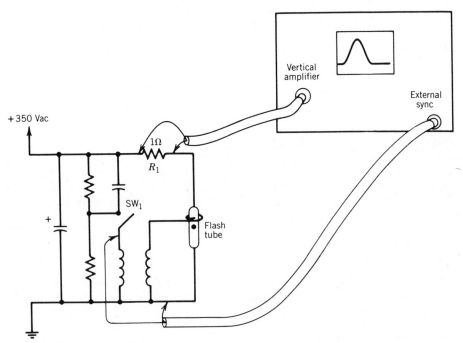

Figure 2-28   External sync use.

sweep will be externally triggered and will start and make only one sweep. As the sweep progresses, current will flow through resistor R_1 and through the flash tube, and the amplitude of the voltage across the resistor will be displayed on the screen. Because the value of R_1 is 1 ohm, the amplitude of the voltage across the resistor is equal to the current flowing through the resistor as

$$I = \frac{E}{R}$$

$$I = \frac{E}{1}$$

Therefore,

$$I = E$$

Thus, if the voltage across the resistor is 5 V, the current through it is 5 A.

To make the measurement, first refer to Figure 2-10 and set the TIME/CM and VOLTS/CM switches to appropriate positions. Set the AC-GND-DC switch to AC, push in the SINGLE trigger mode switch, push in the AC trig coupling switch, and set the TRIGGER SELECT switch to +EXT. Then, before you make a measurement, momentarily push in the RESET trigger mode switch. When the sweep is triggered, the TRIG lamp will light momentarily.

2.5.4 Troubleshooting Example

Figure 2-29 shows the input and output signals from an audio amplifier. Ideally, the only difference between the two signals should be the amplitude when the

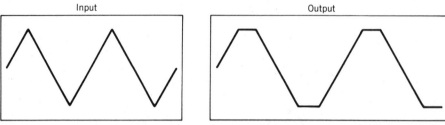

Figure 2-29   Oscilloscope signal displays during troubleshooting.

input amplitude is correctly set. However, because the output waveform is clipped, the amplifier is defective. Then, working from the output to the input and testing each stage should quickly locate the problem.

Oscilloscopes provide a visual picture of voltage versus time. Thus, they allow the troubleshooter to quickly identify improper waveforms at points within malfunctioning equipment, as well as make voltage, phase, and relative time measurements. The visual "picture" of waveforms provided by oscilloscopes makes them valuable troubleshooting tools.

CHAPTER 3

Spectrum Analyzers

Leonard F. Garrett
Tektronix, Inc.

3.1 HOW THEY WORK

The spectrum analyzer is a specialized superheterodyne radio receiver with an electronically swept local oscillator to provide a continuous presentation of amplitude versus frequency on a CRT display. Unlike superheterodyne radio receivers used in communications, where a specific type of modulation must be recovered and all other frequency components as well as noise rejected, the spectrum analyzer must treat all frequency components equally.

Today's modern spectrum analyzer has grown from a complex instrument that is difficult to understand and operate to a fully calibrated instrument with microprocessor-aided controls for simplicity and ease of operation. Such an instrument is capable of providing the same system diagnostics in the frequency domain that the modern oscilloscope provides in the time domain.

The interference or noise itself may be the energy to be measured. Thus, while the spectrum analyzer parallels the basic building blocks of a superheterodyne radio receiver such as those used for communication purposes, the circuits themselves are quite different.

Today's modern spectrum analyzer offers complete calibration in both amplitude and frequency, with information about the various control settings displayed on the CRT along with the signal information of interest. Digital storage and signal processing assist the operator by making measurements more rapidly, more easily, and with greater accuracy.

The CRT display is calibrated in power as well as voltage. The logarithmic scale is calibrated in milliwatts, and the linear scale is calibrated in microvolts and millivolts. A logarithmic scale of 10 dB/div and an eight-division vertical graticule will provide 80 dB of on-screen dynamic range, enabling signal power ratios of $10^8 : 1$ to be observed simultaneously.

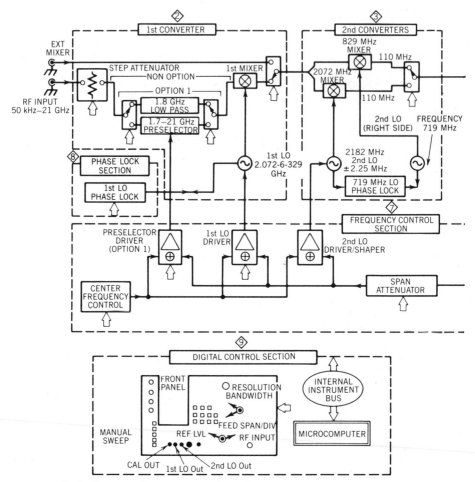

Figure 3-1 Block diagram of modern programmable IEEE 488 bus-compatible spectrum analyzer.

An example of the basic building blocks in today's modern microprocessor-based spectrum analyzer is shown in Figure 3-1.*

The first converter consists of a precision attenuator, optional filters, the first local oscillator, and the first mixer. The precision attenuator reduces the rf input (50 kHz to 21 GHz) signal amplitude to levels that will not damage the first mixer. The optional 1.8-GHz low-pass filter and the 1.7- to 21-GHz preselector (an electronically tuned bandpass filter) are to prevent unwanted mixer products from appearing on the spectrum analyzer CRT screen. When the input signal and the local oscillator (LO) signal are mixed, the sum and difference frequencies, as well as a theoretically infinite number of harmonics, are generated in the mixer. This occurs whether one or several input signals are present. Mixer products of several signals may overlap, producing confusing spectrum

---

*All the illustrations in Chapter 3 are reprinted with the permission of Tektronix, Inc.

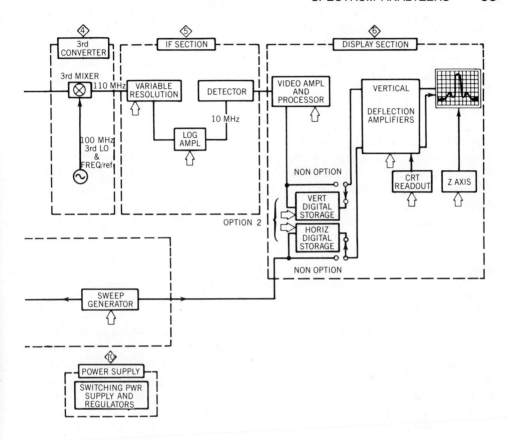

analyzer displays unless unwanted inputs are either filtered out (signals above 1.8 GHz for the lower-frequency bands) or detected one at a time by using the preselector[1] for higher frequencies. The preselector is tuned through the input frequency range as the LO is tuned through its range so that the two track together. These optional filters are valuable for convenient use of the spectrum analyzer.

The local oscillator sweeps from 2.072 to 6.329 GHz in a very linear and precise fashion under the control of the first local oscillator driver. It is phase locked to ensure an accurate sweep. The first detector produces an output each time the LO frequency is 2072 MHz, or 829 MHz away from the input signal frequency depending on the frequency range set on the spectrum analyzer. If the input rf signal is 500 MHz, the first mixer will produce an output when the

---

[1] A preselector is an electronically tuned bandpass filter.

LO has swept to 2572 MHz (2072 MHz + 500 MHz), indicating the presence of a 500-MHz signal on the input.

The frequency control section of Figure 3-1 controls the center frequency of the second LO, drives the first local oscillator through its sweep, and drives the preselector through its sweep under the control of the sweep generator. The span attenuator controls the range of input signals displayed, from a few MHz to several GHz. The center frequency control sets the location of the span within the frequency range of the spectrum analyzer; in other words, whether a 500-MHz sweep is from 250 to 750 MHz or from 1.2 to 1.7 GHz. The sweep generator also starts the horizontal sweep across the CRT so that the horizontal position of the CRT beam corresponds to a particular input frequency.

The second converter produces a 110-MHz intermediate frequency (IF) output when an input is present from the first mixer. The range selected determines whether the 2072-MHz or the 829-MHz mixer is used. The two local oscillators and mixers are used in the second convertor to minimize false indications on the CRT from mixer products. The third converter, the local oscillator, and the mixer convert the 110-MHz first IF signal to the 10-MHz second IF signal.

The IF section contains a precision-adjustable (variable resolution) bandpass filter to set the display resolution. The resolution is the minimum frequency separation between two signals at which they can be distinguished from each other. If the variable-resolution bandpass filter passes too large a range of frequencies, closely spaced signals cannot be resolved and appear as one signal on the spectrum analyzer display. If the bandpass is too narrow, the amplitude of the spectrum analyzer display is very small and incorrect. Time is required for a signal to propagate through the complex circuitry of the spectrum analyzer. If the IF filter bandpass is too narrow, the first LO sweeps through the filter bandpass before appreciable power can propagate through the IF circuitry. An analogy is a radio that is tuned through its receiving band very rapidly. One hears almost no sound from any station in the band; but if the radio is tuned slowly, one can hear each station clearly as it is passed by. The detector in the IF section provides a voltage when the 10-MHz IF signal is present.

The display section provides a horizontal CRT voltage to drive the CRT beam across the screen in synchronism with the first LO sweep. The position of the beam (or spot on the CRT screen) indicates the frequency. The video amplifier provides a vertical deflection on the CRT screen when the 10-MHz IF signal is present. The display section also has optional signal storage capability using digital and analog-to-digital and digital-to-analog conversion circuitry. The control section synchronizes all the other sections so that they are all set correctly for the center frequency and span that the operator has set on the front panel controls.

To visualize the operation of the spectrum analyzer, suppose one has connected the instrument to a circuit which has as an output 250- and 500-MHz signals. The spectrum analyzer has been set to sweep from 50 kHz to 1 GHz to observe the presence of the signals. The first local oscillator will sweep from

2.072 to 3.072 GHz. As each sweep of the LO begins, the CRT beam begins its travel across the CRT face. Each division of the face will represent 100 MHz, since there are 10 divisions. When the LO is 2.322 GHz, the difference between the input and LO frequencies is 2072 MHz, so the second mixer will produce a 110-MHz output. The second converter output is fed to the third converter, which then produces a 10-MHz output. The third converter output is detected, amplified, and fed to the vertical deflection plates of the CRT, causing the beam to be deflected upward. At this time, the CRT beam has traveled 2.5 divisions from the left edge of the screen. The height of the vertical deflection is proportional to the power of the 250-MHz signal. When the first LO frequency reaches 2.572 GHz, the difference between the 500-MHz input and the first LO frequency is 2072 MHz, so another vertical deflection of the beam occurs. At this time, the beam is five divisions from the left edge of the screen. When the first LO reaches 3.072 GHz, the CRT beam has reached the right-hand side of the screen so the beam retraces and the first LO resets to 2072 MHz. The sweeps are rapid and repetitive, so the display will appear steady. The display will be a horizontal line with vertical deflections 2.5 and 5 divisions from the left edge of the CRT screen. Each deflection will appear similar to those in Figure 3-4.

3.2 USE, DO NOT ABUSE, YOUR SPECTRUM ANALYZER

Too much input power will destroy the input mixer and possibly the radio-frequency (rf) attenuator. A continuous-wave (cw) input of 1 watt is typical of most wide-frequency-range spectrum analyzers. When working with potentially high rf power levels, use a directional coupler or power probe to get the power down to acceptable levels. A calibrated directional coupler not only provides a low-level monitoring point but, in addition, allows accurate power measurements of both the fundamental frequency and all harmonics. An important consideration is the coupling coefficient versus frequency. Figure 3-2 shows a typical transmitter installation with transmission line and directional coupler.

It is important to keep in mind when measuring pulsed rf signals, such as those encountered in radar systems, that the peak signal level at the mixer is always greater by $(1.5t_0B)^{-1}$, where t_0 is the pulse width and B is the resolution bandwidth, than the peak level displayed on the CRT. For example, in a spectrum analyzer with a 1-MHz resolution bandwidth, a pulsed rf signal of 100-nsec pulse width would produce a peak level of 6.66 times at the mixer compared to the peak level displayed on the CRT. Figure 3-3 shows a cw signal and pulsed rf signal of the same peak amplitude. Notice the significant difference in amplitude as viewed on the spectrum analyzer.

When viewing a crowded frequency spectrum, be aware of the spectrum analyzer input dynamic range for best linearity (usually -30 dBm at the input mixer). The total energy of all signals that appear at the input mixer must be considered, regardless of whether they appear on the CRT. A simple test for linear operation is to change the rf attenuator by 10 dB and note whether all

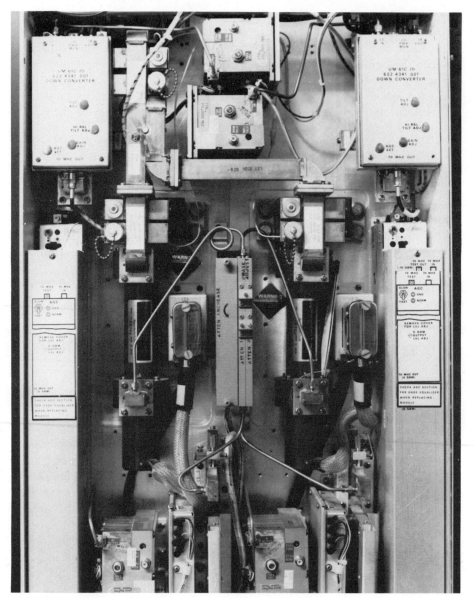

Figure 3-2 Digital microwave radio transmitter showing directional coupler output ports for spectrum analysis and power measurement.

displayed signals follow the attenuator change. Reduce the input power if they don't.

In microwave spectrum analyzers that have a built-in tracking preselector, an auxiliary peaking control is usually provided. This control should be adjusted for maximum displayed signal amplitude in the area of interest. Failure to adjust this control properly may result in a sensitivity loss of many dB.

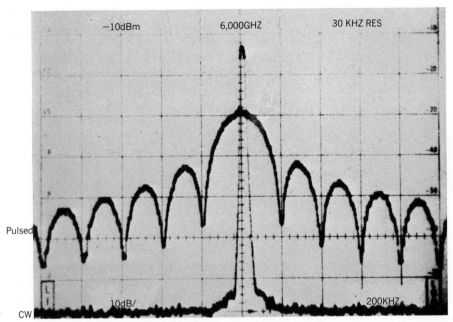

Figure 3-3   Pulsed rf and CW signal with same peak power.

Spectrum analyzers operating the frequency range from 20 GHz to 200 GHz usually employ external waveguide mixers. A bias peaking control on the spectrum analyzer allows the operator to optimize the sensitivity for any frequency. Failure to set this control properly can also result in severe loss of sensitivity.

3.3 MICROWAVE RADIO SYSTEM TESTING

Microwave radio systems use both analog and digital modulation techniques. Performance validation requires the use of a spectrum analyzer to measure such parameters as frequency deviation, signal-to-noise ratio, spurious responses, occupied bandwidth, carrier harmonic levels, power levels, spectrum symmetry, and interference.

3.3.1 Frequency Deviation Measurements

The spectrum analyzer may be used (1) to measure transmitter frequency deviation, (2) as a precise indicator during deviation adjustment, or (3) as a highly accurate means of testing the accuracy of frequency deviation monitors or meters.

The Bessel, or carrier-null, method is often used to measure frequency deviation of transmitters when a single modulating frequency can be used (typical out-of-service testing). Frequency deviation is calculated according to the following relationship:

$$\text{Modulation index (M)} = \frac{\text{peak frequency deviation } (\Delta f)}{\text{modulating frequency } (f_m)}$$

For example, suppose you want to set the deviation of a microwave transmitter to the specified ± 5-MHz limit. First, select a modulating frequency within the normal operating range of the modulator, or a specific frequency in this range specified by the manufacturer. This frequency, multiplied by the modulation index (from a table of Bessel null values) will give the frequency deviation for first, second, third, and so on, carrier nulls. A modulating frequency of 577.8 kHz and the Bessel null number for third-carrier null (8.6531) satisfies the values needed to set a peak deviation of 5 MHz:

$$8.6531 = \frac{5 \text{ MHz}}{0.5778 \text{ MHz}}$$

A null of 60 dB on the spectrum analyzer will provide an accuracy of better than 1%.

Figure 3-4 is a CRT display of the rf carrier unmodulated (left-of-center response) and at carrier null (right-of-center response).

3.3.2 Digital Microwave Radio Measurements

The FCC (Federal Communications Commission) lists spectrum specifications for digital microwave radio in Section 21.106. A graphical representation of

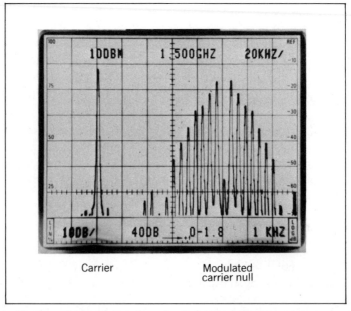

Carrier   Modulated carrier null

Figure 3-4   Display of unmodulated carrier (left side of CRT) and same carrier with FM modulation index of 2.4048.

these specifications for a 6-GHz radio is shown in Figure 3-5. Two points need to be adjusted and/or verified. The spectrum width at a level of 50 dB from the mean transmitted power should not exceed the authorized bandwidth, and the output level should be at least 80 dB down outside the frequencies within the FCC mask.

Besides performing the spectrum occupancy tests, the spectrum analyzer can be used in many other digital radio applications. These applications include such things as checking or adjustment of maximum peak power output, spectrum shape and symmetry, comparison of pre- and postoutput filter performance, spurious emissions far from the carrier, amplitude level variations among several transmissions on the same antenna, interference pick-up, and antenna alignments.

Measurement Needs FCC Regulation 21.106 specifies that for digital modulation transmission below 15 GHz, in any 4-kHz band in which the center frequency is removed from the assigned frequency by more than 50% up to and including 250% of the authorized bandwidth, the attenuation below the mean output power level should be as specified by the following equation, but should never be less than 50 dB:

$$A = 35 + 0.8\,(P - 50)$$
$$+ 10 \log B \quad \text{(attenuation greater than 80 dB not required)}$$

where P is the percent removal from the carrier and B is the authorized bandwidth in MHz. A different equation applies for transmission above 15 GHz. The FCC specifies authorized bandwidths of 30 MHz at 6 GHz and 40 MHz at 11 GHz.

Measuring to FCC Specifications

Occupied bandwidth. The bandwidth is measured at the required down point (50 dB at 6 GHz, 51 dB at 11 GHz), as shown by the FCC mask in Figure 3-5. The 50-dB bandwidth calls for a relative-level measurement in dB rather than an absolute power determination in dBm. What has to be measured is the occupied spectrum width at the point where the spectrum is 50 dB down (51 dB for 11 GHz) from the "mean output power level," as illustrated in Figure 3-6. If the power outside the output filter bandwidth is ignored, then the mean output power is the same as the unmodulated level, and the relative-level formula $10 \log (B_n/f_s)$ holds, where B_n is the measurement noise bandwidth and f_s is the signal frequency.

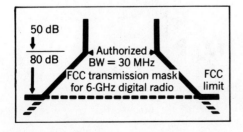

Figure 3-5 FCC occupied bandwidth mask for 6-GHz digital microwave radio.

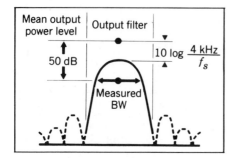

Figure 3-6   Digital microwave radio mean output power level measurement criteria.

The measurement consists of two steps. The first step is to determine the relative level between signal display peak and the mean power level. (Manufacturers will usually specify this number.) Alternately, the user can compute it from 10 log (B_n/f_s). The remainder of the 50 dB or 51 dB is measured with respect to the mainlobe of the signal. It is important to note that the actual measurement bandwidth need not be 4 kHz because this is a relative-level measurement.

Consider the following example: If the specified baud rate is 30.086 MHz, then 10 log (4 kHz/30.086 MHz) = −38.76 dB. With this high baud rate, the output filter truncation error is close to 0.8 dB, and the manufacturer specifies that the mainlobe is 38 dB down from mean output power after the filter. Both numbers meet the intent of the FCC specifications, although referencing to the

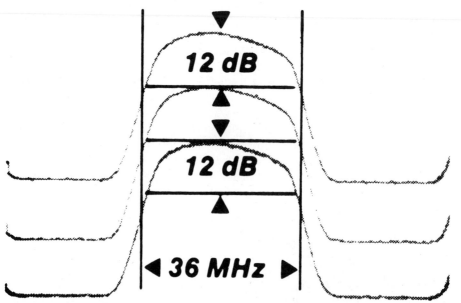

Figure 3-7   Multiple CRT trace shows no change in spectrum shape with three different resolution filters.

output of the filter makes the specification slightly tighter. The remainder of the measurement is illustrated in Figure 3-7, where the spectrum shape is observed using three different resolution bandwidths. In each case, the shape is the same and the bandwidth is 36 MHz at 12 dB down (note 38 + 12 + 0.8 + 51). The change in bandwidth, in each case by a factor of 10 times, moves the spectrum shape by 10 dB (10 log 10 = 10 dB), but this movement has no effect on the bandwidth measurement.

3.3.3 Filter Leakage

The output level in a 4-kHz noise bandwidth must be at least 80 dB down from the mean output level (dBc) outside 250% of specified bandwidth offset. Without an output filter, sidelobes are only about 11 dB down, as shown in Figure 3-8. The function of the output filter is to reduce these sidelobes to the necessary −80 dB level.

Figure 3-9 shows such a measurement. The result can be interpreted in two ways. The simplest technique is to consider the peak of the display as 10 log (4 kHz/f_s) down, and add the relative level of the leakage to this value. Thus, 38 + 53 = 91 dB. The other technique is to consider the effect of the bandwidth used in the measurement. A 300-kHz spectrum analyzer resolution

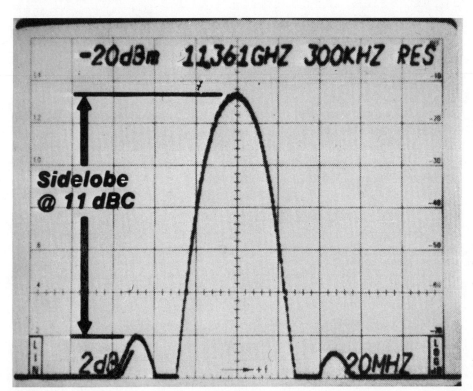

Figure 3-8   Digital microwave radio sidelobe levels prior to output filter.

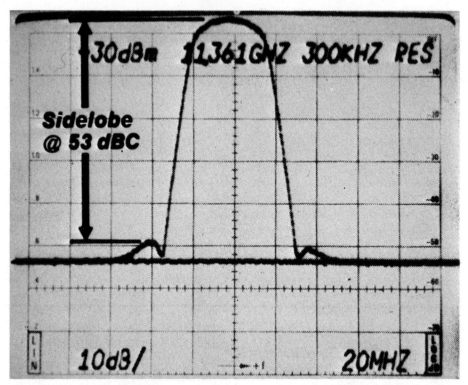

Figure 3-9   Digital microwave radio sidelobe levels following output filter.

bandwidth with a 4 to 1, 60-dB-to-6-dB shape factor equals a (300 × 0.8) = 240-kHz noise bandwidth. Thus, the mainlobe is 10 log (240 kHz/30.086 MHz) = 62 dB. This measurement indicates a level more than 10 dB better than required.

An important point is that the internal noise of the spectrum analyzer be well below the required measurement level.

3.3.4 Other Measurements

Numerous measurements besides those specified by the FCC are possible, as indicated previously. Some examples are given below.

Multiple-carrier level balance. Figure 3-10 shows two digital radio signals transmitted on the same antenna. Although these signals are supposed to be at the same amplitude level, clearly they are not. The use of a vertical display of 2 dB/div, as shown in Figure 3-11, illustrates an amplitude difference of 1 dB.

Spectrum symmetry. A form of symmetry is shown in Figures 3-12 and 3-13, taken before the output filter. Figure 3-12 shows that one sidelobe varies by more than 2 dB in amplitude from the other one. By activating the amplitude difference measurement mode (unique to the Tektronix 492 spectrum analyzer),

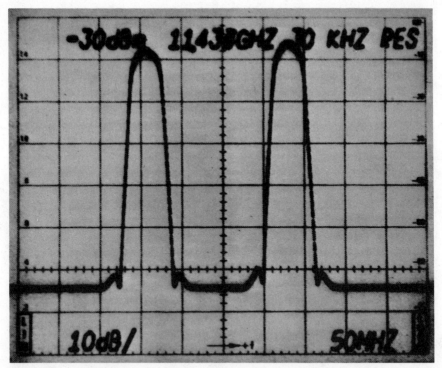

Figure 3-10 Display of two digital microwave signal levels feeding a common antenna.

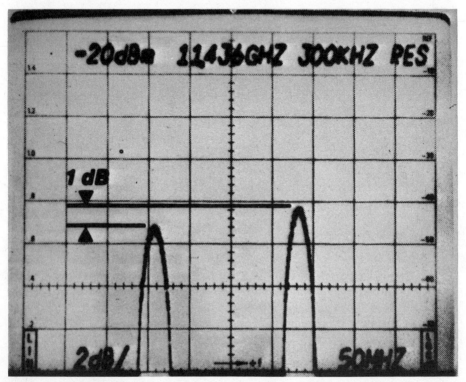

Figure 3-11 Multiple-carrier balance test using the 2 dB/div vertical log mode.

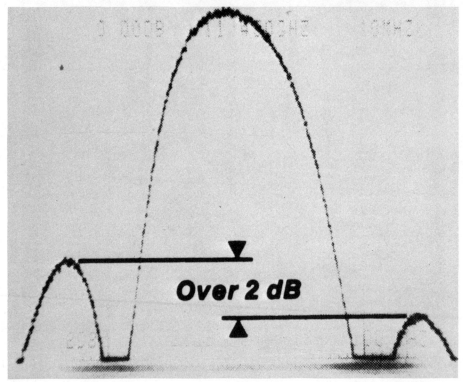

Figure 3-12   Spectrum symmetry measurement using 2 dB/div vertical log mode.

the large sidelobe is adjusted to full screen, and its amplitude relative to the mainlobe is determined at 13.25 dB (upper left readout). Since a perfect sin x/x gives 13.26 dB, it is obvious that the right sidelobe is the one that is incorrect.

Interference. Stray interference can be captured by using digital-storage display with maximum hold function. This technique holds random interference hits even when these hits occur for a short time. Figure 3-14 shows an interference problem by the stray output within the signal nulls on the right side.

Spurious outputs. The spectrum analyzer is an excellent tool in checking for spurious outputs. Figure 3-15 shows a spurious signal offset by the 70-MHz intermediate frequency above the 6-GHz main signal.

3.4 COMMERCIAL TV BROADCAST SYSTEM MEASUREMENTS

Regulatory agencies such as the FCC in the United States impose technical standards on all types of radio emissions. Technical standards are imposed to assure picture quality and to limit interference to other services. TV broadcast stations must meet these standards to maintain their licenses.

Technical standards cover frequency accuracy, frequency response, harmonics, spurious responses, carrier-to-noise ratio, aural carrier deviation, percent modulation of the visual carrier, and field intensity.

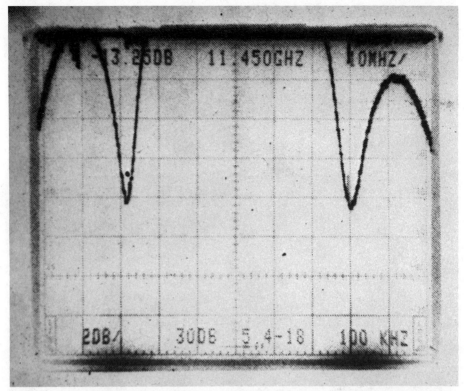

Figure 3-13 Spectrum symmetry measurement using the 0.25-dB step ΔdB mode.

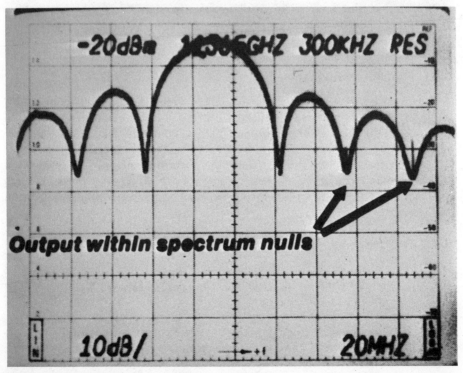

Figure 3-14 Carrier leak measurement showing unwanted output within spectrum nulls.

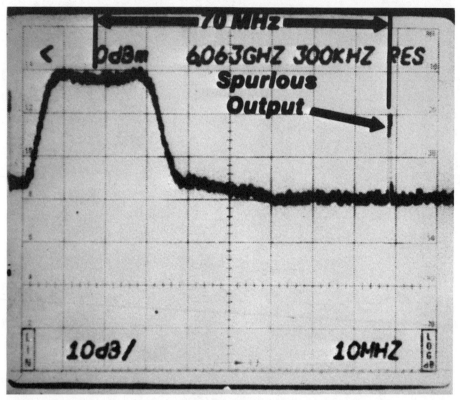

Figure 3-15   Spurious output shown 70 MHz above center of signal spectrum.

3.4.1 Frequency Accuracy

Some analyzers have frequency measurement capability of ± 100 Hz or better. A simple whip antenna may be used when the spectrum analyzer is within a few miles of the transmitter. A resonant dipole antenna should be used at greater distances.

The procedure is simply to tune the spectrum analyzer to the desired frequency and switch to the frequency count mode. Frequency measurements of signal levels as low as -100 dBm or more may be measured with the spectrum analyzer.

3.4.2 Frequency Response

Measurement of frequency response requires the use of a sweep generator or tracking generator with the spectrum analyzer. A specialized tracking generator called a sideband analyzer is frequently used for frequency response testing of TV broadcast transmission systems. The sideband analyzer provides a swept video source when used in conjunction with a companion spectrum analyzer.

The measurement technique is to insert the swept video source (sideband analyzer) into the appropriate video line to the transmitter, tune the sideband analyzer and the spectrum analyzer to the desired TV channel, and finally adjust the sideband analyzer for proper sweep width and output level. Figures 3-16 and 3-17 show the frequency response of a properly adjusted TV transmission system.

3.4.3 Harmonics and Intermodulation Distortion

While there are no actual standards set for U.S. NTSC transmitters for intermodulation distortions, much can be learned about a transmitting system by performing some standard intermodulation tests.

Distortion less than 36 dB down from the visual carrier can cause visible loss of picture fidelity. A two-tone intermodulation distortion test involves driving the TV transmitter with two well-isolated, low-distortion cw signal generators (1 MHz and 1.4 MHz) through a resistive combiner, as shown in Figure 3-18. Possible second- and third-order harmonic and intermodulation components from the 1-MHz and 1.4-MHz test signals are shown in Figure 3-19. Ideally, the TV transmission system should not have visible harmonic

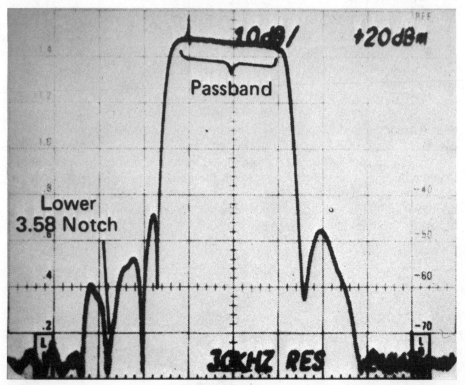

Figure 3-16 Frequency response of TV broadcast transmission system using the spectrum analyzers sideband analyzer system.

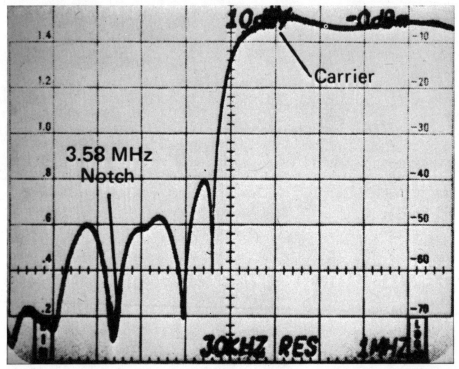

Figure 3-17   Expanded display of TV broadcast transmission showing frequency response and lower-sideband 3.58-MHz notch approximately 70 dB below normal in band response.

and intermodulation components greater than − 50 dBc (50 dB below the visual carrier).

The spectrum analyzer may be coupled to the TV transmission line to the antenna through a suitable coupler (limit the power to the spectrum analyzer to 1 watt), or radiated energy may be picked up with a suitable antenna for harmonic and intermodulation measurements. Figure 3-20 shows the described distortion products on the spectrum analyzer.

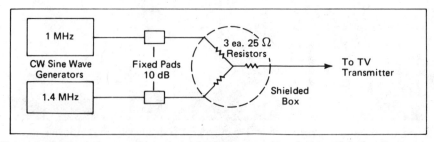

Figure 3-18   Resistive combiner used for TV transmitter intermodulation tests.

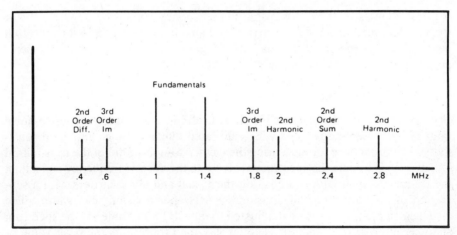

Figure 3-19 Relationship of distortion products produced by 1-MHz and 1.4-MHz signals in a nonlinear system. I_m is intermodulation.

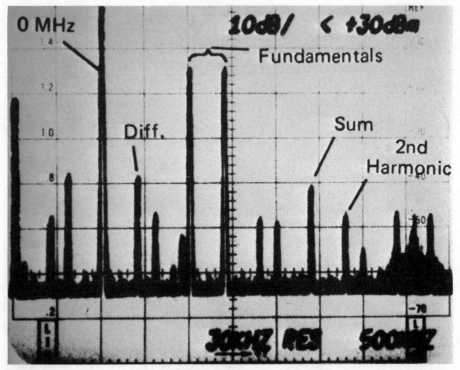

Figure 3-20 CRT photo of fundamental signals and intermodulation products viewed on the spectrum analyzer.

Harmonics of the TV transmitting frequency should be down 50 dB or more, referenced to the visual carrier, and should be measured out to the 10th harmonic using the spectrum analyzer. Figure 3-21 shows the measurement; note that harmonics beyond the 4th are not visible.

3.4.4 Carrier-to-Noise Ratio

The modern spectrum analyzer is a most useful instrument for carrier-to-noise ratio measurements due to its wide calibrated dynamic range and good sensitivity. The procedure for measuring the carrier-to-noise ratio of the transmitted TV channel is to tune the spectrum analyzer to the desired frequency and adjust the frequency span width to position the visual and aural carriers with a separation of four and one-half divisions (1 MHz per division scan width). The visual carrier amplitude is next adjusted to full CRT amplitude via the spectrum analyzer internal rf attenuator. The display then represents the peak value of the carrier-to-noise ratio between the visual and aural carriers. The average value of the noise is obtained by enabling the spectrum analyzer video filter, thus providing a smooth, easier-to-interpret display.

Figure 3-22 shows the resultant display indicating a 60-dB in-channel carrier-to-noise ratio. This value must be modified by two corrections to obtain

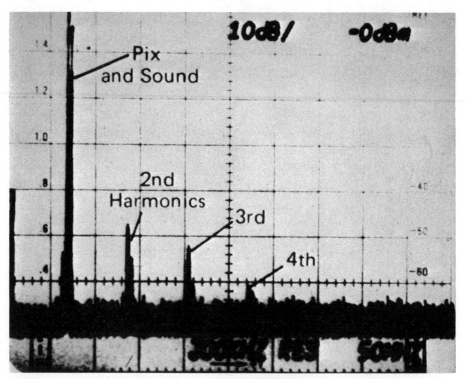

Figure 3-21   CRT photo showing TV transmitter fundamental frequency and harmonics.

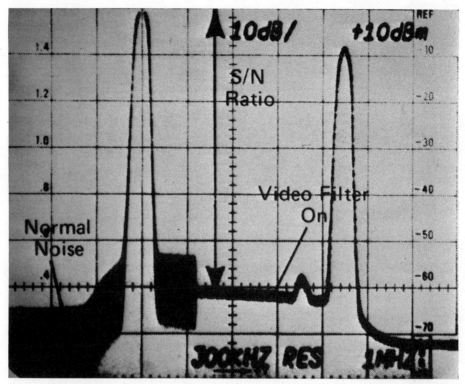

Figure 3-22   Display of carrier-to-noise measurement using the spectrum analyzer with noise-averaging video filter.

the true signal-to-noise ratio. First, the difference between the specified measurement bandwidth (4 MHz for one TV channel) and the measured bandwidth (spectrum analyzer resolution bandwidth used) must be factored in. The spectrum analyzer resolution bandwidth used was 300 kHz, with a calculated noise bandwidth of 240 kHz; thus, the correction becomes

$$-dB = 10 \log_{10} \frac{4 \text{ MHz}}{0.240 \text{ MHz}} \quad \text{or} \quad -12.1 \text{ dB}$$

A second correction factor must be added because of the difference in the way the spectrum analyzer log amplifier and detector process continuous-wave signals and noise (this factor is -2.5 dB and has mathematical verification not covered here). Thus, the true visual carrier-to-noise ratio becomes 60 dB $-$ 12.2 dB $-$ 2.5 dB, or 45.3 dB.

3.4.5 Aural Frequency Deviation Measurement

The spectrum analyzer is used as a wide-dynamic-range-sensitive null indicator capable of measurements to 0.01% accuracy (80 dB dynamic range). The Bessel null technique, as discussed earlier, is used for greatest accuracy due to pure mathematical verification. The basic equation is

$$M = \frac{\Delta f}{f_m}$$

where M is the modulation index, Δf is the peak deviation, and f_m is the modulating frequency.

For TV broadcast, it is convenient to use the first Bessel null (modulation index of 2.405) and a modulation frequency of 10.396 kHz to obtain 25 kHz peak deviation. Figure 3-23 shows the resultant display on the spectrum analyzer.

One word of caution: Any transmitter preemphasis should be bypassed when setting the frequency deviation limits using the above procedure.

3.4.6 Measuring Percent Modulation of the Visual Carrier

The spectrum analyzer in zero span is an accurate receiver with available linear display and adequate demodulation bandwidth to provide modulation depth measurements of the visual carrier. The procedure is to select a resolution

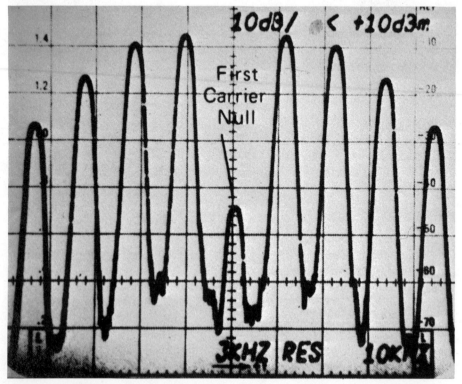

Figure 3-23 Bessel null FM deviation measurement using the spectrum analyzer.

bandwidth of 300 kHz or more (the tip of *sync*, synchronizing or timing pulses, must be resolved), tune the visual carrier to the center of the CRT, and set the frequency span to zero. With the linear vertical mode selected, adjust the input signal level to place the top of sync at the topmost graticule (100% modulation). With an eight-division display, 0% modulation will be at the bottom graticule, and each division will represent 12.5% modulation. Figure 3-24 shows a TV transmitter properly adjusted for 100% modulation.

3.4.7 Field Intensity and Spurious-Signal Measurements

Measurements of field intensity require a calibrated antenna to convert the spectrum analyzer signal level in dBmV (dB referenced to 1 millivolt) to microvolts or millivolts per meter. Several measurements a short distance apart are made at each test site (which should be a clear area), and the results are averaged to account for terrain effects and any surrounding structures that may produce multipath signals.

Spurious signals can originate either within the transmitter or at the antenna. The most logical place to begin testing for spurious signals is at some distance (1 mile or more) from the transmitter, using a resonant dipole antenna of the type used for field intensity measurements. Note any in-channel spurious

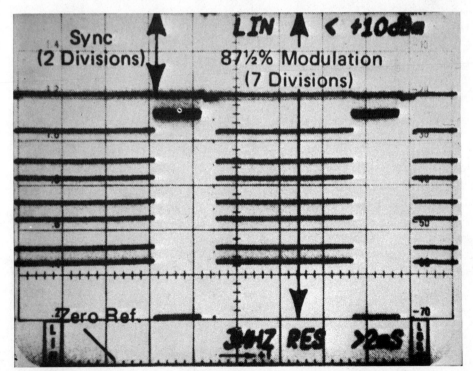

Figure 3-24   Using the spectrum analyzer in the zero span mode for percent modulation measurements.

carriers (Figure 3-25). There should not be any continuous carriers other than the picture, sound, and color subcarriers. If extra spurious signals are noted, momentarily turn off the transmitter to pinpoint whether the problem is internal or off the air. Out-of-channel sum and difference beats are observable in Figure 3-26. See Chapter 9 for further discussion of communications systems.

3.5 BASEBAND MICROWAVE PERFORMANCE EVALUATION

3.5.1 System Description

A microwave transmission system consists of three common frequency groupings generally referred to as baseband, intermediate frequency (IF), and carrier.

Baseband generally refers to a composite signal consisting of voice, teletype, data, or other channels, along with pilots and test tones. This signal is generated by frequency multiplexing the information channels and test tones onto several different subcarriers. The multiplexed composite baseband signal is then modulated onto a microwave carrier for final transmission. Several types of baseband generation schemes and equipment are presently in use. These

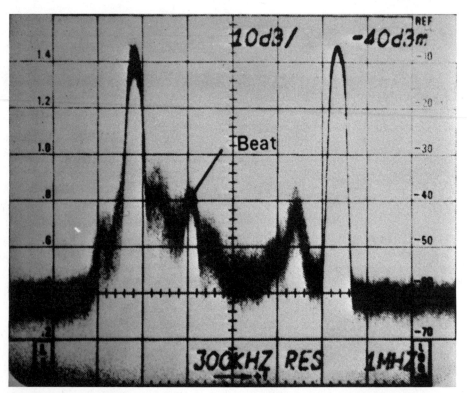

Figure 3-25 CRT display of spurious response approximately 1 MHz above visual carrier.

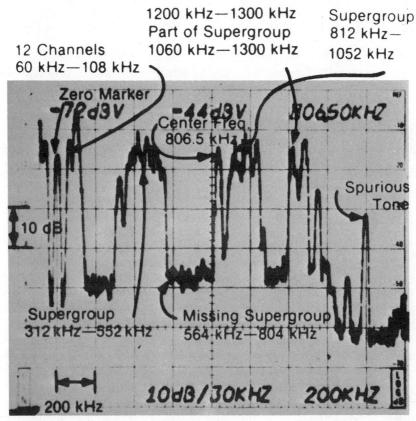

Figure 3-26   Typical baseband radio spectrum.

systems may vary with respect to channel frequency, amplitude level, impedance, and so on. The system described in the following paragraphs is a typical
L-carrier configuration.

A channel bank consists of 12 channel carriers spaced 4 kHz apart from
64 kHz to 108 kHz. They may carry voice information from 300 Hz to 3550
Hz, narrowband teletype signals, and the like. A pilot tone is usually added,
typically at 104.08 kHz.

Five channel banks make up a supergroup band of 60 channels. Supergroup carriers are 420, 468, 516, 564, and 612 kHz. The group bank signal
consists of a 60-channel group from 312 kHz to 552 kHz, in addition to the
carriers and pilot tones (common pilot frequencies are spaced 48 kHz apart
starting at 315.92 kHz).

Ten groups make up a 600-channel mastergroup. Mastergroups, covering
roughly 3 MHz of frequency spectrum, can be combined further in groups of
two, three, or even more. Some installations contain additional test and signaling tones, which contribute to the complexity of the baseband signal. Figure
3-26 is typical of baseband signals viewed on a spectrum analyzer.

While frequencies or signal levels may vary from one installation to another, the basic measurement requirements are quite similar. Basic parameters to be measured include:

1. Amplitude levels of carriers, test tones, signal tones, data, and so on.
2. Spurious signal levels and frequencies due to harmonics and intermodulation.
3. Leakage at channel carrier, group carrier, or other frequencies.
4. System noise levels.
5. Frequency shifts due to changes in degree of modulation or other causes.
6. Identification of random transient noise burst interference.
7. Determination of notch filter shape.

3.5.2 Amplitude Measurements

The easiest measurement to make is that of a signal with an amplitude well above the noise level. First, tune the signal response as close as possible to full screen. At a vertical setting of 2 dB/div and 1-dB reference level increments, the display will be within 1 dB (0.5 div) of full screen. This keeps measurement repeatability to a small fraction of a decibel. It also gives a high-accuracy amplitude level comparison between adjacent signals of similar amplitude. The absolute-level measurement accuracy is essentially equal to reference level accuracy which, in turn, depends on the reference level setting. An accuracy of ± 0.5 dB or better is achievable.

Some spectrum analyzers have the ability to set frequency with a high degree of accuracy, thus making it easy to intercept intermodulation and other spurious signals at frequencies that can be predicted. The high accuracy also means that the frequency of unexpected spurious responses can be pinpointed and the source identified.

If three tones at 57, 2600, and 2714 kHz are fed to the system under test, the following second- and third-order intermodulation responses will occur:

(a) $2(57) = 114$ kHz
     $2714 - 2600 = 114$ kHz

(b) $2600 + 57 = 2657$ kHz
     $2714 - 57 = 2657$ kHz

(c) $2714 + 2(57) = 2828$ kHz
     $2(2714) - 2600 = 2828$ kHz

Theoretically, the pairs of responses fall at precisely the same frequency. However, the actual tones will be displayed as pairs due to slight input signal deviations in frequency.

Figures 3-27, 3-28, and 3-29 display the intermodulation resulting from the three-tone test signal. Figures 3-26 and 3-27 (discussed previously in relation to carrier leak) show the two 114-kHz tones from (a) above. Figure 3-28 shows the two 2657-kHz components computed in (b), and Figure 3-29 shows the output at 2828 kHz.

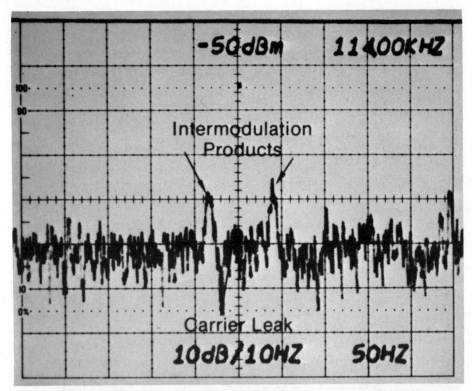

Figure 3-27   CRT photo of baseband intermodulation pair at 114 ± 40 kHz.

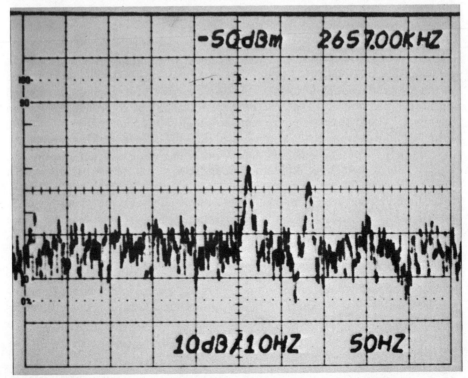

Figure 3-28   CRT photo of baseband intermodulation pair at 2657.015 kHz and
2657.085 kHz.

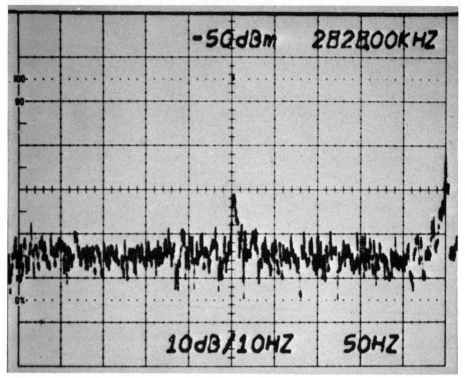

Figure 3-29   CRT display of − 92 dBm intermodulation product.

Figure 3-29 also illustrates the usefulness of averaging functions to pick a low-level signal out of the noise. The peak/average cursor has been moved to the top graticule line to average the displayed noise. In addition, the sweep time is set quite long at 10 sec/div (not shown on readout) to produce maximum averaging. Thus, it is possible to distinguish an intermodulation signal at − 90 dBm from the average channel noise level of − 100 dBm in a 10-Hz bandwidth.

Finally, pilot and signaling tones along with various spurious outputs are illustrated in Figure 3-30. The three strong tones at a level of − 47 dBm are a pilot tone at 1099.4 kHz and two signaling tones at 1097.6 and 1098.6 kHz, respectively. The three quite small signals at about − 85 dBm, and the 1100-kHz tone at − 70 dBm, are caused by harmonics intermodulation and carrier leak.

3.5.3 Noise Measurements

Absolute-noise-level measurement is a highly mathematical, complicated procedure because parameters such as system noise bandwidth, detector characteristics, and logarithmic amplifier effects have to be accounted for. (Those interested in a detailed discussion are referred to "Noise Measurement Using the Spectrum Analyzer—Part One: Random Noise," Tektronix AX-3460.)

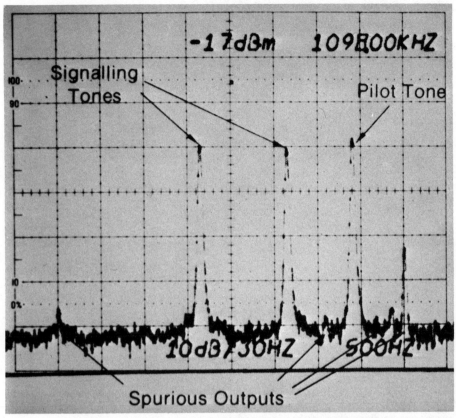

Figure 3-30   CRT display of baseband signaling and pilot tones with low-level spurious tones.

However, relative measurements are easily performed as illustrated in Figure 3-31. This figure shows an input noise level of − 30 dBm at a 3 kHz resolution bandwidth, a 40-dB deep noise slot at 2438 kHz, and baseband system noise at − 65 dBm in a 3 kHz bandwidth.

Figure 3-32 shows another low-frequency noise display. A random noise burst is saved in memory A, while the brighter trace of memory B shows the noisy but coherent spectrum of power supply harmonics.

In Figure 3-33, the lower trace displays the shape of a notch filter at 2.057 kHz. There are also several spurious signals about 300 kHz below the notch frequency. The upper trace shows the update in memory B with the MAX HOLD function activated. The MAX HOLD function causes the display to show the maximum amplitude level that occurs during the observation time. The MAX HOLD feature permits unattended monitoring of random noise burst interference, since the maximum amplitude will be held in memory. In this figure, a noise burst has increased the system noise about 10 dB while totally obliterating the effect of the notch filter.

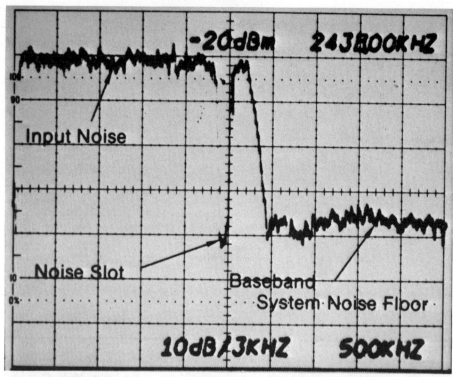

Figure 3-31 Baseband noise measurements using the spectrum analyzer.

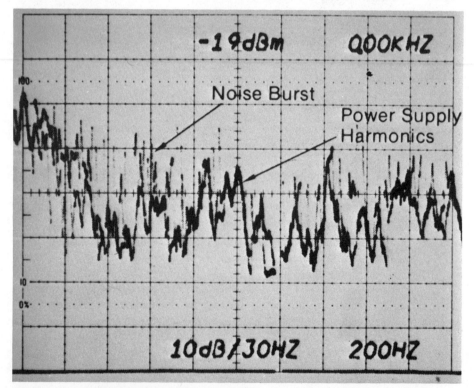

Figure 3-32 Spectrum analyzer low-frequency measurement display showing noise burst and power supply harmonics.

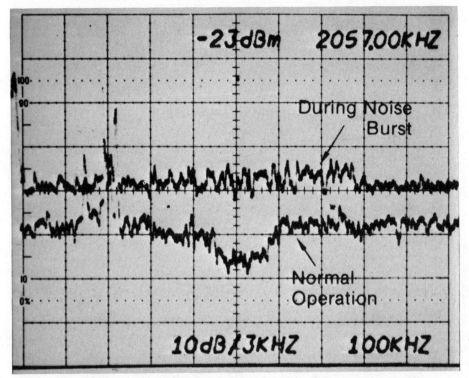

Figure 3-33 Dual-trace display using spectrum analyzer digital storage "save" mode.

3.6 SATELLITE SYSTEM MEASUREMENTS

The spectrum analyzer is practically indispensable in installing and maintaining a satellite up-and-down link. Uses range from initial site selection to daily monitoring of equipment performance.

3.6.1 Initial Site Selection

Of major concern is a site selection that is free from man-made interference. The spectrum analyzer is used in conjunction with a low-noise amplifier (LNA) to evaluate received signal levels and interference at the proposed location.

The spectrum analyzer center frequency and span controls are adjusted to view the 3.7 to 4.2-GHz down-link frequency range with the spectrum analyzer reference level set to −30 dBm. Desired carrier-to-interference levels should be at least 20 dB. It is necessary to slowly scan the horizon a full 360 degrees in azimuth using both antenna feed horn polarizations to discover field variations, obstructions, and local interfering sources.

Locating the satellite is accomplished by searching for the appropriate beacons at 4199 or 3700.5 MHz with the feed horn set to 45 degrees. This

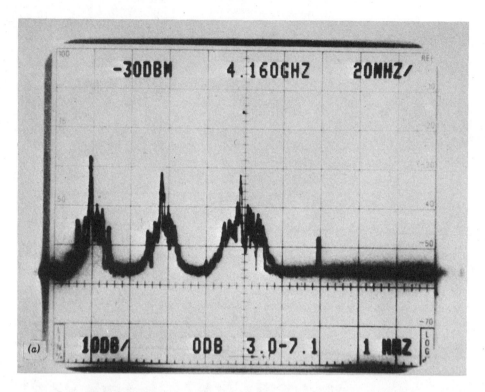

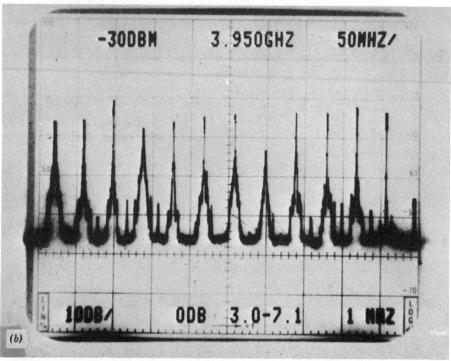

Figure 3-34 (*a*) Satellite TV down-link showing three transponders and beacon.
(*b*) Spectrum of 12 active satellite TV channels and one beacon.

setting will maximize chances of seeing several active channels. A polarization adjustment of 90 degrees is then made by adjusting the antenna feed horn polarization and observing a null of the undesired odd or even transponders on the spectrum analyzer display. Figure 3-34a shows a beacon along with three received transponder signals. Figure 3-34b shows 12 active channels and one beacon for one polarization.

3.6.2 Up-Link Transmitter Spectrum Occupancy

This test is conducted using the spectrum analyzer in MAX HOLD to build up a plot of energy per unit frequency over a period of time. A typical up-link frequency deviation for 100% modulation is ±18 MHz. Spectrum occupancy may be monitored visually by setting the spectrum analyzer scan per division to 5 or 10 MHz and observing any frequency components occurring outside the ±18-MHz limit.

Setting the transmitter frequency deviation requires use of the spectrum analyzer and the Bessel null technique of computing frequency deviation described earlier under microwave system testing. Figure 3-35 shows the spectrum occupancy of a typical satellite channel.

Aural subcarrier injection levels are verified using the spectrum analyzer

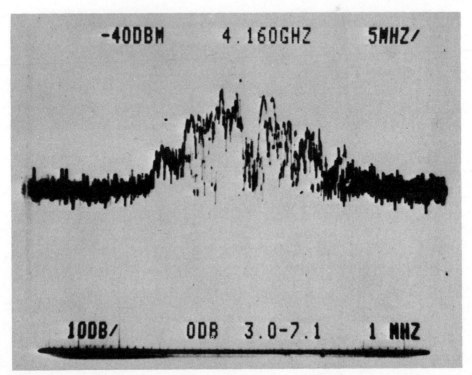

Figure 3-35   Spectrum occupancy measurement of single-satellite TV channel.

to measure the ratio of aural to main carrier levels. Figure 3-36 shows a properly adjusted system.

3.6.3 Measurement of Power into Antenna

First, tabulate all transmission line, directional-coupler, pad, and connecting-cable losses between the transmission line and the spectrum analyzer input connector. For example:

Directional-coupler coupling coefficient	−36.1 dB
Loss in fixed pads	−20.0 dB
Loss in connecting coax to spectrum analyzer	−9.9 dB
Total coupling losses	−66.0 dB

The spectrum analyzer is connected to the directional coupler, and the unmodulated carrier level is read directly in decibels referred to 1 milliwatt (dBm). With an indicated level of −6.0 dBm (Figure 3-37), the total transmission line level is computed to be ±60 dB above 1 milliwatt, or 1000 watts.

3.6.4 Klystron Amplifier Frequency Response Measurements

This measurement requires the use of a leveled microwave sweep generator in conjunction with the spectrum analyzer. This process is simplified by first

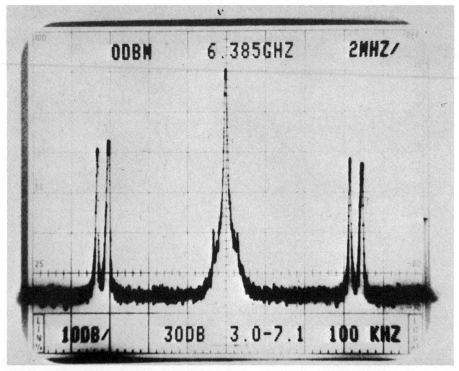

Figure 3-36 Verification of satellite TV aural subcarrier injection levels.

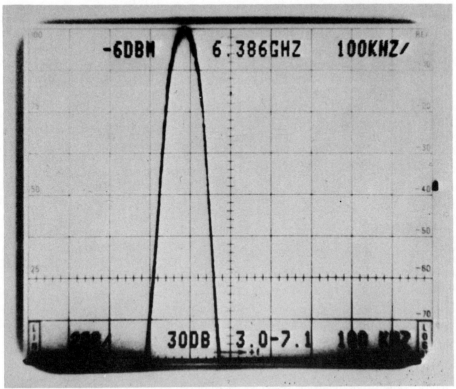

Figure 3-37   Power level measurement of satellite TV up-link at 6.386 GHz.

recording in the spectrum analyzer memory the frequency response of all connecting signal path hardware. The Klystron amplifier is then connected to the dummy load, powered up, and driven by the microwave sweeper.

The frequency response of each cavity can be adjusted for proper operation by monitoring the display shape on the spectrum analyzer CRT. The spectrum analyzer's SAVE A and MAX HOLD functions are used to develop the response accurately even though the sweep generator and spectrum analyzer are not frequency-locked together.

Figure 3-38 shows the response of the connecting hardware (upper trace) and amplifier response (lower trace) in the 10 dB/div log vertical mode. Figure 3-39 shows frequency response in the expanded 2 dB/div log vertical mode.

3.7 PULSED RF (RADAR) TRANSMITTER TESTS

Before describing various tests such as carrier on–off ratio, mainlobe-to-sidelobe ratio, effects of frequency modulation, and so on, a general understanding of how the spectrum analyzer treats pulsed rf signals is required.

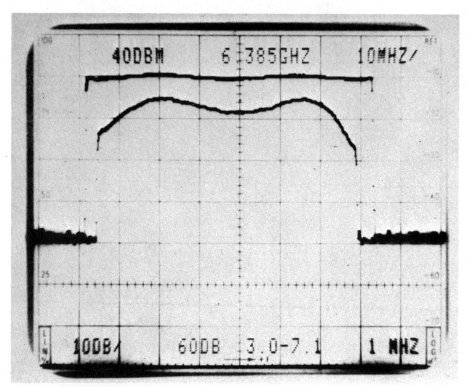

Figure 3-38 Swept frequency response measurements of the individual klystron amplifier cavities.

3.7.1 Continuous Wave

First, consider the spectrum display of a continuous-wave (cw) signal. The spectrum analyzer shows a single spectral line (Figure 3-40). The measured RMS power into the spectrum analyzer is −10 dBm (peak or average). The relationship between peak power and average power is expressed as $P_{ave} = P_{peak}(\tau/T)$, where τ is the pulse width and T is the total "on" time of the source (the reciprocal of the pulse repetition frequency prf). In the continuous-wave display of Figure 3-40, τ is equal to T; therefore, $P_{ave} = P_{peak}$.

3.7.2 Line Spectrum (Resolution Bandwidth < prf)

Now consider the spectrum display of a pulsed rf signal (Figure 3-41). Note that the energy is no longer contained in a single spectral line but in many lines. The indicated power is no longer −10 dBm but −34 dBm; this decrease in amplitude is caused by spreading the available energy over a wider frequency range (the energy per unit bandwidth has decreased). The signal in Figures 3-41 and 3-42 has a pulse width $\tau = 6.67$ μsec. This value is calculated by taking the reciprocal of the sidelobe null-to-null frequency difference or the difference between the central-lobe peak frequency and the first-sidelobe-null

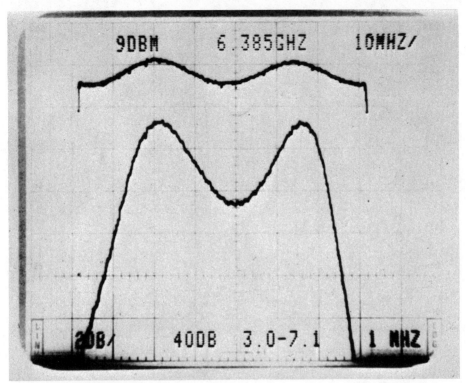

Figure 3-39 Magnified swept response measurements using the spectrum analyzer 2 dB/div vertical log mode.

frequency. The prf is 10 kHz (Figure 3-43). The equation for the *line spectrum* amplitude loss (in dB) is $A_L = 20 \log (\tau/T)$. In Figure 3-43, this loss corresponds to 24 dB.

Figure 3-44 shows the cw-to-pulse dB difference in one display using the spectrum analyzer SAVE A mode. Remember, this relationship is valid only if the resolution bandwidth is equal to or less than 0.3 times the prf. The above examples use a 3-kHz resolution bandwidth, which is 0.3 times the 10-kHz prf.

Figure 3-45 shows the relationship between the spectrum analyzer resolution bandwidth and the spectral-line spacing. Line spectra (as opposed to dense or pulse spectra) can be verified quickly by changing the spectrum analyzer frequency span/div control and noting that the line spacing changes (see Figures 3-41, 3-42, and 3-43).

The average and peak powers during the pulse are dependent on the duty cycle τ/T. These values are expressed in decibels by adding 10 log (τ/T) to the RMS value of the resolved carrier for average power and by adding 20 log (τ/T) to the RMS value for peak power. In Figure 3-41, this addition corresponds to an average power during the pulse of -22 dBm and a peak power during the pulse of -10 dBm. These peak and average values are referenced to the re-solved level of the carrier, Figures 3-42 and 3-43.

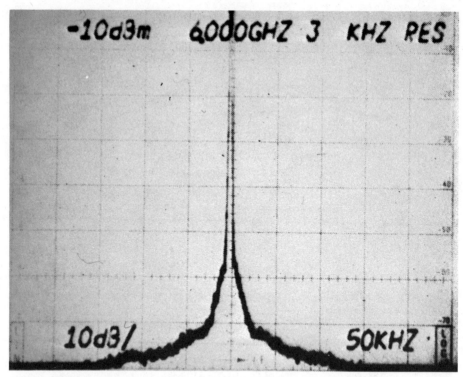

Figure 3-40 Continuous-wave spectrum with excellent spectral purity.

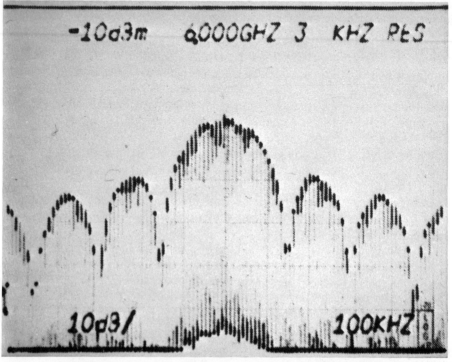

Figure 3-41 Pulsed rf spectrum produced by 6.6-μsec rectangular pulse.

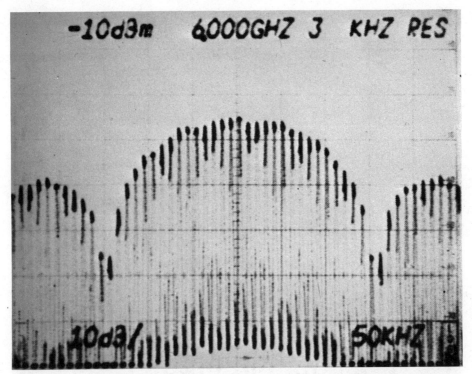

Figure 3-42　Expanded view of pulsed rf spectrum showing repetition rate line structure.

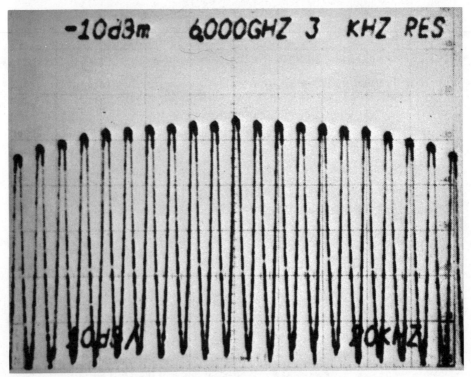

Figure 3-43　Expanded view of main lobe peak amplitude level.

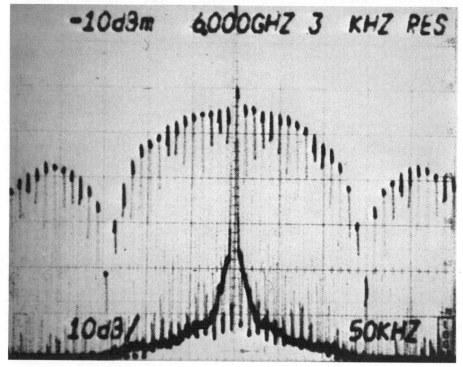

Figure 3-44   Measurement of pulsed rf signal level and cw signal level using spectrum analyzer save "A" mode.

3.7.3 Pulse or Dense Spectra (Resolution Bandwidth > prf)

Individual lines in this display are not resolved (Figure 3-46). In addition, the display is both time- and frequency-related. The envelope is frequency-related and will vary with the spectrum analyzer span/div (Figures 3-47 and 3-48). The lines within the envelope are time-related and will vary with the sweep speed

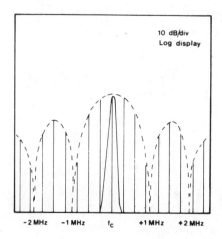

Figure 3-45   Resolution of individual spectral line in a pulsed rf spectrum.

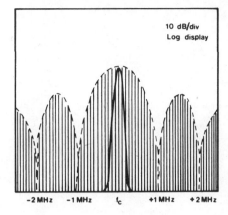

Figure 3-46 Nonresolved repetition frequency lines in a dense pulsed rf spectrum.

or scan time/div of the spectrum analyzer (Figures 3-48 and 3-49). It should be noted that each line represents one complete pulse, whereas in a line spectrum display each line represents a single frequency component within the pulse.

The difference in displayed amplitude between the continuous-wave signal and the pulsed signal is no longer dependent on duty cycle, as in the line spectrum display, but is related to the spectrum analyzer resolution bandwidth/pulse width product. Doubling the resolution bandwidth increases the

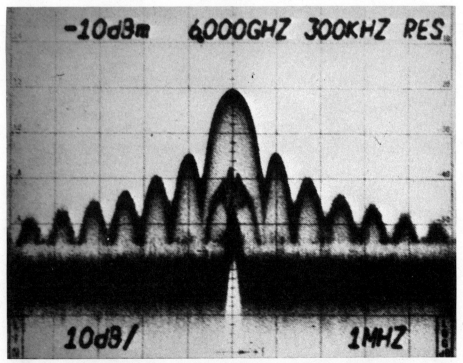

Figure 3-47 Envelope of pulsed or dense rf spectrum with good sidelobe symmetry.

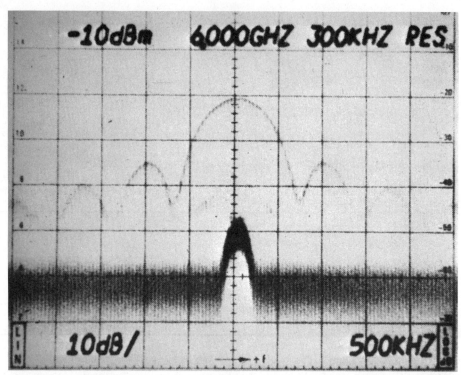

Figure 3-48 CRT display of pulse or dense spectrum showing envelope frequency change with frequency span/division (500 kHz) compared with Figure 3-47 (1 MHz/div).

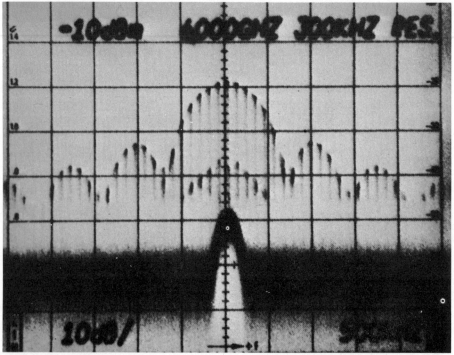

Figure 3-49 CRT display of pulse or dense spectrum showing change in repetition rate line spacing with sweep speed when compared to Figure 3-48.

displayed amplitude by 6 dB, and a 10-fold resolution bandwidth change will produce a 20-dB amplitude change. This linear relationship holds true as long as the pulse width/resolution bandwidth product ($\tau \cdot B$) does not exceed 0.2.

A value of 0.1 yields good null definition consistent with reasonable amplitude loss. The amplitude loss is expressed as

$$A_L \quad (\text{dB}) = 20 \log K \cdot \tau \cdot B$$

where τ is the effective pulse width, B is the spectrum analyzer resolution bandwidth, and K is a correction factor for the difference between the impulse bandwidth and resolution bandwidth. In Figure 3-50, $\tau = 5 \times 10^{-6}$ second, $B = 30$ kHz, and $K = 1$. The calculated amplitude loss is 16.48 dB. For manufacturers who specify the resolution bandwidth as 3 dB and use synchronous filters (Hewlett-Packard), K approximates 1.5. The graphs in Figure 3-51 plot the amplitude loss in dB versus $\tau \cdot B$ for both $K = 1$ and $K = 1.5$. It is important to remember that the power input to the first mixer of the spectrum analyzer should not exceed -10 dBm peak (-1 dB compression) if accurate amplitude measurements are to be obtained.

Reasons for using a pulse spectrum display over a line spectrum are as follows:

1. Better signal-to-noise ratios are obtainable. Display voltage amplitude increases linearly with resolution bandwidth (20 log $K \cdot \tau \cdot B$), while the noise

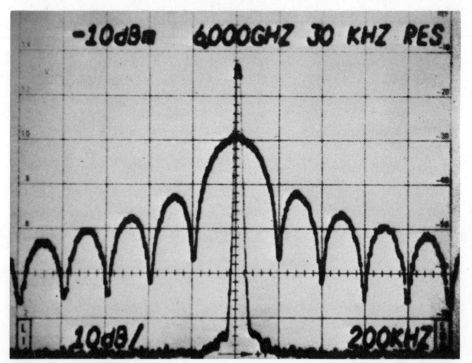

Figure 3-50 CRT display of pulsed rf signal and continuous-wave signal having same peak power level.

Loss in Amplitude, Pulse or Dense Spectrum vs. CW

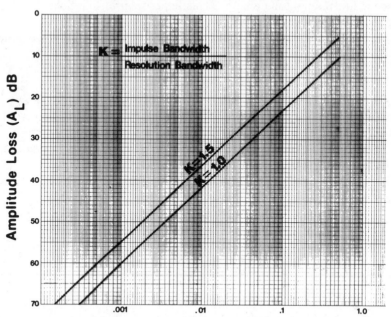

Pulse Width Resolution Bandwidth Product (τ • B)

Figure 3-51   Graph utilized in determining the loss in displayed amplitude on the spectrum analyzer as a function of its resolution bandwidth and the rf signal pulse width.

floor voltage increases as the square root of the resolution bandwidth (10 log B). Figure 3-52 shows a pulse display using 30-kHz resolution bandwidth, and Figure 3-53 shows the same pulse using 300-kHz resolution bandwidth.

2. Faster sweep speeds are possible because wider resolution bandwidth can be used.

3. Good shape and depth of sidelobe nulls are obtainable.

4. No prf jitter to limit line spectrum resolution.

It should be remembered that a sufficient number of pulse lines is required for good envelope definition. The sweep speed should meet the following criterion:

$$\text{sweep speed (sec/div)} \geq \frac{10}{\text{prf (Hz)}}$$

3.7.4 Additional RADAR Transmitter Test Considerations

Connecting the Spectrum Analyzer   Because radar transmitters are generally high-power devices, attenuation in the form of a transmission line directional coupler is required to prevent damaging the spectrum analyzer rf attenuator or

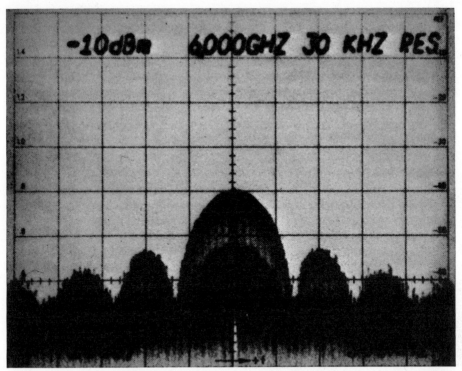

Figure 3-52 CRT display of pulsed rf spectrum utilizing the spectrum analyzer's 30-kHz resolution bandwidth.

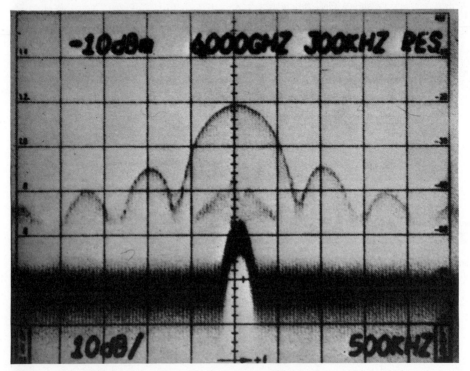

Figure 3-53 CRT display of pulsed rf spectrum utilizing the spectrum analyzer's 300-kHz resolution bandwidth.

input mixer. Alternately, a test antenna can be used to receive a small amount of radiated energy.

Effect of Pulse Shape   A rectangular modulating pulse with good rise and fall times will create an rf energy distribution similar to Figure 3-54, referred to as *SIN X/X*. The mainlobe-to-first-sidelobe ratio is 13.5 dB, which is quite close to the theoretical value of 13.26 dB stated earlier in the section on digital microwave radio measurements. To display the fine detail of the pulsed rf spectrum with sharp lobe-to-lobe null definition, the pulse width/resolution bandwidth product should be less than 1/10.

A trapezoidal pulse (a rectangular pulse with appreciable rise and fall time) will have a large mainlobe-to-sidelobe ratio; 20 dB is indicated in Figure 3-55. A triangular pulse is the extreme rise-and-fall case of the rectangular pulse and has a mainlobe-to-sidelobe ratio of 26 dB (Figure 3-56).

Effect of Frequency Modulation   Pulsing of a magnetron or klystron oscillator is generally accompanied by some frequency shift (frequency modulating). Pulsed rf signals with FM do not have distinct nulls, and the sidelobes are larger than pulsed rf signals without FM. Nonsymmetrical pulse shapes, such as a ramp, will produce an unsymmetrical spectrum (Figure 3-57).

Carrier On-to-Off Ratio   This measurement requires measuring the power level of the continuous feedthrough carrier level and comparing it to the corrected peak value of the mainlobe. This correction for spectrum analyzer cw-to-pulse amplitude loss was described in detail earlier.

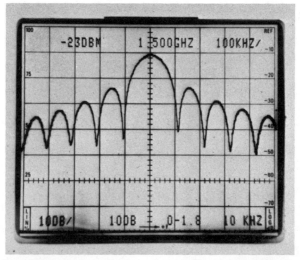

Figure 3-54   CRT digital storage display of a pulsed rf spectrum with good null definition and a rectangular modulating pulse.

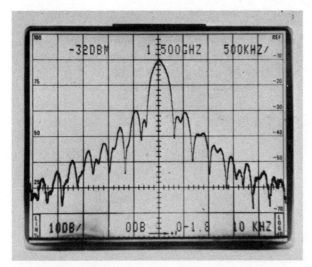

Figure 3-55   CRT digital storage display of a pulsed rf spectrum created by a trapezoidal shape modulating pulse.

For example, consider a 10-kW radar transmitter with a 40-dB directional coupler connected to the spectrum analyzer. The carrier on-to-off ratio is computed to be 20 dB from the CRT display in Figure 3-58, and verified in Figure 3-59 by increasing the resolution bandwidth to 1 MHz (10 times). A 20-dB carrier on-to-off ratio means that 100 watts of continuous power is being transmitted to the antenna during pulse-off conditions.

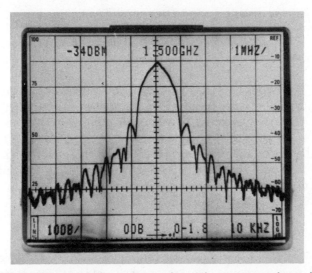

Figure 3-56   Triangular modulating pulse produces the spectrum shown in this display.

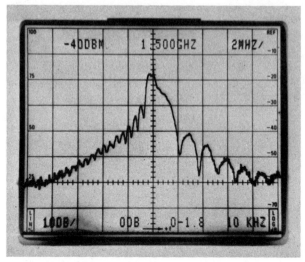

Figure 3-57 The unsymmetrical spectrum shown in this photo is the result of FM'ing of a pulsed rf source.

3.8 SPECTRUM ANALYSIS OF DIGITAL PROCESSING SYSTEMS

3.8.1 General Requirements

Regulatory Agency Requirements The intent of regulations governing interfering emissions from electrical and electronic devices is to optimize the use of the electromagnetic spectrum by limiting unnecessary and undesirable emissions. The FCC in the United States and its counterparts throughout the world govern the level of these unwanted emissions commonly referred to as RFI (radio-frequency interference), or EMI (electromagnetic interference).

Measuring Equipment Spectrum analyzers or radio receivers with appropriate ancillary items, such as antennae and current transformers, are in wide use for RFI/EMI measurements. Testing falls into two basic categories: radiation testing and conducted testing. The spectrum analyzer is well suited for the measurement of radiated emissions because of its wide frequency range and good sensitivity. A calibrated antenna for the appropriate frequency range and an appropriate test range, or a reflectionless room (anechoic chamber), is needed to ensure meaningful and reproducible results.

3.8.2 Terminology Review

Several important terms need to be described, as they are somewhat unique to RFI/EMI measurements. They are: antenna factor, impulse bandwidth, quasi-peak detector, and LISN (line impedance stabilization network).

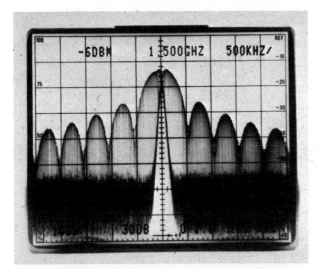

Figure 3-58 The steady-state carrier at the center of this display is cw feed-thru resulting from poor modulator off-conditions.

Antenna Factor An antenna can be thought of as a transducer or coupling device that converts the energy in electrostatic and electromagnetic waves (E) to electrical energy (V) at the input of the spectrum analyzer. This coupling factor is called antenna factor (AF) and is expressed mathematically as the ratio of field intensity (volts per meter) to volts across the antenna terminals.

In field intensity measurements, it is necessary to convert from the spec-

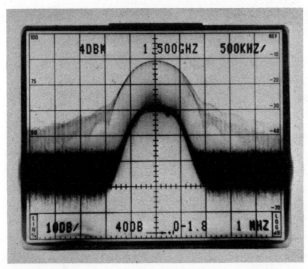

Figure 3-59 Expanded view of cw feed-thru signal used to facilitate measurement of carrier on-to-off ratio.

trum analyzer reading in volts (or microvolts) to field intensity in volts per meter. When making measurements, add 20 log AF to V in dB above a microvolt to get E in dB above a microvolt per meter.

The antenna factor deals with voltages (E/V). Therefore, the dB number is computed as 20 log $AF = AF$ in dB. Antenna gain (G), on the other hand, is a power ratio. Gain, in dB, is given by 10 log $G = G$ in dB. For an ideal matched isotrope ($G = 1$) at 1 MHz, the antenna factor is 29.78 dB. The above relationships yield

$$\text{Antenna Factor } AF = 20 \log f\text{MHz} - G\text{dB} - 29.78 \text{ dB}$$

Typical antenna factors will vary from 7.6 (17.6 dB) for a 300-MHz tuned half-wave dipole to 10 dB for a 30-dB gain horn with reflector at 3 GHz, to 15 dB for an optimum proportioned horn having a 0.1-meter opening, to a linear variation of 26.5 dB at 1 GHz increasing to 47.5 dB at 10 GHz for a log periodic spiral.

The greater the antenna gain (directional properties), the lower the antenna factor and the lesser the requirement for spectrum analyzer sensitivity. Remember that the AF dB is added to the spectrum analyzer reading (V dB) to get the field strength E dB. Conversely, for a given field intensity, there is less input level to the spectrum analyzer as the antenna factor is increased.

The theoretical antenna factor calculations assume a perfectly matched, lossless antenna. Losses add to the antenna factor. Therefore, published numbers may differ slightly from calculations. Figure 3-60 is a graph of antenna factor versus frequency for a specific antenna.

Impulse Bandwidth   Real circuits respond to very narrow pulses as though they were impulses. To measure impulse spectral intensity, one must know the impulse bandwidth of the measuring instrument.

There are several ways to determine a spectrum analyzer's impulse bandwidth. The 6-dB-down bandwidth may be used as a simple approximation for synchronously tuned stages.

For high-accuracy applications, it is useful to measure the actual impulse bandwidth of the individual instrument. This measurement may be made by applying a train of very narrow pulses to the spectrum analyzer input. The impulse bandwidth B_i is then computed from

$$B_i = (V_p/V_{ave}) \text{ prf}$$

where V_p is the peak output voltage, V_{ave} is the average output voltage, and prf is the pulse repetition frequency. The restrictions are that prf $\leq B_i/5$ and that the pulse width be less than about 1/10 the inverse of B_i. V_p and V_{ave} may be most easily measured using the peak and average functions of the spectrum analyzer.

Measurement example: A train of narrow (50 nsec) pulses at a 5000-Hz rate is used to produce the impulse spectral-energy distribution. The spectrum analyzer gain is set for an output peak response of 8 divisions in the linear vertical-display mode. The measured output average response is 1.3 divisions using the spectrum analyzer postdetection (video) filter. Therefore,

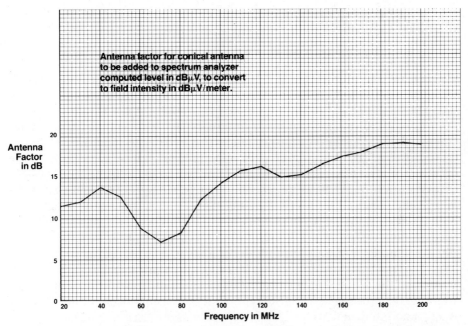

Figure 3-60   Antenna factor versus frequency graph for a typical conical antenna.

$$B_i = \frac{8 \text{ divisions (peak)}}{1.3 \text{ divisions (ave)}} \times 0.5 \text{ kHz} \simeq 3.1 \text{ kHz}$$

A third method is to use a commercially available calibrated impulse generator with known spectral intensity.

Quasi-Peak Detector   This type of detector has long been used in RFI and field intensity meters and has an output proportional to the type of signal measured. An arbitrary weighting function having a charge time of 1 msec and a discharge time of 600 msec is used. The quasi-peak level can be equal to or less than the peak value, but never greater.

Most spectrum analyzers use a peak detector calibrated in RMS units; to obtain quasi-peak values, therefore, a correction factor based on the interference repetition rate and the spectrum analyzer impulse bandwidth must be used. This procedure is covered in detail in Tektronix EMI Application Note 26W-4971. In the United States, the Federal Communications Commission will accept either peak or quasi-peak readings.

LISN (Line Impedance Stabilization Network)   Power-line-conducted emissions refer to the interference fed back into the ac power source via the power cord. Although the long wavelengths (450 kHz) make radiation measurements impractical, it is not difficult to standardize the line impedance over the desired frequency range using an LC network with a 50-ohm voltage divider to the spectrum analyzer. The current standard is a 50-ohm/50-microhenry network installed in series with each current-carrying conductor except the ground conductor. LISNs are available commercially or may be constructed. This subject is treated in greater detail in Tektronix EMI Application Note 26W-4971.

3.8.3 Narrowband and Broadband Measurements

For narrowband measurement, it is only necessary to calibrate the spectrum analyzer in decibels relative to 1 (dbμV). This 1-μV reference is used throughout the industry for this type of measurement.

Broadband implies that the interference being measured has a greater bandwidth than the measurement instrument's bandwidth. The same 1-μV reference is used, but it now becomes important to include the effects of the spectrum analyzer impulse bandwidth. How this bandwidth affects the values obtained is associated with the repetition rate of the signal being measured. As long as the impulse bandwidth is greater than 1/period, the impulse signal will have the same amplitude as the single-pulse signal. When the impulse bandwidth becomes less than 1/period, however, the spectrum analyzer sees the impulse signal as a pure cw signal with a greater amplitude than that of the single pulse.

The spectrum analyzer should be calibrated to include the impulse bandwidth, and the spectral intensity must be constant over the range of this bandwidth. In the broadband (impulse) type of measurement, the spectrum analyzer is calibrated in dBμV/MHz. Calibrated antennas are used in conjunction with the spectrum analyzer.

Figure 3-61 is a graph showing the relationship between dBm, dBμV, and dBμV/MHz for an impulse bandwidth of 100 kHz.

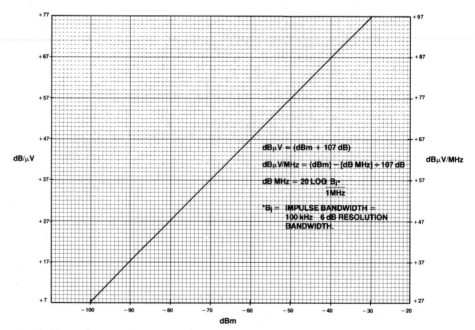

Figure 3-61   Graph used to determine dB/μV or dBμV/MHz from spectrum analyzer level in dBm.

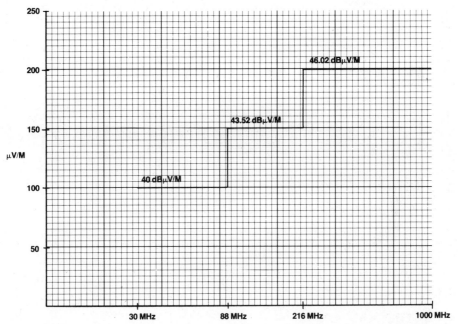

Figure 3-62   Graph of radiated Class B limits versus frequency from FCC Part 15 Subpart J.

3.8.4 Compliance Testing

Figure 3-62 is a graph showing the radiated Class B limits versus frequency. This graph is taken from the FCC Part 15, Subpart J regulations governing all equipment using digital techniques with clock signals or pulses greater than 10 kHz.

The basic test procedures to satisfy limits imposed by regulatory agencies (FCC, VDE, CISPR) or the military (MIL-STD-461B) are very similar. For repeatable results, radiation measurements should be made in an area free from metallic or other reflecting objects. For small distances (3-m spacing), an anechoic chamber is best. For larger distances (30 m), an outdoor site free from power lines and any reflecting objects for a distance of 2 wavelengths or more at the lowest frequency is desirable.

Conductive measurements require attention to grounding detail to avoid or minimize parasitic resonances and ground loops between the device under test and the test equipment. A ground plane of 2 m² or more is desirable.

Additional information on RFI/EMI testing can be found in Tektronix EMI Measurements Note 26W-4971 and in an article entitled "Testing Products Correctly Ensures EMI-Spec Compliance" by Isidor Straus in *Electronic Design News,* November 25, 1981.

CHAPTER 4

Logic and Signature Analyzers

Bernard McIntyre
University of Houston

4.1 INTRODUCTION

The oscilloscope is an ideal instrument for real-time monitoring of repetitive analog voltage waveforms. The trigger system of the oscilloscope allows the beam sweep to begin at a particular input voltage level so that a periodic waveform will appear stationary in a display. A fast (faster than a microprocessor clock) single-beam oscilloscope can be used to accurately monitor microprocessor control lines or check timing relationships between lines. There are several situations involving digital microprocessor systems in which the oscilloscope can be of very limited value. Data and address lines can be floating electrically most of the time, complicating analysis on the oscilloscope and rendering them useless for triggering. Aside from the system clock, the only periodic waveforms are those obtained by programming the microprocessor to go through a short loop routine which repeats indefinitely. Also, typical oscilloscopes display multiple waveforms by chopping the beam or alternating between channels. In the alternating mode, one can observe a waveform on one data channel only by having no data represented on the other data channel. The oscilloscope must then be able to alternate channels at a rate much higher than that of the microprocessor clock.

In the development of digital hardware and software systems, one must be concerned with the simultaneous behavior of a group of voltage lines such as a data bus or an address bus. In these situations, it is advantageous to assimilate the behavior of the group of lines into digital logic words or bytes rather than study the behavior of individual lines. Systematic errors found in

Note: Readers who are not familiar with microprocessors may wish to read Appendix A, Basic Microprocessor Concepts, before reading this chapter.

these bytes or words can be more easily translated into individual line problems, if they exist, or hardware and/or software problems.

Digital data analysis can be greatly facilitated by the use of the logic and signature analyzers. The logic analyzer can be used in the checkout phase of a hardware design problem by tracking and recording the actual flow of data words in the system. Once the system is working properly, a test program is used to generate a stream of data bits. The collective behavior of these bits on each individual line will create a *signature* for that particular line. In the event there is a system failure later, each signature obtained in the test run can be compared with those obtained for the working system. Points at which the two signatures do not match can indicate where possible failures exist. In this chapter, we study the use and theory of operation of the logic and signature analyzers. Examples of their use will be given for a Z80-based microprocessor system.

4.2 INTRODUCTION TO THE LOGIC ANALYZER

4.2.1 Operation

A logic analyzer can be used in a synchronous or asynchronous mode. When used in a synchronous mode, the analyzer data input is strobed in by a clocking signal from the system under test. In the asynchronous mode, the analyzer uses an internal clock to sample the data lines of the test system. For either mode of operation, the data which is clocked in is then treated as shown in the block diagram in Figure 4-1. In this diagram, the data input is clocked into memory and compared with a preset trigger word which is stored in the word recognizer and comparator section. As each word is clocked in, the memory address counter is incremented. When the incoming data match the stored trigger word, a preselected switch determines whether that RAM (random access memory) address will be the first or last one to be displayed.

A sketch of option switches found on any logic analyzer is shown in Figure 4-2. The option switches shown can be used to make the trigger word start or stop the display. In a positive time mode setting, the address of the data word which matches the stored trigger word is effectively made that of the first RAM location for display. Data is then clocked in so that earlier RAM

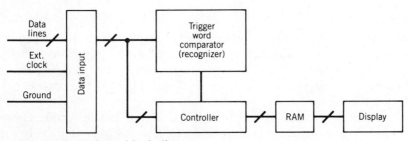

Figure 4-1   Logic analyzer block diagram.

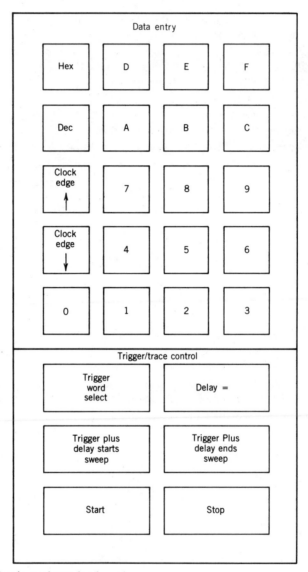

Figure 4-2 Logic analyzer keyboard.

data are replaced by the data that followed the trigger word. In a negative time mode setting, data is clocked into RAM until the trigger word is detected. Further data input is blocked and the trigger word is set to be the last word for display. The display will then show all data in RAM which were clocked in before the trigger word.

All logic analyzers can work in a positive or negative time mode. However, several logic analyzers have additional useful features such as glitch (spurious, short-duration voltage) detection, glitch display, a second RAM for data comparison, the ability to rearrange data words into assembly language format, and other options. The following sections discuss the basic aspects of using a logic analyzer and some of the more common optional features.

4.2.2 Asynchronous Operation

Data Input The data input section of the logic analyzer contains threshold voltage circuits and converts inputs to TTL (transistor–transistor logic) levels, if necessary. The probe dc input impedance is high, usually greater than 1 MΩ, and the parallel capacitance is 5 to 10 pF. For these values, the probes produce no noticeable effects with a microprocessor using a 5-MHz clock. Data on the probes is clocked in with an internal clock that has a sampling rate of at least 20 MHz. The sampling clock speed should be at least three times the microprocessor clock speed. Figure 4-3 shows what a sampled data line could look like when using a 10- and 20-MHz sampling clock. The data input as seen by the 10- and 20-MHz systems in Figure 4-3 is identical as long as the data line is varying periodically. When deviations from this periodicity occur, the system with the faster clock will detect the change in the data line earlier.

Trigger, Memory Systems Word recognition circuits receive a trigger word from the keyboard. While input data are being sampled by the internal clock, the data are stored in consecutive memory locations in RAM. If the user selects the negative time mode, the TRIGGER PLUS DELAY ENDS SWEEP switch is used and the data acquisition will cease when the incoming data match the trigger word. If a trigger delay option is also used, data collection will cease after a set number of sample clock cycles following the trigger word detection. A 256-word RAM may be filled several times over with data, but when the trigger word is recognized, the RAM address of the trigger word is stored and

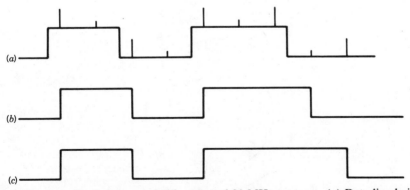

Figure 4-3 Data line being sampled by 10- and 20-MHz systems. (*a*) Data line being sampled. Hash marks indicate when the data is effectively latched by the sampling system. The long hash marks are for the 20- and 10-MHz clocks, while the short marks are for the 20-MHz system only. (*b*) Data line as seen by the 20-MHz system. (*c*) Data line as seen by the 10-MHz system.

the data acquisition ceases. The display will then be set to start with the data word following the trigger word in RAM and end with the trigger word. This is shown in Figure 4-4.

If a trigger delay were also used, the data displayed would depend on the delay count as shown in Figure 4-5. This feature is useful for following the performance of a program containing a loop; after each run with the analyzer, the trigger delay count could be increased by up to 256. Another useful option is the trigger qualifier: an incoming trigger word is not recognized unless a qualifying condition is also true. For example, the trigger word may be the machine code for loading the stack pointer, and the memory address of that code would be the qualifying condition. The qualifying condition may be necessary because the machine code used for the trigger word may be used more than once in the user program, or, the code may also be present in a monitor control program which is used to execute the user program. On some units, there may be only one or two qualifier lines available. In that event, a control line, such as RAM select, is used rather than the address bus. One key address line might also be used to advantage.

When the positive time option is selected, RAM is filled with incoming data until the trigger word is recognized and a counter is started. For the 256-word memory assumed here, data will continue to be clocked in until the eight-bit counter reaches 255. The trigger word becomes the first word displayed, and the word before it in memory becomes the last word displayed.

Display Modes   Once the data lines are sampled and stored in RAM, three modes of display are available to the user. These are the timing, map, and state display modes. Although all three modes may be available in a particular unit, they may not all be available in the asynchronous mode of operation. Figure 4-3 is an example of a timing display for a single data line, although up to eight data lines can usually be displayed in this mode. It is difficult to follow and

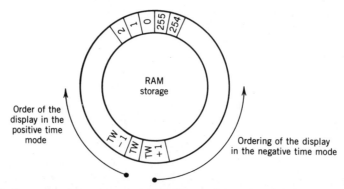

Figure 4-4   Incoming data are stored in RAM starting at 0 and written over earlier data if necessary until the trigger word (TW) is recognized and stored. TW + 1 would be the first word displayed in a negative time mode, and TW would be the first display word in a positive time mode.

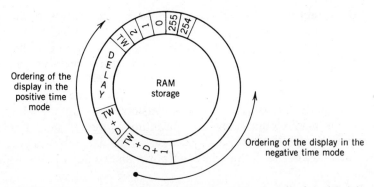

Figure 4-5 The trigger word is stored but not recognized until after the delay. For a delay greater than 255 counts, the trigger word will not appear but will have been written over.

coordinate the data transitions occurring for multiple lines, such as a data bus, so the timing mode is more useful in studying control lines. When displaying a control line, the sampling clock is displayed along with other relevant lines. For example, in a Z-80-based system, there is a control line, M1, which indicates that an op code fetch is in progress. By monitoring the memory read line along with M1, one can determine whether memory is being referenced as part of a machine code fetch operation or a data transfer. This use of the timing mode is similar to the procedure of triggering one channel of a dual-beam oscilloscope off another channel. With the oscilloscope, the number of channels is usually limited to two; with the logic analyzer, there are usually eight.

In the logic state display mode, all data in RAM beginning or ending with the trigger word are displayed in binary, octal, or hexadecimal format. In the positive time mode, the trigger word is the first data word displayed. This mode is useful for following the flow of data on the data or address bus.

In the mapping mode, a grid is formed on the CRT. If the logic analyzer memory is 256 × 8, the CRT might be divided into a 16 × 16 matrix. The low-order nibble of the data byte could be used to order the rows, and the high-order nibble, to order the columns. Each byte of data in RAM will then fit on one of the matrix points. Blank or dark points on the CRT indicate that the corresponding data are not in RAM; bright matrix points indicate that data points are present more than once in RAM. The more frequently the data points occur in RAM, the brighter the corresponding display. In addition, the data or matrix points are vectored. A vector is used to connect consecutive data points, and the vector gets brighter near the new data point. A single data point appearing in the display could be indicative of an error in the data. By using the vectors to and from this data point, one might be able to determine where the data error occurred. Mappings are especially useful when the address bus is being monitored. If the low-order byte of the bus is displayed horizontally and the high-order byte vertically, loops and jumps to subroutines can be made to stand out in the mapping display. As the stored program is being read and executed from consecutive memory locations, the display will move horizontally. When a jump to a subroutine occurs such that the high-order address changes, there is also some vertical displacement. The subroutines can then be made to stand out in an isolated section of the CRT. A glitch on the address bus, resulting in an address far removed from the rest of memory space used, would also stand out in the display. To make use of this mapping feature on a particular logic analyzer model, the programmer should be aware of how the bus is being used for the display. Some models do not use all the address lines to generate the matrix points, and so subroutine addresses should be chosen to be compatible with the logic analyzer model. The data bus would not be as useful as the address bus for mapping purposes since data will not usually be sequential.

Glitches   The data displays discussed in the previous section are available on many models in the asynchronous, or sampling, mode, in which data are strobed on a transition of the sampling clock. If a data line voltage were to rise and fall through the voltage threshold within one sample clock period, the sequence, or glitch, would not be detected in many systems. A glitch detection system can record these rapid voltage fluctuations as they occur. A standard latch system for glitch detection is shown in Figure 4-6. Figure 4-7 shows a data bit stream with a series of glitches G_1–G_4. The leading edge of the clock was used to sample the data line, and it was assumed that both flip-flops were initially reset to zero. The output Q_1 can be set to 1 by NAND gate 1 or by the clocking of input data; it can be reset to 0 by NAND gate 2 or clocked input data. The leading edge of the first glitch, G_1, is detected by NAND gate 1 as it occurs, since $\overline{Q}_2 = 1$, but the falling edge is not detected until the clock edge occurs. Due to this first glitch, the output at Q_2 is seen to be one clock period in width, even though the actual width of G_1 was considerably smaller. Note also that glitch G_3 was not detected because Q_1 and Q_2 had different logic states. The

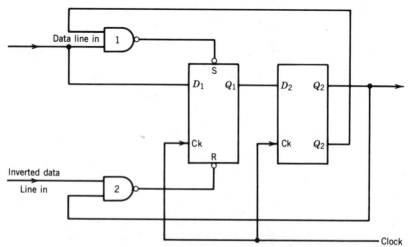

Figure 4-6   Latch system for a logic analyzer.

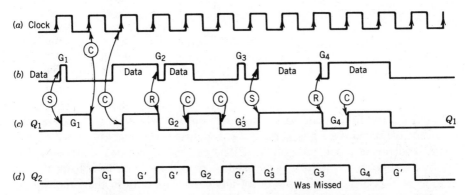

Figure 4-7   The leading edge of the clock in (*a*) is used in the latch system of Figure 4-6. Glitches on the data line in (*b*) are detected and widened to one clock period in (*c*) and are part of the output in (*d*). The arrows coded S, R, and C are used to indicate whether the transition in Q_1 occurred as a result of the first flip-flop being set or reset, or data clocked in.

glitch detection system provides useful information when the data line highs and lows are several clock periods long and glitches are infrequent. However, the glitch detection system cannot sort out data from glitches that occur as frequently as they do relative to the clock frequency in Figure 4-7. In effect, the low data voltage preceding G_3 is so short in duration that it is seen as a glitch. Since the minimum glitch width is one clock period, the data and G_3 are merged into one. The data line in Figure 4-7 is so noisy that the output is very nearly a succession of glitches and would be of very little value to a user.

4.2.3 Synchronous Operation

Data storage in the logic analyzer can also be synchronized to the external system being studied by using the clock input probe line. If the user is not worried about conserving memory in the logic analyzer RAM, the clock line may be connected to the microprocessor clock. Just as in the asynchronous mode, there will be repetitious storing of data because the data or address bus could have stable data for two or more consecutive clock cycles. There could also be a problem with data lines being at some undetermined level (floating) while the clock transition is made. This situation can be easily spotted because it occurs between two groups of valid data words.

A more economical use of RAM is achieved by using a memory I/O request line, chip select line, read, write, or other control line for a clock input. The choice of control line and of leading or trailing edge should be made by referencing timing diagrams for the particular system. One can be seriously misled if the edge of a control line transition is selected when there is no guarantee of valid data being present. Once the data are clocked into RAM and triggered, the display modes available could be the same as in the synchronous mode of operation.

4.2.4 Example Problem

To illustrate the methods of taking data with the logic analyzer, a simple program will be used for a Z80-based system. This computer has a ROM (read only memory) monitor program in memory locations 0000 to $1FFF_H$. The program will be stored in RAM beginning at 2000_H and will result in the data byte 00 being stored in location 2100_H, 01 in 2101_H, and so on. Then, a block transfer of these data will be made to another section of RAM. A logic analyzer will be used to record data bus activity by synchronously clocking on the read, write, and phase clock pins of the CPU (central processing unit). A simple hobby-kit-type logic analyzer was used to record the data flow from the operation of this program. This type of analyzer runs only in the synchronous

Table 4-1 **Example program listing**

Statement number	Memory location	Hex code	Mnemonic	Comments
1	2000	21	START: Ld HL, 210F	Load register pair HL with address 210F
	2001	OF		
	2002	21		
2	2003	75	Loop: LD(HL),L	Load contents of register L into memory location given by HL pair
3	2004	2D	Dec L	Decrement register L; zero flag will be set when L = 0
4	2005	C2	JPNZ LOOP	If zero flag is not set, go back to label LOOP (2003)
	2006	03		
	2007	20		
5	2008	75	LD (HL), L	Load data byte L in memory location (HL); L = 00, HL = 2100
6	2009	11	LD DE, 2200	Load destination address of memory into DE
	200A	00		
	200B	22		
7	200C	01	LD BC, 0010	Load number of data bytes to be transferred into BC
	200D	10		
	200E	00		
8	200F	ED	LDIR;	Contents of memory at (HL) go to memory location (DE); increment DE and HL; decrement BC; repeat until BC = 0
	2010	B0		
9	2011	C3	JP START;	Loop back to beginning
	2012	00		
	2013	20		

mode and has only a state display with no trigger qualifier, but still provides much useful information.

The program listing, in assembly and machine language, appears in Table 4-1 with comments. The hexadecimal numbering system is used throughout the program and its discussion.

The data lines of the analyzer were connected to the data bus of the CPU, and the clock line was connected initially to the CPU read line. The trigger word was set to op code 21_H, and the program was loaded into RAM and executed. The analyzer was not started until after the program was looping. If the analyzer were activated first and waiting for the trigger word, it would most likely trigger from a 21_H data byte or op code used in the monitor program of the Z80 system. (Although the user is only pushing an execute button on the micro to cause the program to execute, a good deal of software is being executed to accomplish this.) We chose to get around that problem by starting the program first, keeping it in a loop, then activating the analyzer. Of course, this is just the situation in which one would want to have features such as trigger delay and qualification. With a trigger delay, one could experiment with the number of clock counts needed to get past the monitor program before data are put into RAM. With data qualification, one could use 21_H as the trigger word only after the address line is 2000_H.

The upward transition of the read line was selected to clock in data, and the positive time mode was used. The upward transition was selected since that is the instant at which the CPU assumes that there is valid data on the bus and strobes it in. The downward transition will work in many cases, but occasionally the data are incorrect because of lines still changing state. The following 16 data bytes were observed and are grouped into four categories for discussion purposes.

CATEGORY	DATA	CATEGORY	DATA
1	21	4	C2
2	75		O3
3	2D		2O
4	C2	2	75
	O3	3	2D
	2O	4	C2
2	75		O3
3	2D		2O

CATEGORY 1. The op code 21_H is observed as part of a data fetch operation because the CPU performs an op code fetch by reading memory. In this case, the analyzer started not when the CPU was at the beginning of the program, but further down, at the address 2100_H. The point is, one cannot predict where the analyzer will interrupt the program. The first data byte, 21_H, read into the CPU after the analyzer is activated initiates the trigger.

CATEGORY 2. The op code 75_H is read into the CPU. The contents of register L will go on the data bus and into RAM. For the CPU to do that, it performs a memory-write operation, so the L register contents do not get stored in the analyzer.

CATEGORY 3. Op code $2D_H$ is read into the CPU.

CATEGORY 4. The op code $C2_H$ and associated address 2003_H are read into the CPU. If the zero flag is not set, the program will loop back. These data illustrate the CPU going through the loop three times. A trigger delay would be useful here if one wished to make sure the CPU was looping in the manner in which the program was designed and exiting properly from the loop. By increasing the trigger delay count each time the program is run, all loops and the exit could be observed. Not having the use of a trigger delay, we assumed that the remainder of the loop is operating properly and selected a trigger word of 11_H, outside the loop. The following data were observed:

CONDITION	DATA	CONDITION	DATA
1	11	3	ED
	00		B0
	22		01
2	01	3	ED
	10		B0
	00		02
3	ED	3	ED
	B0		
	00		

CONDITION 1. The first three bytes are associated with an op code and address being read from memory.

CONDITION 2. Op code LD BC, 0010_H is read from memory.

CONDITION 3. The LDIR op codes are ED and B0. The CPU transfers data from one memory location (HL) to another (DE) and decrements the program counter twice. This results in the LDIR op code being read in again by the CPU. This continues until the counter BC = 0. In order to do all of this, the CPU must perform a read operation at the address given by register HL, and a write operation on the address given by DE. As a result, the contents of HL, 00, also appear here. If the system were clocked on the read and write lines connected for a logical OR function, the byte 00 would appear twice. As the LDIR is repeatedly performed, the data byte transferred is observed to be incrementing. The remainder of this loop can be observed by using trigger words 02 and 07 to look at data deeper into the LDIR loop. Any byte not appearing earlier could be used as a trigger word. In this case, the only byte changing is the integer being stored.

To get a better appreciation of just what the analyzer is doing and what the observed data represent, this program was repeated using the write line to clock the data. The program is read into the CPU using the read line, so it does not appear here. The write line is used only during the execution of the LDIR op code and writing to (HL). The first trigger word used was OF and the following data were observed:

1	0F	
2	0F	0B
	0E	0A
	0D	09
	0C	etc.

1. The program is in the last cycle of LDIR; HL = 210F, BC = 0001, (HL) = 0F; (HL) = (DE) by reading (HL) and writing to (DE). Then, the program loops back to begin again.

2. Register L contains the low-order address byte 0F; the CPU performs a write operation to load 0F into location 210F, then 0E at 210E, then 0D at 210D, and so on.

This program was also run with the CPU clock as the data strobe for the analyzer. We should expect to see a combination of the read/write data runs, some repetition of data, and perhaps some FF data bytes because the bus may have been floating during a clock cycle. The following data were observed when the initial trigger word was 0F:

2	0F	1	C3
	0F		C3
	0F		FF
	0F		00
	0F		00
	0F		20
	0F		20
	0F		20

1. These data bytes are associated with the op codes JMP 2000. The data byte FF occurred when the bus floated high during the op code fetch cycle.

2. The last cycle of the block transfer of data, LDIR, preceded the jump op code. The last byte to be transferred was 0F H. This data byte was read from (HL) and written to (DE).

With only 16 bytes of RAM available in this particular logic analyzer, this last mode of operation is obviously not an economical way to clock data. Because of the repetition of data, one can easily get confused about its meaning, and there may not be enough unique data to determine the part of the program to which the data belong.

4.3 SIGNATURE ANALYSIS

4.3.1 Introduction

In the previous sections of this chapter, the emphasis has been on following the flow of data on a computer bus. In that way, a software/hardware design could be checked by using a test program designed to manipulate the hardware functions. For example, the flow of data through a parallel input/output (PIO) interface to a parallel input printer could be verified by clocking the PIO data into the logic analyzer. Conditional jumps, loops, and loop exits in the software could be directly observed in a logic analyzer with sufficient memory. But once the complete system is operating properly, one would like to be able to check or correct a hardware failure without having to work through a long series of op codes and data from a logic analyzer. In this situation, it would be faster to use a signature analyzer (SA). An SA compresses a large amount of data from a single data or control line into a *signature*, characterized by the contents of a 16-bit shift register. When the system performs properly, the correct signature is left behind. The correct signature must have been determined previously.

4.3.2 The Meaning of a Signature

To determine how an SA compresses a large amount of data, consider the shift register in Figure 4-8. This is a 16-bit shift register with feedback loops; the input to bit 1 consists of a data input stream exclusive-ORed with the contents of bits 7, 9, 12, and 16. The data in the register are clocked to the right simultaneously for all bits. If the data input line is not utilized, the contents of the register must be one of 64,000 possible combinations. Because of the feedback loops, there will be a sequencing through the entire range of possible bit combinations, giving the appearance a random-number sequence. As an ex-

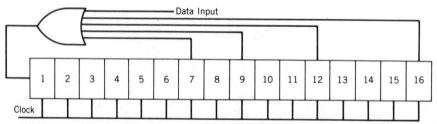

Figure 4-8   Sixteen-bit shift register with feedback loops.

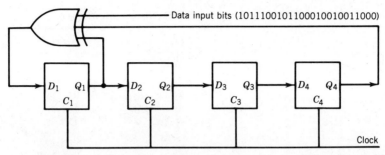

Figure 4-9 Four-bit pseudorandom number generator using a shift register with feedback.

ample, a four-bit pseudo-random-number generator is shown in Figure 4-9. When the input data bits are zero in Figure 4-9, the contents of the four-bit shift register are as shown in Table 4-2. All 15 nonzero combinations of the four bits are generated in this configuration, and then the pattern begins to repeat. If a data or address bus line is also fed into the exclusive-OR input, this stream of bits will be superimposed on the random-number sequence, and result in the register contents shown in Table 4-3. Let us now assume that the register is reset to zero each time prior to use. As an input data bit stream is clocked into the system the register contents change in an apparently random way. But each time the register is reset and has clocked in the same bit pattern, the register contents will be the same. In summary, a long bit stream clocked into a previously reset register will leave behind a unique combination of bits,

Table 4-2 **Four-bit shift register contents when the input bits are zero**

Cycle	Q_1	Q_2	Q_3	Q_4	$D_1 = Q_1 + Q_4$
0	1	1	1	0	1
1	1	1	1	1	0
2	0	1	1	1	0
3	1	0	1	1	0
4	0	1	0	1	1
5	1	0	1	0	1
6	1	1	0	1	0
7	0	1	1	0	0
8	0	0	1	1	1
9	1	0	0	1	0
10	0	1	0	0	0
11	0	0	1	0	0
12	0	0	0	1	1
13	1	0	0	0	1
14	1	1	0	0	1
15	1	1	1	0	1
16	1	1	1	1	0

Table 4-3 **Four-bit shift register contents when input bits are nonzero**

Cycle	Q_1	Q_2	Q_3	Q_4	$D_1 = Q_1 + Q_4 + D_i$	D_i
0	0	0	0	0	1	1
1	1	0	0	0	1	0
2	1	1	0	0	0	1
3	0	1	1	0	1	1
4	1	0	1	1	1	1
5	1	1	0	1	0	0
6	0	1	1	0	0	0
7	0	0	1	1	0	1
8	0	0	0	1	1	0
9	1	0	0	0	0	1
10	0	1	0	0	1	1
11	1	0	1	0	1	0
12	1	1	0	1	0	0
13	0	1	1	0	0	0
14	0	0	1	1	0	1
15	0	0	0	1	1	0
16	1	0	0	0	1	0
17	1	1	0	0	0	1
18	0	1	1	0	0	0
19	0	0	1	1	1	0
20	1	0	0	1	1	1
21	1	1	0	0	0	1
22	0	1	1	0	0	0
23	0	0	1	1	1	0
24	1	0	0	1	0	0

or signature, if the bit stream is always clocked in, started, and stopped in precisely the same way each time.

4.3.3 Use of a Signature

To use the properties of a signature on a particular system, one must either have signatures provided by the manufacturer or obtain the signatures from a properly operating system. In either situation, the test conditions under which the correct signatures were obtained must be clearly stated and reproducible to allow comparisons with signatures at a later time.

When the signatures are available from the manufacturer and no design changes are later made on the system, the signatures given for each IC pin should be valid as long as there are no hardware malfunctions. The software used in the test would usually be provided in a test ROM and instructions would be given for clocking, starting, and stopping the data flow. When an invalid signature is encountered, the IC outputting the signature is bad or it received incorrect data from another IC. Familiarity with the circuit is essential for tracing the signatures back to the source of incorrect data. Of course, the user must be confident that the invalid signature is not the result of an incorrect test setup. When the faulty ICs are found and replaced, one should then obtain the correct signatures. It is assumed here that it would be cost effective to look for defective ICs rather than replace complete circuit boards or the entire system.

When signatures are not available from the original designer or manufacturer, they must be generated before the system becomes defective or they must be obtained from an identical, properly operating system. This is no simple task. It is much more difficult to devise a test capable of generating valid signatures than it is simply to follow the directions for repeating a previously designed test. The test designer must have detailed knowledge of the software and hardware aspects of the system and also of the SA. The emphasis in the remainder of this chapter will be on how the SA clocks in data from a probe point to help a user design a valid test. Some examples of tests for a Z80-based system will also be discussed.

4.3.4 Timing Considerations

For the signature analyzer shown in Figure 4-10, the external system must provide five signals to the SA: ground, data, clock, start, and stop. They will be discussed in that order. The SA and the test system should have a common ground to eliminate ground loops. (When the actual ground reference of the two units differs, the voltage difference will create a current between the two systems.) The data probe is a single probe connected to an IC or bus pin. (Note

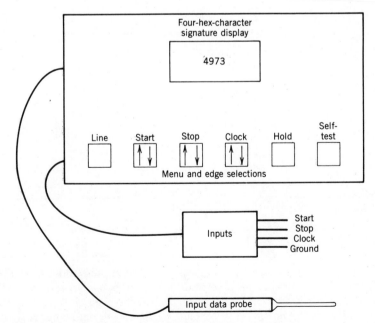

Figure 4-10 Input and input controls for the signature analyzer.

that the SA uses a single data line while the logic analyzer in general uses at least an eight-bit bus.) If properly clocked, each data bit flowing to the particular pin will be clocked into the shift register. The clocking signal will probably come from a control line rather than the test system clock. The problem with using the test system clock is that it may oscillate several times while there is no pertinent data flowing. Even worse, the data lines may be changing while being clocked into the SA. If even one bit out of the entire bit stream varies, a different signature will be observed. Because the shift register clock strobe does not have to be a continuously varying signal, it is best to use a control line, which signifies the presence of stable data to the register. This point will be illustrated more in the examples used later in the chapter.

Once rising or falling edges are selected for the clock and both start and stop signals selected, the signature analyzer logic checks the start line at each clock edge to detect a transition in the start line. If the rising edge of a control line is selected for a start signal, the analyzer logic must detect the low-to-high transition on successive clock pulses. The first data bit will be clocked in from the data probe at the second clock pulse, as illustrated in Figure 4-11. If the rising edge of the start line were to occur but then fall before the clock edge, the analyzer would not recognize the transition. This requirement on the start line lowers, but does not eliminate, the probability of a glitch triggering the start of the analyzer. There is a similar requirement on the selection of the stop line. If the falling edge is selected, then at one clock edge that signal must be high, and at the next clock edge it must be low. No data are shifted in at this second clock edge.

4.3.5 The Free-Running Mode

A good way to gain familiarity with a signature analyzer and its capabilities is to run a microprocessor in the free-running mode. Consider what happens when a microprocessor is powered up. Registers in the microprocessor are initialized

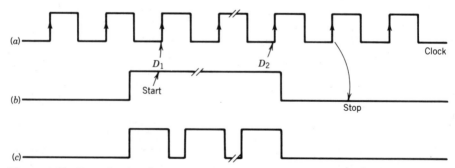

Figure 4-11   Data for the signature analyzer is latched at the positive transition of the signal in (*a*). When the control line in (*b*) is used to both start and stop the data taking, the data line is first sampled at point D_1. The last data bit is taken at the instant D_L. The control line in (*c*) is used just as that in (*b*), but note that signal (*b*) is just the envelope of (*c*).

to zero, and the CPU is put into the op code fetch mode. The program counter contents, assumed to be zero here, are put on the address bus. The CPU then accesses that memory location for a data byte and routes it to the op code decoding register. The program counter is incremented, and the microprocessor is ready to address the next op code or operand. If all the memory locations contain 00_H and the CPU is a Z80 or 8080, the op code fetch each time will be the NO OP (No operation) code.

When this code is decoded, the CPU does nothing but fetch the next byte of code. The only effect of this code is to introduce a small delay time equal to that of the op code fetch cycle. When all memory locations contain 00_H, the CPU will repeatedly fetch the code, increment the program counter, and fetch the next code. For a 64K memory system having 16 address lines, locations 0000_H through $FFFF_H$ will be read, and this process will begin again. In other microprocessors, the coding for a NO OP may be EA_H or some other expression, but if all memory locations contain the byte corresponding to that op code, the result will be the same as with OO_H on a Z80.

A computer memory consisting of NO OP codes is of no use to a CPU, but it can be used to check the CPU with an SA because the address bus of this system will behave like the output of a 16-bit counter. If the SA were used in the same test situation (start, stop, and clock) on a 16-bit counter as on this address bus, the respective pins will have the same signatures. If the signatures are valid, we can assume that the program counter, incrementer, and internal bus of the CPU are operating properly. As long as the CPU continues to properly decode the NO OP code, we will assume that it will also decode other codes correctly. While the CPU is in this free-running mode of an op code fetch loop, an oscilloscope can be used to check the control lines.

This free-running mode can be used to check several registers of the CPU if there were a practical method for setting it up. This can be done by disconnecting the data pins of the CPU from the data bus and grounding them. The CPU data pins may be disconnected from the bus by bending them out away from the socket or by constructing a wire harness between the circuit board socket and the CPU. Each pin of the CPU is connected to its respective socket pin via the harness, except for the CPU data pins, which are grounded. Data can still flow on the data bus, but the only data word clocked into the CPU during an op code fetch operation is 00_H. The system now operates as though all memory contains 00_H and it is in the free-running mode.

The next step is to select control lines and edge transitions for the SA. Consider the timing diagram of Figure 4-11. The clock signal shown there could be the read line in the present situation, while the other could be the highest ordered bit of the address bus, A_{15}. As the program counter cycles between 0000_H and $FFFF_H$ in the free-running mode, the A_{15} line will be low for addresses up to 32K ($7FFF_H$). From $7FFF_H$ to $FFFF_H$, A_{15} will be high every time the address bus is active, as indicated in Figure 4-12. The read line, an active low line, is active once for each of the 64K code fetches. If A_{15} is active, the read line activity period will be contained within that of A_{15}. The CPU assumes that valid data are on the bus at the end of the read strobe and latches

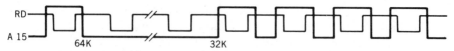

Figure 4-12 The superposition of the read line, active low, and the address line, A 15. For the first 32K addresses in RAM A 15 remains low; for the next 32K op code fetch cycles, A 15 goes active high long enough for memory to be accessed.

data on the upward transition of the read (Rd $\uparrow$). Data may be on the bus earlier than that, but it may not be stable. For instance, the signal Rd $\downarrow$ is used as an enable line for memory, but if it is also used to clock data into the SA, the signatures in general would be expected to vary randomly. Of course, in the free-running mode, that will not be a problem, since there are always valid data being latched into the CPU: the no op code. A good start signal for the SA is $A_{15}\downarrow$.

This can be seen by visualizing the program counter as it approaches $FFFF_H$. The line A_{15} will go high each time memory is accessed and will be sampled by the SA at every clock edge, Rd $\uparrow$. As far as the logic of the SA it concerned, then, the line A_{15} does not make a downward transition because it is always high when sampled by the clock. When memory location 0000_H is accessed, though, A_{15} will be low at the clock edge, the transition of A_{15} will be recognized, and the data bit on the data probe line will be clocked into the SA. The first data bit used by the SA, then, is associated with address 0000_H; a new data bit will be clocked in at every clock edge until a stop signal is recognized by the SA. It may appear confusing, but $A_{15}\downarrow$ is also used as a stop strobe for the SA. For address values from zero to 32K, A_{15} will be low at every clock edge. From 32K to 64K, the line A_{15} will be high at every clock edge. The SA will continue clocking a data bit at each Rd $\uparrow$ until the transition $A_{15}\downarrow$ is recognized. The transition is recognized during the sequencing of addresses $FFFF_H$ to 0000_H. The last data bit clocked in occurred before A_{15} went low, at $FFFF_H$. The set of SA control lines, Rd $\uparrow$, $A_{15}\downarrow$, and $A_{15}\downarrow$, ensures that the data to the SA probe are associated with the read operation of memory locations 0000_H to $FFFF_H$ inclusive.

While in the free-running mode, a system ROM can also be checked for valid signatures, assuming that the signatures were recorded earlier while the system was performing properly. Again, the appropriate clock signal is Rd $\uparrow$. The start and stop signals to the SA are a function of the position of the ROM in the memory map of the system. For a 2K ROM at the low end of memory, the locations to be read into the SA are 0000_H to $07FF_H$, and A_{15} is again an appropriate start line. Address line A_{11} will be low until the last byte is read from $07FF_H$; then, it will be high each time memory is read, until A_{12} goes high. If $A_{11}\uparrow$ is chosen for the SA stop strobe, the last data byte to be accessed from the ROM will be from $07FF_H$. As each ROM location is accessed by the CPU in the free-running mode, the ROM data will go to the data bus but not to the CPU data pins. Signatures can be obtained for bus pins D_0 through D_7 and recorded or compared with previously obtained values. Each of the eight signatures is a compact representation of 2K bits in ROM; if any of these bits are changed, the signature will change.

The situation for RAM is different from that of the ROM, because its data are not permanent. Still, if a frequently used program is run in RAM, and the remainder of RAM is loaded with zeros, signatures can be obtained for the program in the event that it is not loaded or stored properly. For a 1K RAM occupying memory locations 2000_H to $23FF_H$, it should be clear that A_{15} will not be a good control line. In binary, the RAM locations are 0010 0000 0000 0000 to 0010 0011 1111 1111. The start and stop control lines must be chosen so that all 1K locations are read, but no others. As the first RAM location is accessed, A_{13} will go high, but this A_{13} will go high when A_{14} is high. The appropriate combination of control lines is Rd $\uparrow$, $(A_{13}{\cdot}\overline{A}_{14}{\cdot}\overline{A}_{15})\uparrow$, and $A_{10}\uparrow$ for clock start and stop.

CHAPTER 5

TROUBLESHOOTING METHODS

Luces M. Faulkenberry
University of Houston

5.1 TROUBLESHOOTING

Troubleshooting can be defined as the logical art of identifying and repairing the cause of a malfunction. In our field, the term implies a malfunction in electronic systems, though all fields from automotive mechanics to foreign diplomacy employ troubleshooters. The process of identifying malfunctions is primarily one of using logical mental processes. Many of the logical mental steps are done subsconsciously by experienced troubleshooters, and that is where the art comes in. We stress that the process of troubleshooting is mostly mental and that the measurements that provide the clues for logical deduction occupy only a fraction of the troubleshooting time.

Troubleshooting is an important task. The economy and life style in a developed country depend on keeping the electronic industrial control systems, communications systems, and computer systems operating. To illustrate the importance of troubleshooting and repair to modern organizations, one may note that about one-third of the Department of Defense electronics budget is allocated for maintenance.

Figure 5-1 illustrates the typical failure rate versus time for electronic equipment. The early failure rate is very high compared to the midlife failure rate. For this reason, many companies use *burn-in*, the operation of components or equipment at elevated temperatures, to accelerate the early failures. Early failures are then eliminated before the equipment leaves the factory, and the equipment is delivered to the customer entering the bottom of the "bathtub curve" shown in Figure 5-1. The early failures are referred to as infant mortality and are usually caused by the initial stress of operation on marginal components.

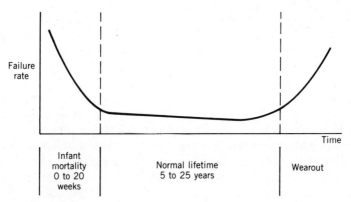

Figure 5-1   Failure rate curve.

After the infant mortality period, the equipment enters the normal lifetime period which will be anywhere from 5 to 25 years depending upon the reliability designed and built into the equipment. Failures during the normal lifetime are often associated with drift because of component aging, overstress of the equipment by misuse, as well as component malfunction. Like humans, the end of the equipment life cycle is characterized by a marked increase in the number of malfunctions. Everything that can fail seems to do so, and it is more economical to replace the equipment than to continue to maintain it. Note that there is work for the troubleshooter throughout the equipment life cycle since the failure rate of sophisticated equipment is almost never zero for any length of time.

The demands on the skill of the troubleshooter become more exacting as systems become more complex. Only a thorough knowledge of the system to be maintained and considerable experience can equip a person to become an expert troubleshooter. There are, however, some guidelines which can aid the beginning troubleshooter in approaching the task. The purpose of this chapter is to present these guidelines.

5.2 APPROACHING THE TASK

The first and foremost thing a troubleshooter must do is *stay calm*. Turbulent emotion and anxiety interfere with the logical thought processes required for successful troubleshooting. The only effective troubleshooting is that done calmly. So *stay calm*!

The more information we have about a system, the better we can troubleshoot it. Many systems such as large computer mainframes and spectrum analyzers are so complex that special training is required to competently troubleshoot many types of malfunctions. With a good general knowledge of electronics and basic troubleshooting procedure, one can find many kinds of malfunctions in most equipment. Before troubleshooting any equipment, spend some time learning about it. Even when time is very short, the troubleshooter should know the function of each major subsystem of the failed equipment.

The best place to learn a great deal about equipment in a short time is in the equipment service manual. The service manual usually contains the equipment specifications, description, theory of operation, schematics, parts layout, and a troubleshooting guide. The troubleshooting guide contains a list of common malfunction symptoms and causes. The service manual should be studied as well as time permits before repair is attempted. The equipment manufacturer is also a good source of troubleshooting information. A telephone call to the manufacturer's service shop can save hours of looking for an obscure problem. One should have a very concise description of the malfunction symptoms and any measurements already made prior to calling the manufacturer. If a service manual is not available, obtain one for future reference, and do the best you can. Often, if the problem is not simple, the best you can do is obtain a manual as quickly as possible.

A good troubleshooter uses all the resources available to find a malfunction, including any diagnostic programs available for equipment with computing power. There is no disgrace in using the equipment to find its own problem. Doctors ask patients where it hurts.

Use the right test equipment for the job. A spectrum analyzer can be valuable for troubleshooting communications equipment but worthless for digital equipment. A simple VOM is hard to beat for continuity checks. For some digital troubleshooting, a logic analyzer must be used. Use the equipment needed and appropriate for the job. A little thought on the test equipment required can prevent a lot of wasted time. Most test equipment can be rented if it is only needed occasionally.

The troubleshooter must also know the operation of the test equipment. The operating manual supplied with the test equipment is the best place to find operating instructions in concise form. Don't spend an hour finding out that the 50-MHz signal on the spectrum analyzer screen is a marker, not a spurious signal in the malfunctioning equipment.

In summary, a troubleshooter should stay calm, know the equipment to be repaired, know the test equipment to be used, and use the right test equipment for the job.

5.3 SOME TROUBLESHOOTING METHODOLOGY

There are some accepted steps and methods to use when troubleshooting. This section presents the steps and common methods used in isolating faults.

5.3.1 Visual Inspection

The first step to take in locating a malfunction is to look the equipment over thoroughly. Check for blown fuses, broken or discolored components, broken wires, broken printed circuit conductors, cold solder joints, smelly transformers, corrosion, components that are too hot, and leaking electrolytic capacitors. In other words, look for anything that is not quite right. A thorough visual inspection will reveal the malfunction often enough to be worthwhile. If nothing else, the troubleshooter learns the location of all the parts from a thorough visual inspection.

5.3.2 Power and Quiescent Conditions

The first electrical check should be for proper voltage levels from all power supplies operating under normal load. Even if the supply output must be disconnected and a dummy load inserted, the power supply should be checked under load. A lot of time has been wasted looking for obscure troubles when a malfunctioning power supply was the problem.

The quiescent conditions of analog circuitry should be checked in analog equipment or analog sections of digital equipment. If the static conditions are incorrect, signals will not be processed properly. Digital equipment often can be checked in the reset state for correct initial conditions.

5.3.3 Back to Front

The back-to-front methodology simply means starting with the output section when making dynamic measurements, as shown in Figure 5-2. One then progresses toward the input until the proper signal (or code for digital equipment) is found; the stage immediately toward the output is then suspected of malfunction. The troubleshooter must know the proper output of each stage for any troubleshooting method to be effective.

5.3.4 Signal Injection

When a malfunctioning stage affects the output of a previous stage or the normal input of the equipment is not available, then the troubleshooter must use a function generator to "inject a signal into the equipment." This signal should be as close to the normal signal as possible. Often, a stage will need an ac signal superimposed on a dc level. The offset control on most function generators can provide a limited dc level. When the offset adjust lacks sufficient range, a voltage divider and capacitor can be used as shown in Figure 5-3.

5.3.5 Half-Splitting

Half-splitting refers to checking the output of a complex circuit of many stages at the halfway point and successively in the middle of each remaining set of

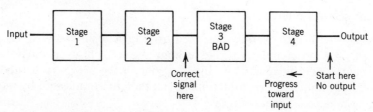

Figure 5-2   Back-to-front fault isolation.

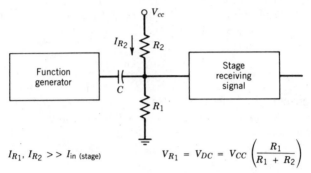

$$I_{R1}, I_{R2} >> I_{\text{in (stage)}} \qquad V_{R1} = V_{DC} = V_{CC}\left(\frac{R_1}{R_1 + R_2}\right)$$

Figure 5-3   Offset for signal injection.

stages. For example, suppose an eight-stage circuit has a malfunction in stage 5. To find that stage 5 is malfunctioning, one would use the following procedure: After checking the power supplies, one would check the output of stage 4 (halfway through the circuit). If the proper output is found at stage 4 (as it would be in our example), one would check the output of stage 6, halfway between stage 4 and the final stage. In our example, the stage 6 output would be wrong. One would now check halfway in the circuit between the last two checks, or the stage 5 output. It would be wrong in our example, so stage 5 would be the malfunctioning stage. This method works best for equipment with independent stages in series, such as radio receivers and transmitters.

5.3.6 Breaking the Loop

Electronic systems or subsystems with feedback loops are difficult to troubleshoot unless the feedback loop is broken. The appropriate dc level or signal must be injected at the point where the feedback loop is broken. The voltages and signals throughout the circuit should then be monitored for error. The voltage or signal injected at the loop break can be varied to check for appropriate response to the change throughout the circuit. Normally, the loop is broken at the point at which a low-power signal can be conveniently injected. The technique is shown for a phase lock loop block diagram in Figure 5-4. The power supplies and reference oscillator output would be checked prior to breaking the loop. In this case, f_o would be incorrect or unstable rather than missing, or we would know the VCO was bad.

5.3.7 Compartmentalize

Complex systems are generally designed from logical subsystems. The entire system may be too complex to troubleshoot at once, but each subsystem can usually be checked independently by using one of the previous methods. The subsystem(s) found to be malfunctioning can then be troubleshot by using the previously mentioned methods.

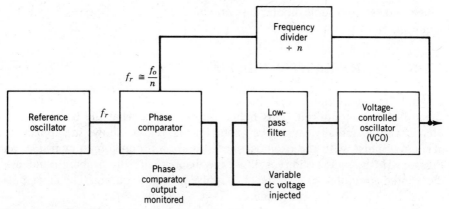

Figure 5-4   Breaking-the-loop, phase-locked loop example.

5.3.8 Known Good Comparison

To recognize an erroneous output, we compare it with the output of a properly functioning circuit. We either know the circuit well enough to recognize what the output should be, or compare it with a waveform (bit pattern, state sequence, memory map, or timing diagram for digital equipment) in the service manual. If the documentation giving proper outputs is not available and the troubleshooter lacks enough experience with the equipment to recognize the proper outputs, then the outputs obtained should be compared with those of a similar, properly functioning unit to determine whether or not they are good. This technique is particularly valuable with digital and microprocessor-based equipment.

5.3.9 Swapping

The swapping of good printed circuit (pc) boards into a system to isolate faults is called board swapping. The practice is also used with components. For quick, in-the-field repairs, this practice is acceptable and used by many large companies, but, as a primary troubleshooting procedure, the method is costly. Many good components and pc boards have been ruined by putting them into a system with the power on or into a system with a fault that affected the substituted part fatally. When equipment must be operating again quickly, the replacement of the pc board or module that has failed is fine. It is an effective technique to minimize down time of equipment in the field. The bad component(s) of the malfunctioning pc board or module can then be found at a more leisurely pace at a later time. The troubleshooter should be quite sure that the trouble is in the module to be swapped before swapping, and the module should usually be swapped with the power off. There is equipment in which no part can be damaged by swapping, even with the power on, but one should be sure before proceeding. The worst problem with board swapping as a troubleshooting procedure, as opposed to other repair procedures, is that it can become a habit, which makes the troubleshooter less effective than others who use procedures requiring more thought.

5.4 SOME FREQUENT PROBLEMS

Ground the test equipment probe as close as possible to the point where the signal is being monitored. This helps to assure that what the troubleshooter sees is what is actually there. For example, Figure 5-5a shows the appearance of a clock pulse with the oscilloscope ground on a minicomputer frame, and Figure 5-5b shows the same oscilloscope display with the probe grounded on the printed circuit board near the clock pulse generator output. The clock pulse is fine, but the waveform of Figure 5-5a is misleading.

When working with metal–oxide–semiconductor (MOS) integrated circuits, be careful not to touch the leads with static electricity on your body or the ICs will be damaged. Use special handling tools, or, lacking them, ground the work surface, or equipment frame, and your wrist through wires running to a high-resistance (typically 1 MΩ) ground before replacing an MOS device. Do not use the wires when making operational checks on the equipment, but

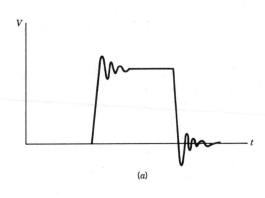

(a)

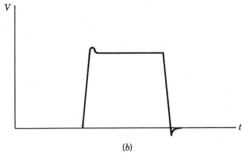

(b)

Figure 5-5   Effect of oscilloscope probe ground placement on observed clock pulse. (a) Ground on minicomputer frame. (b) Ground on pc board common near clock pulse generator.

only when changing an MOS device in unplugged equipment. At the very least, touch a good ground immediately prior to touching an MOS device. The static electricity on a person's body will rupture the thin, insulated gates of MOS devices. Protected MOS devices are less prone to static voltage breakdown but should still be handled carefully. If the level of static charge on a person's body is high enough, it will damage even a protected device.

The order in which power supplies are turned on is critical in many pieces of equipment, especially those using complementary metal–oxide–semiconductor (CMOS) integrated circuits. The n- and p-channel MOSFETs in CMOS have a parasitic SCR (silicon controlled rectifier) built in which can latch on if the power supply turn-on sequence is wrong. When a device latches up, it will destroy itself and possibly other equipment. Don't forget to check the power supply turn-on sequence in malfunctioning CMOS equipment.

When working with digital equipment, step the equipment manually from one state to the next, if possible. The single-step mode is a great aid to the troubleshooter in finding problems because the state (high or low) of various points can be compared to the desired state with a VOM or oscilloscope. Otherwise, the states would change too fast to be monitored without a logic analyzer.

Noise problems in both analog and digital equipment can be the result of a poor ground return path in the equipment. Check the ground resistance and the ground impedance; both should be very low. Conducted electromagnetic interference filters sometimes place enough inductance in the ground line to cause noise problems. Figure 5-6 shows the appearance of a signal on an oscilloscope when the ground between two subsystems is missing.

A final note: Dirty or oxidized printed circuit board contacts cause many problems, and a routine cleaning of the contacts can prevent problems. The tried and true method of cleaning pc board contacts with a pencil eraser should be used only as a last resort. The very thin platings now used on pc board contacts can be removed by a pencil eraser and cause the contacts to rapidly corrode after an eraser cleaning. One of the contact cleaning solvents should be used whenever possible.

5.5 TRENDS

The microprocessor and inexpensive digital-to-analog and analog-to-digital converters have enabled self-test and diagnostic capabilities to be designed into equipment that will locate and display (or print out) common failures when the equipment malfunctions. Automatic test equipment that can both test and diagnose troubles in complex subsystems and systems is being developed rapidly. Some automatic test equipment is programmed to tell the troubleshooter where to measure in the malfunctioning equipment; this is called guided probing. Many problems in some computers can already be diagnosed by letting the ailing computer talk, via a telephone data link, to a computer at the service center. The service center computer remotely troubleshoots the malfunctioning one and determines which board to change. Equipment will be built with increasing

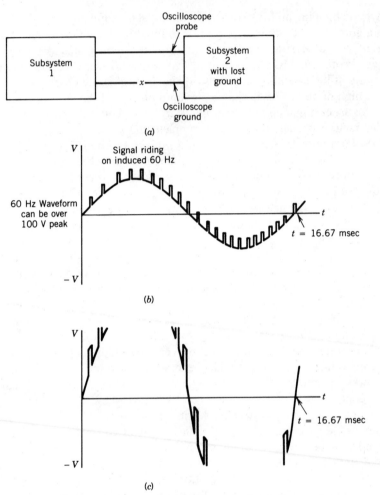

Figure 5-6   Oscilloscope indication of a missing ground. (*a*) Connection. (*b*) Oscilloscope display. (*c*) Oscilloscope display on low-voltage scale.

self-diagnostic power. Work will continue in the area of modelling malfunctions mathematically so that the probability of certain types of failures can be predicted and found by machine. Automatic test equipment will become even more sophisticated. Test instruments will be able to perform more functions once left to the operator and even tell the operator where to check next. These trends do not mean the end of the electronic troubleshooter. Instead, the job of troubleshooting will become even more challenging because the human troubleshooter will be called upon to find the cause of malfunctions the machines cannot handle.

5.6 SUMMARY

A good troubleshooter will remain calm, know the malfunctioning system and test equipment, visually inspect the failed equipment, check the power supplies and quiescent conditions, and use one or more of the fault isolation techniques to identify the trouble. One more trait rounds out the requirements a troubleshooter must have: the tenacity to continue until the trouble is found.

5.7 REFERENCES

Allen, L., *Practical Electronic Servicing Techniques*. Blue Ridge Summit, PA: Tab Books, 1970.

Anderson, R. E., Board Testing in the 80's. *IEEE 1979 Semiconductor Test Conference*, Cherry Hill, NJ: IEEE Computer Society Press, p. 7.

Bedrosian, S. D., Fault Analysis of Analog Circuits Using Graph Theory. In Saeks, R., and Leberty, S. R., (eds.), *Rational Fault Analysis*. New York, NY: Marcel Dekker, Inc., 1977.

Craig, R. S., Alternative Strategies for Incoming Inspection. *IEEE 1979 Semiconductor Test Conference*, Cherry Hill, NJ: p. 244.

Gardner, H., *Handbook of Solid-State Troubleshooting*. Reston, VA: Reston Publishing Co., Inc., 1976.

Herrick, C. N., *Electronic Troubleshooting; A Manual for Engineers and Technicians*, 2nd ed., Reston, VA: Reston Publishing, 1977.

Kochan, S., Landis, N., and Monson, D., Computer-Guided Probing Techniques. *IEEE 1981 Test Conference*. Philadelphia, PA: IEEE Computer Society Press, p. 253.

Liberty, S. R., Tung, L., and Saeks, R., Fault Prediction—Toward a Mathematical Theory. In Saeks, R., and Liberty, S. R., eds. *Rational Fault Analysis*. New York, NY: Marcel Dekker, Inc., 1977.

Schultz, J. J., *Electronic Test and Measurement Handbook*. Blue Ridge Summit, PA: Tab Books, 1969.

Zender, T. J., Driving Forces Behind Field-Based System Testing. *IEEE 1981 Test Conference*, Philadelphia, PA: p. 407.

PART 2

ANALOG SYSTEM TROUBLESHOOTING

CHAPTER 6

Linear Power Supplies

Lowell L. Winans

Heath Company

6.1 INTRODUCTION

The dc power supplies described in this chapter are electronic circuits that convert alternating current to direct current.

Linear voltage regulation is the producing of a constant dc voltage at a predetermined level by continuously controlling a variable resistance that is either in series or in parallel with the load. This is usually achieved by sensing the output voltage and then changing a variable series resistance, as necessary, to keep the output voltage at the desired level even though the load or input voltage may vary.

6.2 UNREGULATED POWER SUPPLIES

6.2.1 Half-Wave Rectifier

Rectifier circuits convert alternating current to pulsating direct current. Figure 6-1a shows a half-wave rectifier. When an ac signal is applied to the circuit, the diode only conducts on the positive half-cycle of the ac signal, and thus current flows through the load resistor and diode at this time and produces a pulsating positive dc voltage. Of course, to obtain a negative voltage, we can just turn the diode around as shown in Figure 6-1b.

The average value of the ac signal applied to the input of Figure 6-1 is zero volts, as the positive half-cycle cancels the negative half-cycle.

Note: Portions of this Chapter are reprinted with the permission of Heath Company, Benton Harbor, MI, from the following document: SS-9000 (595–2303) HF Synthesized Transceiver Owner's Manual Copyright 1982.

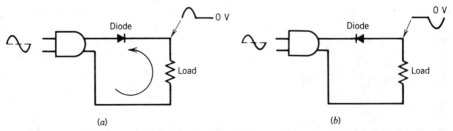

Figure 6-1   Half-wave rectifier. (*a*) Positive half-wave rectifier. (*b*) Negative half-wave rectifier.

Another voltage value is the RMS (or effective) value. This is the amount of ac voltage required to produce an equal amount of heat in a resistor as a known dc voltage. As it turns out, 120 RMS ac is equivalent to 120 V dc. Because meters are calibrated to display RMS voltages, it is usually the ac peak voltage that is being sought, and this is found by the following formula (see Figure 6-2):

$$E_{peak} = E_{RMS} \times 1.414$$
$$= 120 \text{ V} \times 1.414$$
$$= 169.7 \text{ V}$$

The average value of the half-wave output voltage can also be found as shown in Figure 6-3. It is the peak value times 0.318. If the peak value is 169.7 volts, then the average voltage is:

$$E_{ave} = E_{peak} \times 0.318$$
$$= 169.7 \text{ V} \times 0.318$$
$$= 53.9 \text{ V}$$

6.2.2 Transformers

Power transformers can step up or step down line voltage to the desired ac value. They can also provide electrical isolation from the power line for the protection of both people and equipment.

As shown in Figure 6-4, equipment with line-operated power supplies can be dangerous. A severe electrical shock can result from line voltage present between two pieces of equipment, when each chassis is connected to one side of the power line.

If both pieces of equipment are plugged into an ac outlet such that each

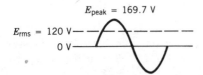

Figure 6-2   Finding the peak valtage.

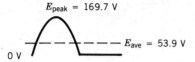

$E_{\text{peak}} = 169.7 \text{ V}$

$E_{\text{ave}} = 53.9 \text{ V}$

0 V

Figure 6-3   Finding the average voltage.

chassis is connected to the same side of the ac line, there is no problem. However, the safest practice is to always use one or more isolation transformers, as shown in Figure 6-5. This isolates the equipment from the ac line and protects against accidental shock.

6.2.3 Full-Wave Rectifier

Full-wave rectification can be achieved by using two diodes and a center-tapped transformer secondary, as shown in Figure 6-6, or by using a bridge rectifier, as shown in Figure 6-7.

As shown in Figure 6-6*a*, diode D_1 conducts on the positive half-cycle of the incoming waveform, and diode D_2 (see Figure 6-6*b*) conducts on the negative half-cycle. The result is that the load is supplied a pulsating dc that has twice the ripple frequency of the half-wave power supply, meaning that more power is applied to the load. This rectification method has the drawback that the peak rectified voltage can be no more than half of the peak voltage at the secondary of the transformer. Because the transformer is center tapped, only half of the transformer voltage is being used effectively at any one time.

After measuring the RMS voltage at the load, the peak voltage can be calculated by

$$E_{\text{peak}} = E_{\text{RMS}} \times 1.414$$

The average voltage is twice that of the half-wave rectifier and can be found by

$$E_{\text{ave}} = E_{\text{peak}} \times 0.636$$

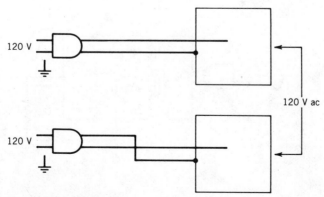

120 V

120 V

120 V ac

Figure 6-4   AC line shock hazard.

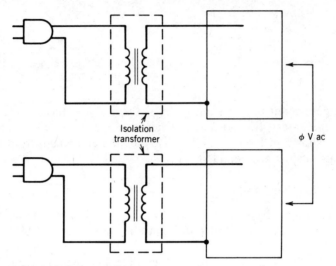

Figure 6-5   Using isolation transformers.

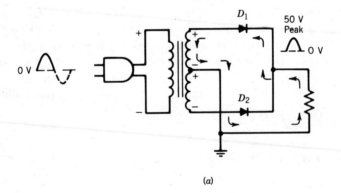

(a)

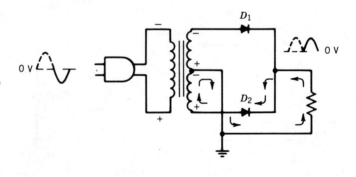

(b)

Figure 6-6   Full-wave rectifier. (a) First half-cycle. (b) Second half-cycle.

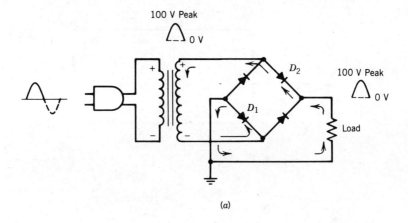

(a)

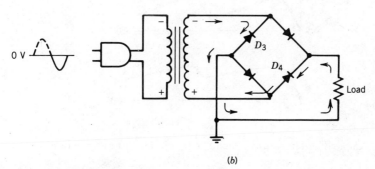

(b)

Figure 6-7   Bridge rectifier. (*a*) First half-cycle. (*b*) Second half-cycle.

6.2.4 Bridge Rectifier

This rectifier produces the same peak voltage as the transformer secondary. As shown in Figure 6-7*a*, diodes D_1 and D_2 conduct on the first half-cycle, and diodes D_3 and D_4 conduct on the next half-cycle. The result is a full-wave-rectified voltage whose average value can again be found by

$$E_{ave} = E_{peak} \times 0.636$$

6.2.5 Filter Circuits

Filter circuits smooth out the pulsating dc created by the rectifiers to form a smooth dc voltage required by most electronic circuits.

Capacitor Filters   A simple filter can be a capacitor connected to the output of either a half-wave or full-wave rectifier (see Figure 6-8*a*). As shown in Figure 6-8*b*, the capacitor will charge as the input voltage rises and will attempt to

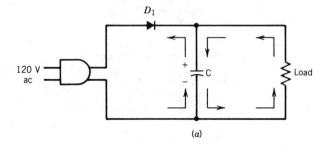

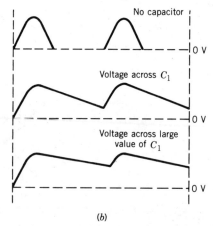

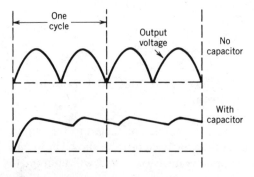

Figure 6-8   Capacitive filter. (*a*) Capacitor filter circuit. (*b*) Filter action.

supply operating voltage to the load during the times when the pulsations are not at their peak value. In doing this, the capacitor will charge and discharge with each positive pulse. How much the capacitor discharges depends on how great a load is connected to the circuit, on the value of capacitor C_1, and on how much current the transformer can supply. Figure 6-9 shows a capacitor filter connected to a full-wave power supply.

Figure 6-9   Filtering a full-wave rectifier.

Any filter that has a capacitor immediately following the rectifiers places an additional load on the rectifiers. When the circuit is first turned on, the capacitor is completely discharged and appears as a short circuit to the rectifiers until it charges. This causes the transformer and rectifier current to be very high until the capacitor is charged.

Also, after the capacitor has charged and the secondary voltage goes to the peak of its negative half-cycle, there is twice the voltage across the diode than there is across the capacitor. This is true for either half-wave or full-wave rectifiers. Therefore, diodes should be selected that can withstand this higher voltage. This situation does not exist with the bridge rectifier. These diodes are never exposed to voltages greater than the peak of the secondary voltage.

RC Filters   The RC (resistor–capacitor) filter shown in Figure 6-10 provides the same filtering as a very large (and costly) capacitor, but it uses two smaller capacitors and a resistor.

Capacitor C_1 performs the same function that it did in the single-capacitor filter. However, capacitor C_2 represents a very high impedance to the dc that is passed to the load. C_2 represents a very low impedance to any ac ripple that gets through the R_1 and C_1 network, and the ripple is reduced for a smooth dc. If needed, another stage of RC filtering can be added.

LC Filters   The LC filter in Figure 6-11 is similar to the RC filter in Figure 6-10, except that it uses an inductor (choke) in the place of resistor R_1. The inductor has a lower impedance to the dc flowing through it and is therefore more efficient than a resistor. However, inductors are more expensive than resistors, and they are bulky. Inductor-type filters work well because they react to current changes, while the capacitors react to voltage changes.

CAUTION: Do not remove and reconnect the loads from operating power supplies that use LC filters. This can cause inductive spikes that may destroy the rectifier diodes.

6.2.6 Voltage Doublers

To produce dc voltages that are higher than the peak value of the line voltage, either a step-up transformer or a voltage multiplier circuit must be used. Voltage doublers, to produce twice the input voltage, are common. Higher-order voltage multipliers are used only in low-current applications, where the value of the voltage is not critical.

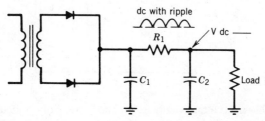

Figure 6-10   RC filter.

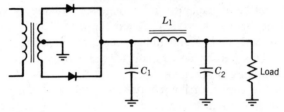

Figure 6-11   LC filter.

Half-Wave Voltage Doublers   Figure 6-12a shows a half-wave voltage doubler. As shown in Figure 6-12b, capacitor C_1 charges to the peak voltage on the
negative half-cycle of the input waveform. Then, on the positive half-cycle,
the peak value of the input waveform is added to the voltage on C_1. This voltage
causes diode D_2 to conduct and charges C_2 to a peak voltage that is twice as

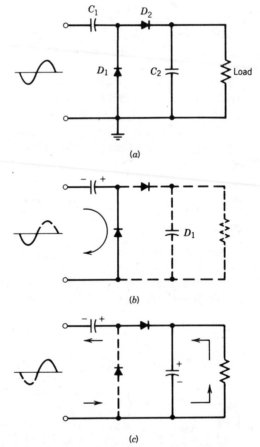

Figure 6-12   Half-wave voltage doubler. (*a*) Half-wave voltage doubler circuit.
(*b*) First half-cycle of doubler. (*c*) Second half-cycle of doubler.

high as the input signal. Because capacitor C_2 is charged only on one of the half-cycles of the input signal, it is a half-wave voltage doubler.

Full-Wave Voltage Doubler   Figure 6-13*a* shows a full-wave voltage doubler. On the positive half-cycle of the input signal, capacitor C_1 charges through diode D_1 as shown in Figure 6-13*b*. Then, on the negative half-cycle, capacitor C_2 charges through diode D_2, and the voltages on the two capacitors add to form a total peak voltage that is twice that of the input peak voltage. The total output voltage is divided between capacitors C_1 and C_2 so that neither capacitor is subjected to the full output voltage as it is with the half-wave voltage doubler.

6.3 SERIES-REGULATED POWER SUPPLIES

Voltage regulator circuits attempt to produce a constant dc voltage by overcoming changes such as a varying line voltage or a changing load resistance. How well a circuit performs this task is called **percent regulation** and can be determined by the following formula:

$$\% \text{ reg} = [(F_{\text{no-load}} - E_{\text{full-load}})/E_{\text{full-load}}] \times 100$$

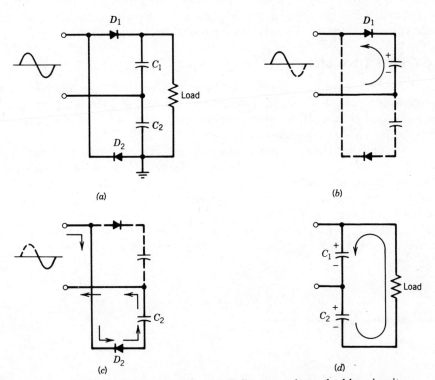

Figure 6-13   Full-wave voltage doubler. (*a*) Full-wave voltage doubler circuit. (*b*) First half-cycle of doubler. (*c*) Second half-cycle of doubler. (*d*) Split capacitors supply load.

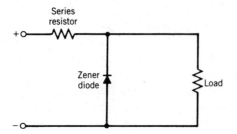

Figure 6-14   Zener diode regulator.

Zero percent regulation is ideal and unattainable. It would mean that the output voltage is constant under all load conditions.

6.3.1 Zener Diode

A simple voltage regulator is a zener diode and a resistor, as shown in Figure 6-14. The diode will zener (break down) at a specific voltage and maintain a constant dc voltage across itself. Of course, the input voltage to the circuit must be greater than the zener voltage for any regulation to occur. The series resistor limits the circuit current to keep the zener diode from being damaged by excessive current. This is the most basic of shunt regulators because the zener diode is connected in parallel (or shunt) with the load.

6.3.2 Pass Transistor

As shown in Figure 6-15, a series regulator is a variable resistance that is in series with the load. If the load decreases, the R_{var} decreases to maintain a constant load voltage. If, however, the input voltage to the regulator increases, then the R_{var} will also increase.

A simple but effective series regulator can be made from the zener diode regulator by adding a transistor as shown in Figure 6-16. Because the transistor is connected as an emitter follower, the only difference between the output voltage and the zener voltage is approximately 0.6 volts, the drop across the emitter–base junction of the transistor. Using a transistor in this way allows the regulator to supply more current to the load. For even greater current-handling ability, a second transistor can be connected as a Darlington pair as

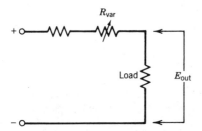

Figure 6-15   Series regulator.

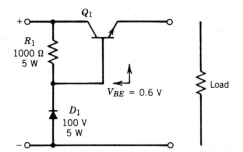

Figure 6-16   Pass transistor.

shown in Figure 6-17. This configuration has the advantage that a lower-wattage resistor and zener diode can be used.

6.3.3 Feedback Regulator

The feedback regulator is very popular with both discrete component and integrated circuit voltage regulators. As shown in Figure 6-18, the output voltage is sampled. Error is detected by comparing the sample voltage against a known voltage, and the error signal (the difference between the sample and known voltages) is amplified and then used to control the variable series resistance (called the pass transistor). A basic transistor circuit is shown in Figure 6-19, and a circuit that uses an operational amplifier is shown in Figure 6-20.

In Figure 6-19, transistor Q_1 operates as both the error detector and the error amplifier. If the output voltage should increase, E_f will increase, causing Q_1 to conduct more and increased current to flow through R_1. This reduces the voltage at the base of Q_2, which increases the resistance of Q_2 and returns the output voltage to its proper value.

The regulator in Figure 6-20 operates in much the same way as the one in Figure 6-19, except that the operational amplifier has much more gain than a single transistor. The amplifier amplifies the difference between the reference voltage of D_1 and the sampled voltage of R_3, and drives transistor Q_1 accordingly. If the voltage of R_3 goes up, then U_1 starts to turn off Q_1.

A feature of feedback regulators is that the output voltage is easily changed by varying control R_2.

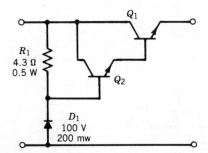

Figure 6-17   Darlington pair.

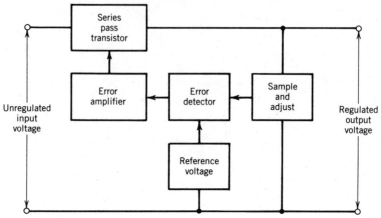

Figure 6-18   Feedback regulator block diagram.

6.3.4 Thermal Sensing

Thermal sensing is necessary to protect both the voltage regulator and the load. Should the pass transistor get too hot, the regulator circuitry could be damaged, and this could damage the load circuitry.

Integrated circuit voltage regulators usually have internal thermal protection that causes them to shut down when they get too hot. This is done by a thermal-sensing transistor that is physically located beside the pass transistor. Its base is held cut off at approximately 0.4 V. As the pass transistor heats up, the voltage required to turn on the thermal-sensing transistor decreases. If excessively high temperature turns the transistor on, the thermal transistor removes the base drive from the pass transistor and shuts down the voltage regulator. The thermal circuits usually have a hysteresis built into them so that the regulator does not turn back on too quickly. This reduces the chance of low-frequency thermal oscillation.

Other thermal protection devices include fuses and circuit breakers. A

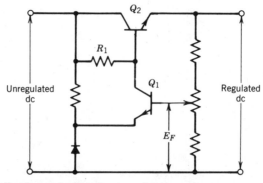

Figure 6-19   Feedback regulator using transistors.

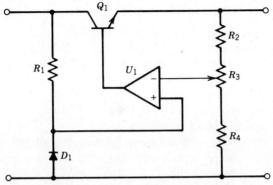

Figure 6-20 Feedback regulator using an operational amplifier.

fuse, of course, is a thin piece of metal (usually in a glass case) that melts when it gets too hot. Then, the fuse must be replaced before the circuit can function again. The circuit breaker is a strip of two dissimilar metals sandwiched together. As the strip heats, the metals expand at different rates causing it to bend and separate the electrical contacts. This opens the circuit and shuts down the regulator.

6.3.5 Current Limiting

Current limiting protects the voltage regulator and load circuits by limiting the output current to a safe value.

A current-limiting device can be a transistor and a resistor, as shown in Figure 6-21. Under normal operation, the emitter–base voltage of current-limiting transistor Q_3 is not sufficient to turn the transistor on. However, when the current from the regulator becomes excessive, the increased voltage drop across R_1 turns transistor Q_3 on. This reduces the voltage at the base of Q_1, which causes Q_1 to conduct less and hold the output current to a safe level.

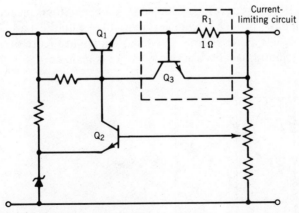

Figure 6-21 Regulator with current limiting.

6.3.6 Crow Bar Circuit

Another protective circuit is the *crow bar*. This circuit, shown in Figure 6-22, does not protect the power supply, but it does protect the load. The crow bar consists of an SCR connected directly across the load. If the output voltage should go too high, the SCR is fired, and the output voltage is quickly brought to a very low value. This normally blows a fuse or opens a circuit breaker because the power supply becomes shorted. The circuit must then be turned off and back on again to reset the SCR.

The circuit senses the overvoltage as follows: Normally, the power supply puts out +5 V. Therefore, the 6.2 V zener diode does not conduct, and there is no gate current to turn on the SCR. However, if the output voltage goes above 6.2 V (due to a power supply component failure), then the SCR will receive gate current, and the SCR will turn on. This will bring the output voltage down to a very low level.

6.3.7 IC Regulators

There are a variety of integrated circuit voltage regulators. They come in many positive or negative fixed voltages, both positive and negative fixed voltages, as well as variable voltages. Figure 6-23 shows how simple a well-regulated power supply can be when one of these ICs is used. The IC will probably also contain current-limiting and thermal shutdown features.

One popular type of IC regulator is the 78 hundred series. These ICs have an output current in excess of 1 A, with current limiting and automatic thermal shutdown. The ICs come in fixed values of 5, 6, 8, 12, 15, 18, and 24 V and are very easy to use because they have only three terminals as shown in Figure 6-23.

6.4 SHUNT-REGULATED POWER SUPPLIES

Shunt regulators are less efficient than series regulators because the power dissipated in them is largely wasted. However, the shunt regulators do have an advantage. If a short develops in the load, the regulator is protected, because no current will pass through it since it is in parallel with the load. A very simple resistor and a zener diode shunt regulator is shown in Figure 6-14. More complex shunt regulators usually use one or more transistors.

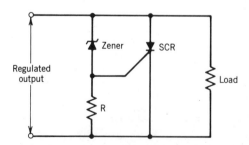

Figure 6-22   Crow bar.

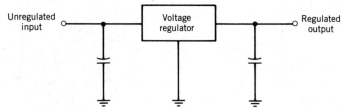

Figure 6-23   IC voltage regulator.

Figure 6-24 shows a shunt regulator that uses an op amp for an error detector and amplifier, and a transistor for the control device (Q_1). How it works: R_1, P, and R_2 are a voltage divider. If the voltage across R_1 increases, transistor Q_1 turns on more and lowers the output voltage by conducting more current to ground. Transistor Q_1 is controlled by the op amp whose inverting input is held constant by the zener diode and whose noninverting input detects the voltage increase through voltage divider R_2, P, and R_1. Should the load resistance decrease, lowering the load voltage, the change is again detected by the R_2, P, and R_1 voltage divider. Now, the op amp causes Q_1 to conduct less, thereby decreasing the total load so that the output voltage remains constant. R_3 is a current-limiting resistor that prevents damage to the regulator transistor if the load is removed.

6.5 HIGH-CURRENT VOLTAGE REGULATORS

Integrated circuit voltage regulators that can operate at several amperes are common. However, to keep cost down, normally a lower-current voltage regulator IC is used to control one or more pass transistors. An example of this is shown in Figure 6-25, which is a 13.8-V dc power supply that is capable of producing 15 A continuously or 25 A intermittently.

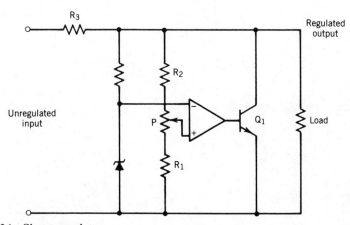

Figure 6-24   Shunt regulator.

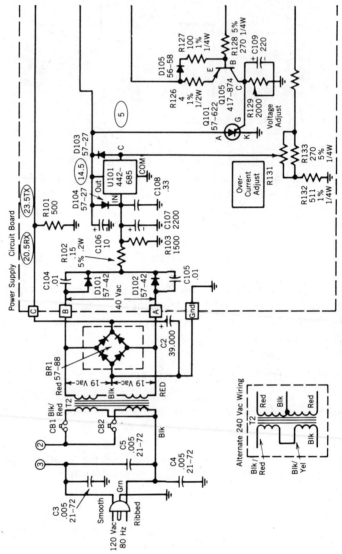

Figure 6-25 Schematic of the Heathkit power supply model PS-9000. (Courtesy of Health, Inc.)

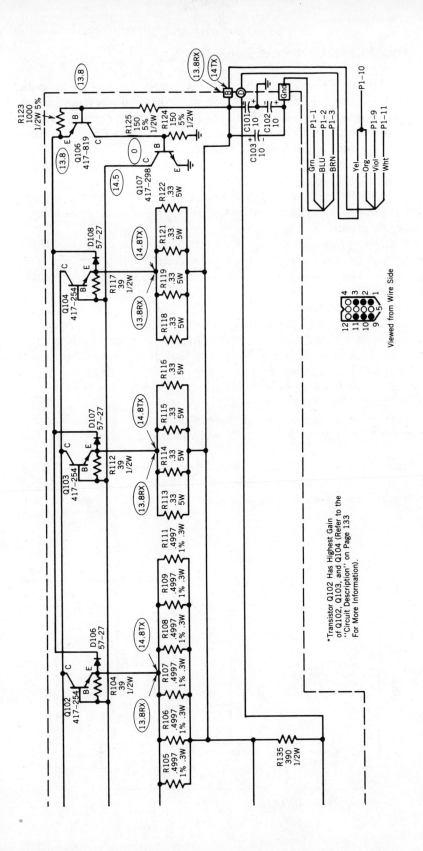

*Transistor Q102 Has Highest Gain of Q102, Q103, and Q104 (Refer to the "Circuit Description" on Page 133 For More Information).

Viewed from Wire Side

As shown by the block diagram in Figure 6-26, the power supply schematic in Figure 6-25 has current limiting, short-circuit protection, thermal sensing, and a control for adjusting the output voltage. These functions operate similarly to those of the simpler circuits explained earlier, except that this circuit uses more parts for greater current-handling capability.

In Figure 6-25, full-wave rectifiers D_{101} and D_{102} and bridge rectifier BR_1 supply unfiltered dc to voltage regulator U_{101} and pass transistors Q_{102}, Q_{103}, and Q_{104}. A common heat sink ensures that a rise in temperature in any one of the pass transistors will be detected by the voltage regulator to provide high-temperature protection.

Surge current protection is provided by transistors Q_{106} and Q_{107}. If the base-to-emitter current of the pass transistors exceeds the predetermined limit, then Q_{106} will turn on Q_{107} to rebias the pass transistors during the surge and limit the current to a safe value.

Short-circuit protection is provided by transistor Q_{105} and SCR Q_{101}. If a short circuit occurs in the load, current flowing through the pass transistor emitter resistors will create a high enough voltage to turn on transistor Q_{105}, which will fire SCR Q_{101}. This grounds the output of regulator U_{101}, which turns off the pass transistors and protects the power supply. Regulator U_{101} is self-protecting and will not be damaged. The power supply must be turned off and back on again to reset the SCR Q_{101}.

If a short circuit should develop in the power supply, the circuit breakers in the primary circuit of transformer T_2 will open and disconnect the line voltage.

A portion of the output voltage is coupled from voltage divider R_{131} through R_{134} and applied to the voltage regulator. This is the voltage regulator output voltage sample. The voltage divider also allows the output voltage to be adjusted.

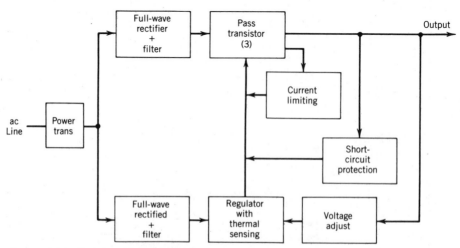

Figure 6-26 Power supply block diagram.

6.6 TROUBLESHOOTING TECHNIQUES

When troubleshooting, be sure to locate and repair the problem and not just replace a bad part. As a blown fuse is usually a sign that another circuit component is defective, a burned resistor can be the sign that a transistor or capacitor is faulty. Look beyond the obvious.

6.6.1 Visual Checks

Troubleshooting should start with a good visual check of the power supply. Check the fuse or reset the circuit breaker and look for burnt, discolored, cracked, or broken parts. These must be replaced before proceeding.

If the power supply was just on, touch the pass transistors, voltage regulator, or other active devices to see if one is hotter than it should be. Some pass devices will normally run very warm, however. Be careful doing this step or a blistered finger will result. Use a temperature-measuring device, if possible.

6.6.2 Voltage Measurements

If practical, remove the load from the power supply and then measure the output voltage. If the voltage is at the proper value, then the problem may be in the load circuit and not in the power supply at all. The following troubleshooting technique is called *divide and conquer*. Start at the output of the questionable circuit and, when you find the proper circuit voltage, proceed toward the beginning by dividing the circuit into logical sections. The problem is probably in the previous section.

For example, if the primary fuse of the power supply blows, you may want to disconnect the regulator portion of the power supply from the rectifier section and then see if the circuit still blows the fuse. This will tell you if the problem is in the regulator section or not.

Oscilloscope measurements can be very useful if the power supply is oscillating. This type of problem can be caused by a bad bypass capacitor near the regulator IC or the error amplifier, depending on the type of regulator circuit being used.

6.6.3 Current Measurements

Current measurements can tell you if the current-limiting feature is working or not, and it can tell you if the pass transistors are each supplying their share of the load properly and not just one transistor doing all the work.

If a suitable ammeter is not available, you can place a fractional-ohm resistor (0.1 Ω or so, with a large wattage value) in the current path, measure the voltage across the resistor, and then solve for the current using Ohm's law:

$$I = \frac{E}{R}$$

where

I = current, in amperes

E = voltage, in volts

R = resistance, in ohms

6.6.4 Common Problems

The most common problems are:

Components. Open or shorted rectifier diodes, regulator ICs, pass transistors, or filter capacitors. Replace them as required, but be sure to determine the source of the problem before restoring power.

Improper voltage regulation. Check the regulator, voltage reference device, or error amplifier. If the output voltage is zero after you remove the load, check the crow bar circuit.

Power supply oscillation. Check the IC bypass capacitors if an IC voltage regulator is used. If operational amplifiers or transistors are used, check other bypass or stabilization capacitors of the error detector or error amplifier.

Overheating pass transistor(s). Check the pass transistor. If more than one pass transistor is used in parallel, make sure they are properly matched. (One transistor could be supplying more than its share of current and overheating.) Also, thermal runaway can be caused by the emitter resistor of the pass transistor changing value. The increased current will flow through the pass transistor, the current-limiting feature will not work, and the pass transistor will overheat. It could eventually fail.

If a pass transistor is being driven by a regulator IC, the pass transistor could overheat if the thermal sensing feature of the IC has failed.

6.6.5 Replacing Components

When you replace components, be sure to replace them with components of the proper value. For example, be sure that capacitors are not just the correct value in microfarads but that they also meet the voltage requirements of the circuit.

Other factors to be aware of are current, power, and tolerance. For example, transistors have separate voltage and current specifications. They also have a power specification that is usually less than the product of the maximum voltage and current specifications.

Never replace protective components such as fuses with anything other than proper replacement parts. Installing a fuse with too high a current rating could endanger the equipment and is a potential fire hazard.

When you consider a part on a circuit board, be sure to use enough heat to melt the solder, but not too much heat, or you could damage the board. Foils on a multilayer computer circuit board may require more than the usual amount of heat because of supply and ground planes inside the circuit board. In such a case, be sure the foils are all unsoldered, or you may accidentally raise a foil off the board or damage a foil inside the board when removing the component. To be on the safe side, cut off the bad part and solder the new part to the length of lead left protruding from the board.

CHAPTER 7

Switching Power Supply Converters

Luces M. Faulkenberry
University of Houston

7.1 INTRODUCTION

Switching power supplies are in common use and are continuously replacing linear power supplies. In this chapter, we will discuss the principle of operation of switching power supplies and their advantages over linear power supplies. We then look at the configurations (topologies) of the most commonly used switching power supplies and the most used regulation scheme for them. Finally, we look at some starting points for troubleshooting malfunctioning switching power supplies. This chapter discusses driven dc-to-dc switching power supply converters.

Switching power supplies, also called switched-mode power supplies, differ from linear power supplies in that, as the name implies, they operate by turning the power-regulating device on and off. The basic switching power supply is shown in block diagram form in Figure 7-1a. It consists of an unregulated supply voltage V_s (which may be any of several rectifier–filter combinations), a switching element, a low-pass filter to turn the switching waveform into dc, and a regulator to hold the output constant as the load power requirement or the input voltage varies. The regulator compares the output voltage sample to the reference voltage and adjusts the duty cycle of the switching element to compensate for any difference, as shown in Figure 7-1b. The process normally used is called *pulse width modulation* (PWM).

Pulse width modulation in our basic supply of Figure 7-1 operates as follows. The output voltage is filtered so that it is the average voltage of the switching element. The longer the switching element is on, the higher the average voltage, and vice versa. If the power requirement of the load goes from

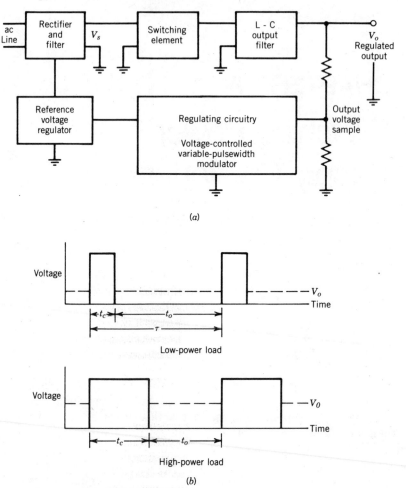

Figure 7-1 Šwitching power supply. (*a*) Block diagram. (*b*) Pulse width modulation.

low to high, as shown in Figure 7-1*b*, the output voltage will drop as increased load current causes higher voltage drops across the source resistance, switch resistance, and filter resistance. The output voltage drop is sensed by the regulator which increases the on-time of the switch to hold the output terminal voltage constant in spite of the additional power (therefore current) drawn by the load.

The ideal switching power supply would act as a dc transformer converting with 100% efficiency one dc level on the input to another dc level on the output. The switching frequency f_s will normally be about 10 times the corner frequency of the output filter f_o.

The advantages of switched-mode converters are many. Typical efficiencies of switching power supplies are in the 60 to 90% range as opposed to 30 to 60% for linear power supplies. By using a high switching frequency, anywhere from 20 to 200 kHz, the size of the magnetic and capacitive filtering components (as well as isolation transformers, if used) are much smaller, thereby reducing the size, weight, and material usage of the power supply.

Switched-mode converters are more complex than linear power supplies, generally respond to transient or sudden load changes slower (though as f_s increases, the transient response difference narrows), and because of the switching produce more line and electromagnetic interference, which must be carefully filtered out. These disadvantages are usually more than offset by the switched-mode supply advantages in most applications.

7.2 THE BASIC FOUR SINGLE-SWITCH CONVERTER CONFIGURATIONS

There are four major single-switch converter configurations in common use. They are the *buck* (step down) and its dc isolated configuration called the forward converter, the *boost* (step up) and its dc isolated version called the reverse converter (which is seldom used today, so we will skip it), the *buck–boost* (voltage inverting) and its derivative called the flyback, and the $\overline{C}uk$ and its dc isolated version called the dc isolated $\overline{C}$uk. The $\overline{C}$uk converter is sometimes called the boost–buck.

7.2.1 Buck

Figure 7-2a shows the buck configuration, and Figure 7-2b shows the input and output waveforms. The buck converter requires only an inductor L, a capacitor C, a diode D_1, and a transistor for the power stage. The buck, boost, and buck–boost are simply different arrangements of the same components.

The transistor switch may be either a bipolar transistor or a power MOSFET. The faster switching times and lack of storage time of the MOSFET provide lower switching losses at higher operating frequencies, and the lower drive power needed to switch it reduces circuit complexity. The lower on-resistance of bipolar transistors causes them to be favored at lower frequencies where switching losses are less significant. The tradeoff frequency at which bipolar transistors are favored over power FETs on loss considerations is changing almost monthly. If an FET can handle the voltage and current, it is usually the favored power switch above 30 kHz at this time. A *pnp* (or *p* channel) rather than an *npn* used as the power switch in the buck avoids the problem of having to use a floating drive circuit.

The circuit operates as follows: When the transistor is on (t_c), the inductor stores energy in its magnetic field. The capacitor charging and load currents are provided by the inductor current as the inductor is charged by the supply voltage V_s. The diode is reverse-biased and is off. Current flows from the supply only when the transistor is on. When the transistor is off, the collapsing field of the inductor causes its polarity to reverse, forward-biasing the diode. The capacitor and inductor provide the load current during t_o. Since the volt-seconds

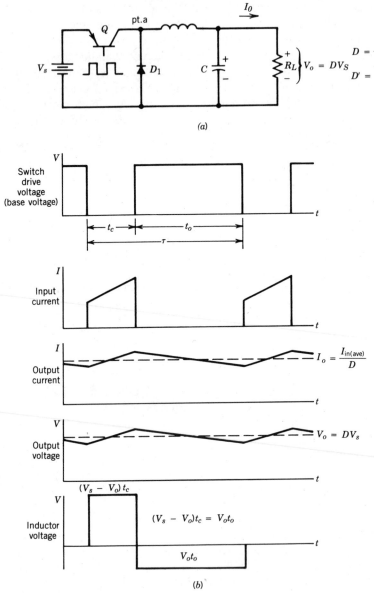

Figure 7-2   Buck converter. (*a*) Circuit configuration. (*b*) Waveforms.

stored on the inductor must equal the volt-seconds released, neglecting diode and transistor voltage drops:

$$V_o = V_s \frac{t_c}{\tau} = V_s D$$

where $D = t_c/\tau$, the duty cycle of the switching signal. Since power is conserved (neglecting losses),

$$I_o = \frac{I_{in(ave)}}{D}$$

From these two relations, one sees why the buck is also called step down since $D \leq 1$. The buck converter can only step down. The major disadvantages of the buck are its pulsating input current, which often causes it to require an input filter, and its ability to only step down the voltage. Buck converters are typically used for higher-current, moderate-power (to around 800 W) step-down applications.

7.2.2 Boost

The boost, or step-up, power circuit configuration is shown in Figure 7-3a. Some applicable waveforms are shown in Figure 7-3b. The boost converter can use an *npn* switch without a floating drive circuit since the switch connects the inductor directly across the input supply for energy storage.

The circuit operates as follows: When the transistor is on (t_c), energy is stored in the inductor. The diode is off, so the load voltage and current during t_c are supplied by the capacitor. During t_o, the transistor is off, and the energy stored in the inductor is delivered to the load and capacitor through the now forward-biased diode. The inductor discharge voltage is of the same polarity and in series with V_s and thus provides a boost function.

If the losses and switching component voltage drops are neglected,

$$V_o = \frac{V_s}{1 - D} = \frac{V_s}{D'}$$

and

$$I_o = I_{in}(1 - D) = I_{in}D'$$

Note that V_o cannot go to infinity as the equation might suggest. The various resistive loss elements will cause a limit in output voltage to be reached at a step-up ratio somewhere between 5 and 10 typically. The boost converter is normally used in lower-power (to about 500 W), lower-current applications because of the high output ripple. The output current is delivered to R_L and C in pulses, which cause noise problems. The boost converter can only step up voltage.

7.2.3 Buck–Boost

The buck–boost converter, shown in Figure 7-4, is also called the voltage inverter because the output voltage is opposite in polarity with respect to the input. It can be used to step up or step down the input voltage. If $D > 0.5$, it functions as a step-up converter; and if $D < 0.5$, it functions as a step-down converter.

When Q is on (t_c), the inductor is magnetized because it is connected across V_s. The diode is off and the load current is supplied by C. When Q is off, the energy stored in L is delivered to C and R_L through D_1, because the collapsing magnetic field around L inverts the polarity of the inductor voltage.

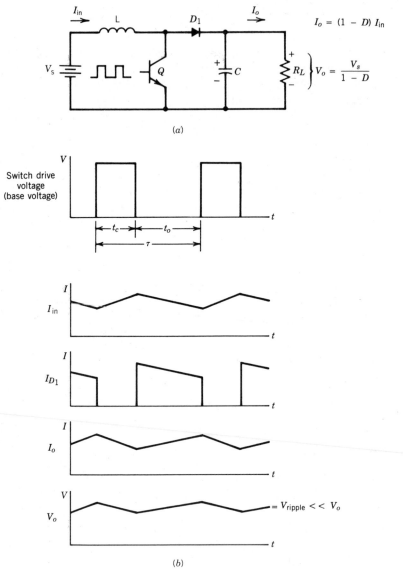

Figure 7-3   Boost converter. (*a*) Circuit. (*b*) Waveforms.

The ideal input and output voltage and current relationships are shown in Figure 7-4.

The currents drawn from the voltage source (V_s) and delivered to the output section via D_1 are both pulsating, making the control of conducted electromagnetic interference (EMI) in the buck–boost converter difficult. Conducted EMI consists of switching transients which travel through the power cord and interfere with other equipment connected to the same power circuit. The buck–boost is used in moderate-power (to about 500 W) conversion applications.

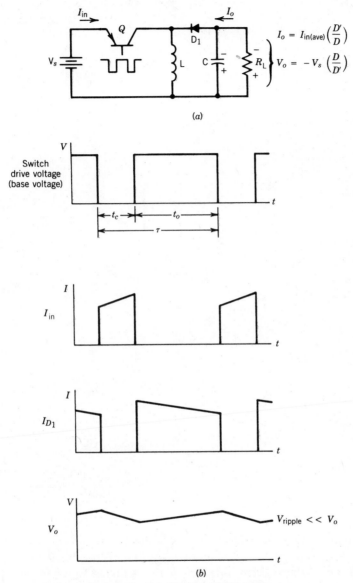

Figure 7-4   Buck–boost converter. (*a*) Circuit. (*b*) Waveforms.

7.2.4 Čuk

In 1977, Dr. Slodoban Čuk invented the converter topology that bears his name. It is derived from cascaded boost and buck converter stages. Both the input and output circuits contain inductors as shown in Figure 7-5*a*, thus neither the input nor output current is pulsating. This greatly reduces conducted EMI problems, allowing smaller filtering components.

The circuit of Figure 7-5*a* works as follows: During the transistor on-time, the diode is off since C_1 is charged as shown in Figure 7-5*a*. The inductor

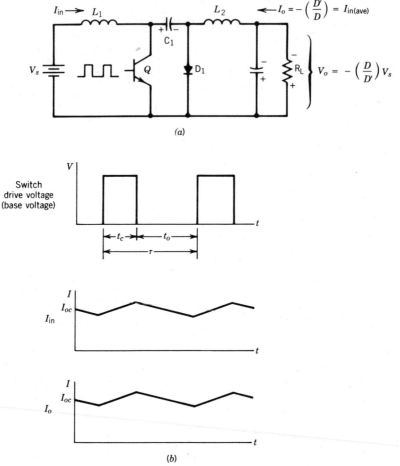

Figure 7-5   Čuk converter. (*a*) Circuit. (*b*) Waveforms.

L_1 is receiving energy from V_s via Q. C_1 is discharging through L_2, the load, and through the output filter capacitor C_2, so C_1 is transferring energy to L_2 during this time. When Q turns off, the collapsing flux of L_2 reverses L_2 polarity and turns the diode on. L_2 then transfers its energy to C_2 and R_L. C_1 charges through D_1 to $V_s + V_{L1}$ during the time the transistor is off. Note that C_1 is a coupling capacitor and is critical in the energy transfer process. L_1 delivers its energy to C_1 during t_o, and C_1 delivers its energy to L_2 as well as C_2 and R_L during t_c.

The ideal input-to-output relationships of the $\overline{\text{C}}$uk converter are the same as the buck–boost,

$$V_o = - \left(\frac{D}{D'} \right) V_s$$

and

$$I_o = -\left(\frac{D'}{D}\right) I_{in}$$

so it can function as a step-up converter when $D > 0.5$ and as a step-down converter when $D < 0.5$.

The $\overline{C}$uk converter is somewhat more complex than the other configurations, but the improved ripple performance makes the additional cost worthwhile in many applications. The $\overline{C}$uk can be used for converters up to the kilowatt range and possibly beyond.

The $\overline{C}$uk converter has some additional desirable properties. If L_1 and L_2 are coupled as shown in Figure 7-6a and the turns ratio and the coupling coefficient are properly set, the input or the ouput current ripple can be set to zero. (The output ripple is chosen to be zero, naturally.) The $\overline{C}$uk converter can also be operated as a two-quadrant converter (reversible input and output), as shown in Figure 7-6b. If Q_1 is switched and Q_2 is off, the power flow is to the right. If $-V_s$ replaces R_L, or if the load can deliver energy (a decelerating dc motor with a high inertia load), and Q_2 is switched with Q_1 off, the power flow is to the left. The transistor–diode pairs can be replaced by a suitable power MOSFET which has an inherent reverse diode built in. The $\overline{C}$uk can also easily be operated as a four-quadrant converter connected as shown in Figure 7-6c. In this configuration, it can operate as a push–pull converter (to be discussed later) or a switched-mode amplifier in which the output is a high-power amplified replica of the pulse width modulator reference, which is the input signal rather than a dc reference in this mode of operation.

In this configuration, both the top and the bottom converters must be capable of bidirectional current flow, because the top converter must sink all of the current the bottom converter provides, and vice versa. To have the top of R_L negative, Q_1 and Q_4 must be switched with equal duty ratios; and to reverse the output polarity, Q_2 and Q_3 must operate with equal D. For this configuration,

$$V_o = V_s \left(\frac{D - D'}{D\,D'}\right)$$

The configurations of Figures 7-6b and 7-6c have not yet been widely used.

7.3 DC ISOLATED SINGLE-SWITCH CONVERTERS

Each of the single-switch converters we have discussed has a derived configuration that uses an isolation transformer for dc isolation and, if needed, voltage step-up, step-down, or inversion. The derived versions can also, by use of multiple secondary windings, provide multiple dc output voltages, though the regulation of each secondary supply is dependent on the one chosen for the PWM regulator sample. Since it is very seldom used, we shall not discuss the boost-derived reverse converter.

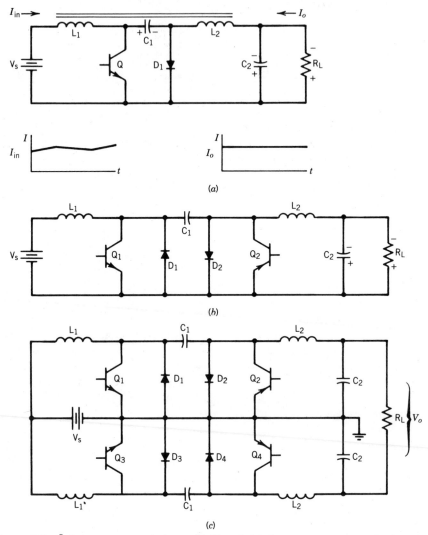

Figure 7-6 Čuk converter variations. (*a*) Coupled inductor–zero output ripple connection. (*b*) Two-quadrant (bidirectional) connection. (*c*) Bridge or push–pull connection.

7.3.1 Forward Converter

The forward converter is the buck-derived converter. The forward converter is shown in Figure 7-7. Note that L_1, D_2, C, and R_L are arranged as a buck converter with Q and V_s missing. The transformer secondary and D_2 act as the switch. The forward converter is, then, just a dc-isolated buck.

When Q is on, the primary of T_1 induces a voltage in the middle winding that turns D_1 off. During t_c, voltage in the right-hand secondary of T_1 in Figure 7-7 turns D_2 on and provides charging current for L_1 and C and load current.

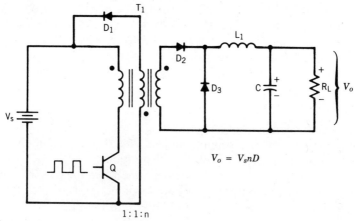

Figure 7-7   Forward (buck-derived) converter.

When Q turns off, the collapsing transformer flux reverses the polarity on all the windings of T_1 and on L_1. D_2 is thus turned off and D_3 is turned on. During t_o, L_1 provides energy to the output circuit, as in a nonisolated buck. The middle winding of T_1 turns D_1 on and provides a path for the transformer energy back into the supply, during t_o. This resets the core in the opposite direction, preventing core saturation and preventing the collector voltage of the transistor from exceeding $2V_s$.

The forward converter is used in the same power range as a buck when dc isolation is needed. The ideal output voltage equation is shown in Figure 7-7.

7.3.2 Flyback

The flyback, or buck–boost-derived, converter is the simplest dc isolated switched-mode power circuit configuration. Its use is in higher-voltage, low-current or low-power applications where its comparatively high output ripple can be tolerated.

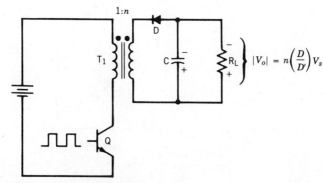

Figure 7-8   Flyback (buck–boost-derived) converter.

Referring to Figure 7-8, when Q is on, magnetizing energy is stored in the transformer. D_1 is off and C provides the load current. When the transistor is off (t_o), the collapsing flux reverses the transformer winding polarity, D_1 turns on, and the transformer secondary provides charging current to C and load current to R_L. The output voltage relation is shown in the figure.

7.3.3 Isolated C̄uk

The isolated C̄uk, shown in Figure 7-9a, is the newest member of the isolated dc-to-dc converter family. It is more complex than the others but provides less input and output ripple than the other configurations.

Figures 7-9b and 7-9c show the current flow when the transistor is on (t_c) and off (t_o), respectively, for $D < 0.5$ (step-down mode). During t_c, the current from the supply magnetizes L_1. C_1 discharges into the primary of T_1 causing C_2 to discharge through L_2, charging it; through C_3, charging it; and through the load. The diode is off because of the reverse bias applied by the isolation transformer secondary; $i_2 > i_1$ because $D < 0.5$, and the circuit is operating as a step-down converter. During t_o, the discharge of L_1 in series with the source

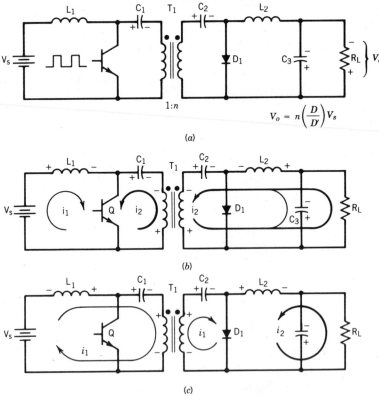

$$V_o = n\left(\frac{D}{D'}\right)V_s$$

(a)

(b)

(c)

Figure 7-9 Isolated C̄uk converter. (a) The circuit. (b) Current during t_c, $D < 0.5$. (c) Current during t_o, $D < 0.5$.

charges C_1. The current coupled through the isolation transformer charges C_2 through D_1, which is now forward-biased. L_2 flux collapse forces current through D_1, providing load current. The relative magnitudes of i_1 and i_2 as indicated by the light and heavy lines reverse when the output voltage is stepped up ($D >$ 0.5).

Figure 7-10 shows a $\overline{C}$uk in which all the magnetics are coupled and the coupling coefficients of L_1 and L_2 to T_1 are adjusted by changing the core gaps g_1 and g_2. If g_1 and g_2 are set properly, the input and output currents both have *zero* ripple over a wide range of operating conditions! This configuration will probably find wide use in applications where EMI is critical.

The two quadrant and bridge $\overline{C}$uk connections shown in Figures 7-6*b* and 7-6*c* can be implemented in the same fashion with the dc isolated versions. The dc isolated $\overline{C}$uk bridge is shown in Figure 7-11.

7.4 CONTINUOUS- AND DISCONTINUOUS-CONDUCTION MODES

Most switched-mode dc-to-dc converters are operated in a continuous-inductor-current mode, in which the inductor is either charging or discharging at all

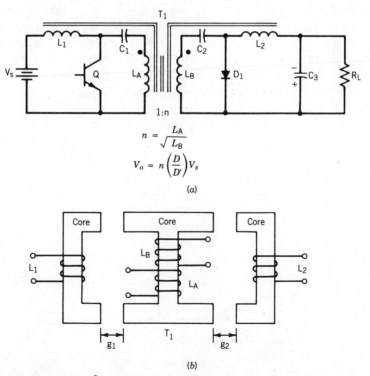

$$n = \sqrt{\frac{L_A}{L_B}}$$

$$V_o = n\left(\frac{D}{D'}\right)V_s$$

(*a*)

(*b*)

Figure 7-10   Zero ripple $\overline{C}$uk. (*a*) The zero ripple showing the coupled magnetics. (*b*) A coupled magnetics structure.

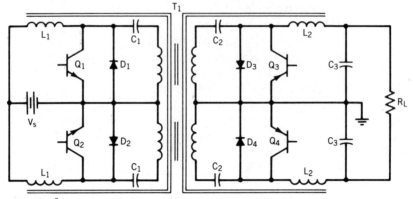

Figure 7-11 Čuk isolated bridge with coupled magnetics.

times, as shown in Figure 7-12*a*. The inductor current never goes to zero, and the inductor is receiving or transferring energy continuously. If during the transistor off-time an inductor current goes to zero, the converter is operating in the discontinuous mode, as shown in Figure 7-12*b*, in which no inductor current is flowing for a portion of t_o. The discontinuous mode is entered by a decrease in the switching frequency f_s, a reduction in duty cycle D for a given load, a drop in the filter inductor L, or a reduction in the load current, which causes the average current to drop while the ac ripple current stays the same. The discontinuous-conduction mode is normally entered when operating a switching power supply at light loads. When the inductor current in the converter reaches zero, the diode turns off, preventing the inductor current from reversing and thus causing zero inductor current during t_o'.

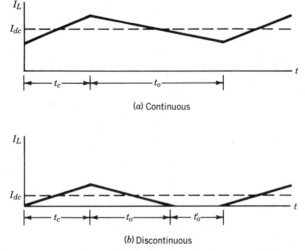

Figure 7-12 Inductor current waveforms for the continuous and discontinuous inductor current modes. (*a*) Continuous. (*b*) Discontinuous.

Entering the discontinuous-conduction mode can actually increase a switching power supply's stability; and if entered because of an increase of R_L, which is usually the case, the voltage ripple does not worsen. However, the gain equations for the converters are no longer valid in the discontinuous mode. The discontinuous mode causes a slight shift in the output voltage when it is entered. R. E. Middlebrook and Slodoban Čuk have derived a set of conditions that predict the R_L for which the discontinuous mode is entered. If $R_L > R$ critical (R_{CR}), the converter operates in the discontinuous mode; and if $R_L < R_{CR}$, the converter operates in the continuous mode.

If a nominal resistance (R_{nom}) is defined as

$$R_{nom} = 2Lf_s$$

where L is the inductor value and f_s is the switching frequency,

then

$$R_{CR} = \frac{R_{nom}}{1 - D} \quad \text{for the buck}$$

$$R_{CR} = \frac{R_{nom}}{D(1 - D)^2} \quad \text{for the boost}$$

$$R_{CR} = \frac{R_{nom}}{(1 - D)^2} \quad \text{for the buck–boost}$$

The Čuk converter has the discontinuous mode defined in terms of a dimensionless k value such that if $k > k_{CR}$, the continuous mode is obtained and if $k < k_{CR}$, the discontinuous mode is obtained. For the Čuk, $k_{CR} = (1 - D)^2$. If $R_{nom} = 2(L_1 \parallel L_2)f_s$, then $R_{CR} = R_{nom}/(1 - D)^2$ for the Čuk. The k_{CR} values for the buck, boost, and buck–boost converters are $1 - D$, $D(1 - D)^2$, and $(1 - D)^2$, respectively.

7.5 MULTIPLE-SWITCH CONVERTERS

At power levels over about a kilowatt, high efficiency and reduced power component stress dictate the coupling of power to the load throughout a greater portion of the switching period. This results in the use of two or more switches and increased circuit complexity. Three major converter configurations are in common use: the push–pull, half-bridge, and bridge. Others exist but are not in such common usage. The Čuk bridge has been discussed.

7.5.1 Push–Pull

The basic push–pull dc-to-dc converter is shown in Figure 7-13a. When Q_1 is on, D_1 turns on as energy is magnetically coupled through T_1. When Q_1 turns off, the collapsing magnetic flux causes all the transformer polarities to reverse. The positive voltage on the anode of D_2 causes it to turn on, and the energy from the transformer flux collapse is coupled to the output. During t_o, the collector voltage of Q_1 is $2V_s$. When Q_2 turns on, the process repeats, with D_2

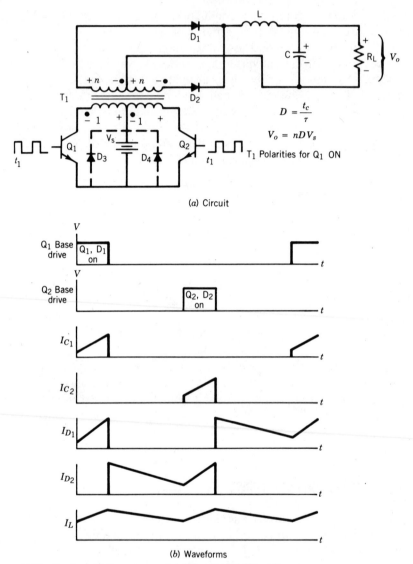

(a) Circuit

(b) Waveforms

Figure 7-13   Push–pull converter. (a) Circuit. (b) Waveforms.

conducting while Q_2 is on, and D_1 conducting when Q_2 turns off. The output voltage is ideally

$$V_0 = nDV_s$$

Diodes D_3 and D_4, when used, prevent a negative voltage from being applied to the transistor that has just turned off.

It is important that the on-time be identical for Q_1 and Q_2, that the forward voltages of the diodes be fairly well matched, and that the on voltages of the transistors be closely matched to prevent more magnetization of the transformer core in one direction than the other. Otherwise, over a few periods of operation, the core will saturate in one direction, which will result in the failure of the converter.

The push–pull converter can power several supplies on the secondary of the power transformer by the use of multiple secondary windings, as can the half-bridge and the bridge. Only one secondary supply can be used as a sample for the voltage-regulating circuitry.

If, when the push–pull converter is under a heavy load or when it is first turned on, Q_1 receives its turn on signal at the same time that Q_2 is turned off, $D = 1$. Then, during the storage time of the transistor just turned off, very high collector current can flow, with $V_{CE} = 2V_s$. This causes very high switching losses and can overstress the transistors. Therefore, the control and regulating circuitry must provide a delay between turning the on-transistor off and the off-transistor on; a delay equal to the turn-off time of the transistors. The delay, called dead time, prevents the condition.

7.5.2 Half-Bridge

The half-bridge circuit shown in Figure 7-14 eliminates the necessity of transistors having to block $2V_s$ as in the push–pull. Each transistor sees a maximum voltage of V_s. As in the push–pull converter, several voltages can be provided by the use of multiple secondary windings on T_1. The two capacitors C_1 and C_2 provide a voltage of $V_s/2$ to one side of the transformer primary, limiting its voltage to $V_s/2$. The output voltage is then ideally

$$V_o = \frac{nD}{2}$$

When Q_1 is on and Q_2 is off, the transformer primary is attached to $V_s/2$ and V_s. Energy is magnetically coupled to the secondary and on through D_3 to the load. When Q_1 turns off (Q_2 is still off), the collapsing field reverses the trans-

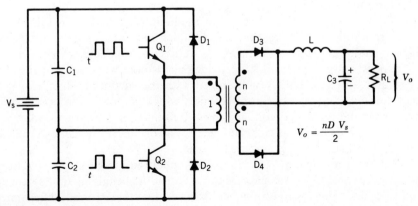

Figure 7-14   Half-bridge converter.

former polarities. Energy is coupled from T_1 through D_4 to the load. Free-wheeling diode D_2 prevents the collector of Q_2 from going negative. When Q_2 turns on, D_4 couples energy to the load. Between the time Q_2 turns off and Q_1 turns on again, D_3 couples the energy from the magnetic flux decay of T_1 to the load, and D_1 prevents the collector voltage of Q_1 from rising above V_s.

Dead time is essential for the half-bridge circuit because, if there is any time overlap between one transistor turning off and the other turning on, very large switching currents can flow and damage the transistors.

7.5.3 Bridge

The bridge (also called the full-bridge) converter shown in Figure 7-15 combines the most power delivered to the load with the least blocking voltage requirements for the transistors. The bridge converter operates much like the half-bridge. Diodes D_1 through D_4 prevent any collector voltage from rising above V_s or below zero volts when all transistors are off. The transistors are turned on in pairs, Q_1 and Q_3 or Q_2 and Q_4. If Q_1 and Q_3 are on, D_5 is forward-biased by the transformer secondary and provides load current. When Q_1 and Q_3 are turned off, the decaying magnetic flux reverses the transformer polarities and D_6 provides current to the load until transistors Q_2 and Q_4 turn on.

As in the half-bridge and for the same reason, dead time must be provided in the control signals to assure there will never be any conduction overlap between the time one set of transistors is turned on and the other set is turned off.

Note that both the half-bridge and bridge need floating-drive sources for the supply side transistors.

7.6 REGULATION OF SWITCHED-MODE DC-TO-DC CONVERTERS

Numerous integrated circuit manufacturers make switching power supply PWM regulators. Most manufacturers of switched-mode power supplies use IC regulators in their supplies.

The basic parts of a pulse width modulator are shown in Figure 7-16. These are a reference voltage regulator, which provides a stable voltage with which the output voltage is compared, as well as a voltage supply for the internal IC circuitry; an oscillator, which provides a sawtooth waveform for the PWM comparator and an output synchronized with the sawtooth which drives the pulse-steering circuitry; an error amplifier, which compares the power supply output with the reference voltage; and a pulse-steering circuit that turns the output transistors on in the right sequence. The oscillator frequency is set with an external capacitor (C_{ext}) and resistor (R_{ext}).

Referring to Figure 7-16b, the basic operation is as follows: An output transistor, assume Q_1 in this case, is turned on at the beginning of the sawtooth ramp (t_1). Since the sawtooth voltage is lower than the error amplifier output, the PWM comparator output is high. The pulse steering circuit steers this high output to Q_1 because the synchronizing signal has enabled the steering circuitry

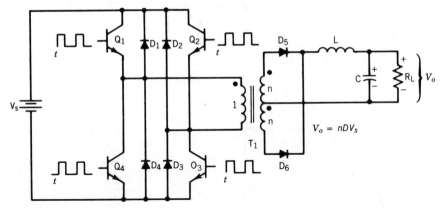

Figure 7-15 Bridge converter.

at the start of the ramp. When the ramp becomes more positive than the error amplifier output, the PWM comparator output goes low, turning Q_1 off via the pulse-steering circuit. The process repeats the next ramp cycle, except that the pulse-steering circuit now couples the PWM high output to Q_2. If the power supply voltage should drop, the output voltage sample will drop lower than the reference voltage so that the error amplifier voltage will rise. This will cause a longer time between the beginning of the ramp and when the ramp voltage becomes more positive than the error amplifier output, thus increasing the on-time of each transistor. Most PWM regulators incorporate a minimum dead time of about 5% of the cycle and also have provision for setting the dead time higher by an external resistor or voltage divider.

A current high-performance pulse width modulator is shown in Figure 7-17a. It is the Silicon General 1526 designed by Stan Dendinger of Silicon General. The SG 1526 integrated circuit can operate with an input voltage from 8 to 40 V, be set to switch at frequencies from 1 Hz to 400 kHz, and provide 100 mA of drive current to the converter switches.

The IC provides a soft-start option so that the addition of an external capacitor will force the error amplifier output to rise slowly. This causes D to become progressively larger, reducing the inrush current to filter capacitors at turn-on. To allow normal operation to proceed, a diode disconnects the soft-start voltage from the error amplifier output when the soft-start capacitor voltage exceeds the error amplifier output. It provides a low-voltage lockout to the output transistors and prevents the soft-start capacitor from charging until the on-chip reference has risen to $+4.5$ V. This occurs with approximately $+7.5$ V supplied to the SG 1526.

The totem pole (current sink/source) output transistors have low on impedance and provide considerable flexibility in driving the power switches. The output is high during D (t_c/τ). The regulator has a built-in current-limiting comparator that terminates the output pulse by disabling the AND gate at the output of the PWM comparator. The current-limiting comparator thus allows current limiting on a pulse-by-pulse basis. An active low shutdown pin is provided at the same AND gate terminal for other monitoring functions to stop

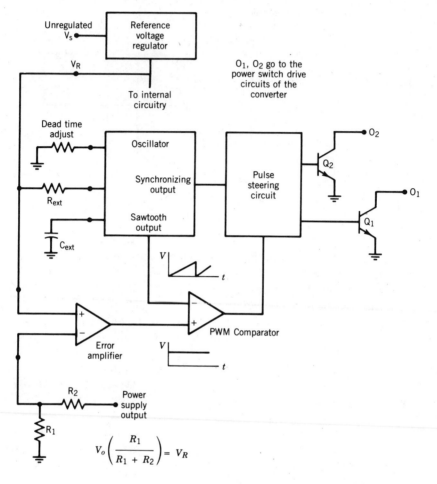

(a) Key parts

Figure 7-16   Basic pulse width modulator block diagram. (*a*) Key parts. (*b*) Waveforms.

the supply. Both the current-limiting and PWM comparators have a small amount of hysteresis to prevent state changes from small amounts of noise. The oscillator capacitor is charged and discharged by current sources providing a linear ramp. The discharge time is also the dead time and can be lengthened from 3% of the cycle to 50% by an external resistor that lowers the discharge current.

Operation of the pulse-steering circuit can be seen by examining the circuit and the waveforms in Figure 7-17*b*. Starting with the output, the upper transistor is on and the lower transistor is off during D. The $\overline{\text{SYNC}}$ low disables both output gates during the dead time, and the $\overline{\text{SYNC}}$ high enables them. When $\overline{\text{SYNC}}$ is high, $\overline{\text{Q}}$ of the metering F/F and one output of the toggle F/F are low, and one gate is enabled. Then, the upper transistor turns on and the lower

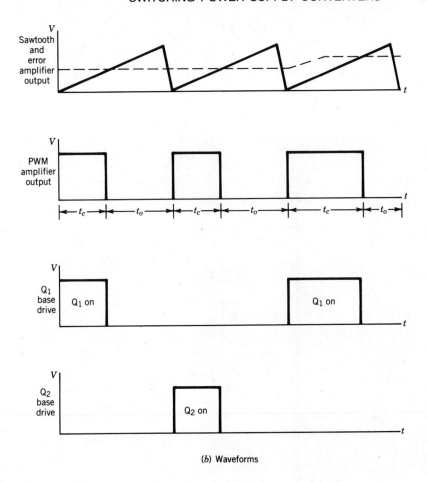

(b) Waveforms

transistor turns off. Assume that at the beginning of a cycle, Q of the toggle F/F is low. The $\overline{\text{SYNC}}$ will set the metering F/F; its $\overline{Q}$ output will then go high when its D input goes high. The $\overline{\text{SYNC}}$ also sets the memory F/F, driving its $\overline{Q}$ output low, toggling the toggle F/F, and setting its $\overline{Q}$ output low. Since the error amplifier output is higher than the sawtooth, the PWM comparator output is high. With the current-limiting comparator output high and no low on the $\overline{\text{SHUTDOWN}}$ pin, the PWM signal passes through the AND gate to the D input of the metering F/F. The metering F/F Q output then resets the memory F/F, causing its $\overline{Q}$ output to go high again. The metering F/F $\overline{Q}$ output goes low, causing the bottom transistor pair of Figure 7-17a to operate, that is, it causes the top transistor to turn on and the bottom transistor to turn off.

During the next cycle, the toggle F/F Q output will be driven low by the memory F/F when the $\overline{\text{SYNC}}$ low sets the memory and the metering flip-flops. The circuit operation will then be the same, except that the upper output transistor pair will operate during the PWM comparator output high. As the error amplifier output voltage goes lower, the duty cycle of the PWM is reduced.

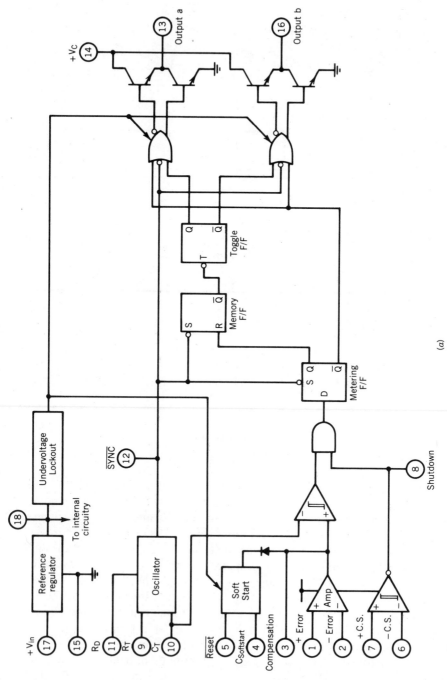

Figure 7-17 Silicon General 1526 pulse width modulator (courtesy of Silicon General). (*a*) SG 1526 block diagram (*b*) Waveforms.

(*a*)

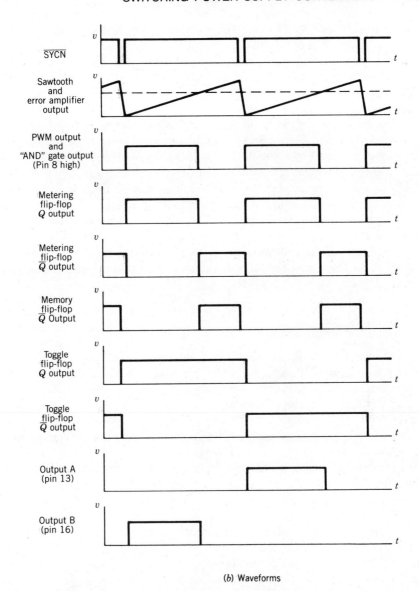

(b) Waveforms

When the PWM comparator drops low during a period, or the current limiter operates, or a $\overline{\text{SHUTDOWN}}$ signal appears, the output pulse terminates. The metering F/F cannot accept another input until it is set by the next $\overline{\text{SYNC}}$ low. This is how pulse-by-pulse current limiting occurs. If a PWM high output does not appear during a period, the memory F/F $\overline{\text{Q}}$ output will not be reset high, so the next $\overline{\text{SYNC}}$ signal will not toggle the toggle F/F. This prevents a transistor in a push–pull converter from receiving two successive pulses.

A voltage divider from the reference regulator output provides the reference voltage for the noninverting input of the current limit (CS for current shutdown) comparator. The inverting input then senses the voltage drop across a small resistor (R_{cs}) or, if the current feedback must be floating, the rectified output of a current transformer. If the $I_o R_{cs}$ voltage drop exceeds the current-limiting comparator reference voltage, the comparator output drops low, disabling the AND gate at the output of the PWM comparator terminating the output pulse.

The SG 1526 can easily be used for either single- or multiple-switch converters as shown in Figure 7-18. Switched-mode-converter pulse width modulators are produced by several IC manufacturers, and most work on principles similar to the SG 1526, although circuit details and performance vary considerably. Some, designed only to drive single-switch dc-to-dc converters, have a power diode and transistor switch built into the IC.

Some single-switch IC regulators terminate the transistor on-time (D) by sensing when the switch current has reached a predetermined level. They also have a voltage comparison amplifier, such as that in the regulator we discussed, that will shut the transistor off when the sampled output voltage exceeds the reference voltage.

7.7 TROUBLES

As with most electronic systems, the troubles that may occur in a switched-mode power supply are almost infinite in variety. We will discuss only a few of the most common: loss of output voltage, loss of regulation, failure of power switch transistors, excessive output ripple or output switching spikes, conducted EMI, oscillations with step change of input voltage or load, and $V_o = V_s$ in buck-and-boost converters. We will also discuss possible causes for each complaint.

The loss of output in a dc-to-dc converter can be caused by many things. One should immediately check for overload shutdown. Reduce the load and start the supply again. If it doesn't start, then check the rectifier power supply from the line that feeds the converter. A blown rectifier or shorted filter capacitor might be the trouble. If the unregulated power supply is okay, then check the power switches for shorts or loss of base drive. If the base drive is lost, the feedback signal may be too high, the soft start may be hung low, the regulation chip may be bad, or the output filter capacitor may be shorted.

Do not hesitate to break the feedback loop and inject the appropriate voltage or signal at the break to aid the fault isolation process.

In push–pull converters, check for transformer saturation by monitoring a few cycles of the switch as the supply is turned on. If the duty cycles vary for the transistors or the pulses stop coupling to the transformer in a few periods, the transformer core is *walking* or moving along the BH curve into saturation. This check is normally done after replacing a bad transistor. Any duty cycle imbalance must be corrected, and replacement power diodes and switches must match those already there.

If the power supply provides an output and the power switches are func-

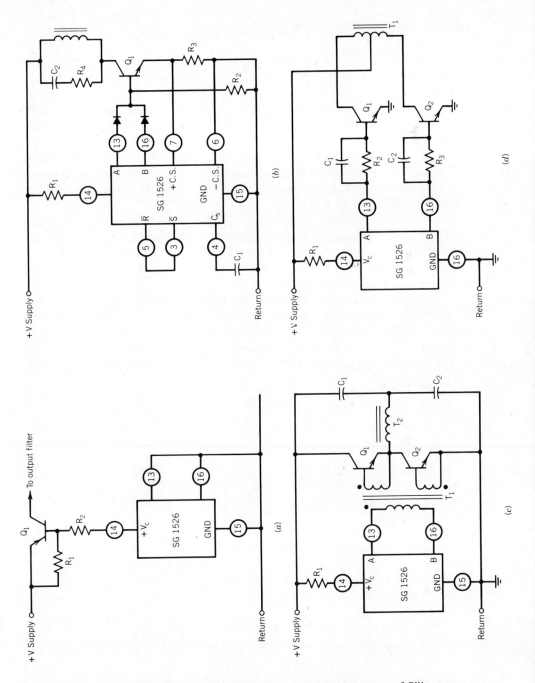

Figure 7-18 SG 1526 connections for four convertor types (courtesy of Silicon General). (*a*) For single-ended configurations, the V_c terminal is alternately switched to ground by the driver pull-up transistors. (*b*) Using the SG 1526 in a flyback converter with current limiting. (*c*) Low-power transformers are driven directly by the output terminals. (*d*) Basic connections for a push–pull grounded-emitter configuration.

tioning but the output voltage varies with the load and input voltage, the regulation is gone. Either the feedback signal is wrong or the regulator's IC has failed.

Equal output and input voltages are a problem usually associated with buck, boost, or buck–boost converters. In the buck and boost converters, this problem is caused by the power switches being locked on because of switch failure or switch drive circuitry malfunction. In the buck–boost, this condition will occur if the power switch loses its drive so that it is off. An emitter-base short, usually caused by excessive drive, can cause a bipolar transistor to appear locked in the off-state.

If the power switches fail periodically, the problem is usually excessive switching loss. Huskier power switches or a snubbing circuit such as the one shown in Figure 7-19 must then be used. If a snubbing circuit is already in use, check to see that it is okay. The snubbing circuit works as follows: As the transistor turns off, its rising collector voltage turns the diode on and shunts some of the turn-off current away from the collector. When the transistor turns on again, the capacitor discharges through the resistor and transistor. The snubbing circuit does lengthen the transistor turn-on time. The resistor should be wire wound so that its small inductance will help delay the discharge of the capacitor until the transistor is on. A starting point for selecting the capacitor is

$$C = \frac{I_{\text{div}}t_s}{V_{CE}}$$

where

I_{div} = amount of turn-off current to be diverted

t_s = transistor storage time

V_{CE} = transistor collector-to-emitter voltage when off

The resistor should be selected so that the capacitor discharge current added to the load current will not exceed the transistor maximum collector current but still discharge almost completely during t_c. Because the snubbing circuit lengthens the power switch turn-on time, C should be as small as possible.

When excessive output switching spikes are found, one should first check that the supply has a good low-impedance ground. The capacitors used for

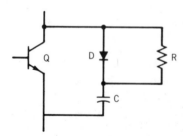

Figure 7-19   Power switch snubbing circuit.

switching power supply filters are usually electrolytic. The filter capacitors must have a low equivalent series resistance (ESR) to successfully filter the high current ripple frequency at f_s and to filter the switching spikes. Any replacement filter capacitors should be of the low-ESR type. If output switching spikes or ripple worsen over time, one of the filter capacitors, probably the output, is failing. If the switching spikes are higher than desired, shunting the output filter capacitor with a good polystyrene or silver mica capacitor will bring them under control. A larger output filter capacitor will help the output ripple.

Conducted EMI will interfere with other equipment on the same power circuit as the switching-power supply. Converter topologies with pulsating input current are the worst offenders. The buck, buck–boost, and all their derived circuits such as the push–pull are prone to conducted EMI. An input filter with an impedance lower than the supply input impedance is the only sure cure. Note that the input filter *must* have an output impedance lower than the power supply effective input impedance, or the supply regulation and output impedance will be affected. The regulation will get worse and the supply output impedance will get larger. Referring to the buck supply shown in Figure 7-20, the reflected input impedance is

$$Z_{ei} = \frac{\sqrt{\dfrac{L_e}{C_o}}}{M^2}$$

where M is the gain function V_o/V_s. For a buck supply, $M = D$; for a boost, $M = 1/(1 - D)$; for a $\overline{C}$uk and a buck–boost, $M = D/(1 - D)$. Commercial line filters are available that reduce conducted EMI to the level required by the FCC Docket 20780 rules of 1979. The purchase of a commercial filter with the proper output impedance will often be the most economical conducted-EMI repair.

The oscillation of a switching-power supply when subjected to a line or load change can damage a delicate load of digital integrated circuits and render the function of an analog circuit load meaningless. The condition usually arises

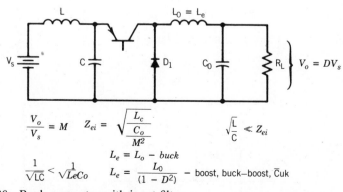

Figure 7-20   Buck converter with input filter.

from inadequate damping of one of the power supply filters. The power supply filter inductors and capacitors must have low series resistance to filter well and hold losses to a minimum. This means they will resonate at some frequency, at least for a while, if excited. Oscillatory output response should cause one to check the filter damping circuits for proper component functioning; and if no damping exists—install it.

Input filter damping is shown in Figure 7-21. A discussion of input filter and damping calculations follows.

7.8 INPUT FILTER AND DAMPING CALCULATIONS

7.8.1 Operation

As was mentioned earlier, the addition of an input filter can cause the regulator loop gain to decrease, the filter output impedance to increase, and the input voltage change coupling to the output voltage to increase, unless the input filter output impedance (Z_o) is small compared to the reflected input impedance of the output filter (Z_{ei}). Additionally, since the input impedance of most converters is negative (at a given output power, the input current drops as input voltage rises), proper damping is required to prevent oscillations as well as to assure that resonant peaking of the filter circuits does not cause a rise in Z_o at ω_o. The resonant rise of the input filter output impedance and the resonant drop of the input impedance of the output filter can be severe. This is because the parasitic resistances of the filter inductors and capacitors must be kept low for high electrical efficiency. Figure 7-22 shows a graph of frequency versus the input impedances of interest and output impedance of the input filter. If the output impedance of the input filter is lower than the lowest input impedance factor (the short-circuit output filter impedance) and $\omega_o < \omega_{OF}$, so that the resonant peaks of the output and input filters do not overlap, the input filter will not interfere with normal converter operation. So,

$$\sqrt{\frac{L_1}{C_1}} \ll \frac{R_E}{M^2}$$

and

$$\omega_o < \omega_{OF}$$

is sufficient for satisfactory converter operation.

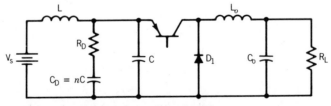

Figure 7-21   Buck converter with input filter damping.

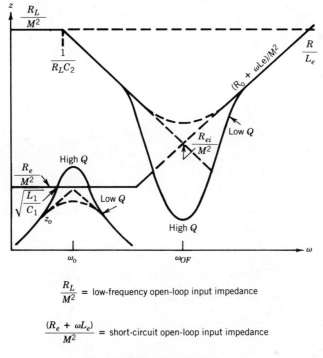

$\dfrac{R_L}{M^2}$ = low-frequency open-loop input impedance

$\dfrac{(R_e + \omega L_e)}{M^2}$ = short-circuit open-loop input impedance

Z_o = input filter output impedance

Figure 7-22   Filter impedance versus frequency.

The input filter damping components R_D and C_D shown in Figure 7-21 are necessary, as mentioned, because R_1, the internal resistance of the inductor, must be kept very low to avoid degrading the converter efficiency. The choice of R_D is not arbitrary; R_D should be chosen so that the output impedance is at its minimum, that is, maximum $|Z_o|_{mm}$ for optimal damping. There is a relationship between the ratio of $|Z_o|_{mm}$ and $R_D = \sqrt{L_1/C_1}$, expressed by n, that allows the smallest capacitor to be used in series with R_D. As shown in Figure 7-23, if R_D is too large, insufficient damping occurs; and if R_D is too small, ω_o shifts to $1/\sqrt{L_1(C_1 + C_D)}$, and damping is decreased as R_o becomes even smaller. A reasonable choice for the $|Z_o|_{mm}$ and R_o ratio is

$$\frac{|Z_o|_{mm}}{R_o} = 1$$

Then, using the equations derived by R. E. Middlebrook, who devised a design technique for quickly calculating the damping circuit for optimum filter Q,

$$\frac{|Z_o|_{mm}}{R_o} = \sqrt{\frac{2(2 + n)}{n^2}}$$

$$Q_{o(\text{opt})} = \sqrt{\frac{(4 + 3n)(2 + n)}{2n^2(4 + n)}}$$

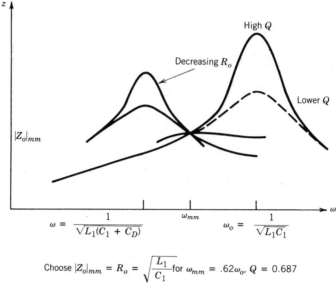

$$\text{Choose } |Z_o|_{mm} = R_o = \sqrt{\frac{L_1}{C_1}} \text{ for } \omega_{mm} = .62\omega_o, \ Q = 0.687$$

Figure 7-23 Damping circuit effect on input filter.

and

$$R_D = Q_{opt}R_o$$

at which

$$\frac{\omega_{mm}}{\omega_o} = \sqrt{\frac{2}{2 + n}}$$

The output filter damping is handled similarly.

A dc isolated converter input filter damping is the same, except that if the primary to secondary transformer winding ratio is 1:n, then

$$Z_{ei} = \frac{\sqrt{\dfrac{L_e}{C_2}}}{n^2}$$

7.8.2 Buck Example

The buck converter of Figure 7-24 will be used for an example calculation. Only the components directly related to the input filter are calculated. The pulse width modulation feedback loop and protective circuits are omitted. Calculations are for nominal criteria, with losses from parasitic resistances and switching element voltage drops neglected.

Let $V_o = 48$ V at $I_o = 4.2$ A (200 W), and $V_s = 165$ V at $I_s = 1.72$ A (V_s = rectified 117 V RMS). Thus, $D = V_o/V_s = 0.29$. If $f_s = 50$ kHz, then $t_c = 5.8$ μsec and $t_o = 14.2$ μsec. If the inductor current ripple is $\pm 20\%$ of I_o, then L_2 is calculated from

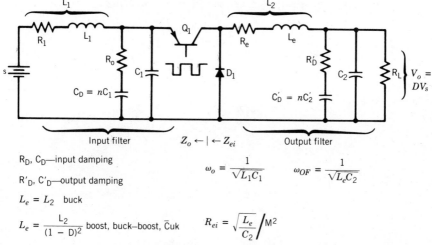

R_D, C_D—input damping

R'_D, C'_D—output damping

$L_e = L_2$ buck

$L_e = \dfrac{L_2}{(1 - D)^2}$ boost, buck–boost, Ćuk

$\omega_o = \dfrac{1}{\sqrt{L_1 C_1}}$ $\omega_{OF} = \dfrac{1}{\sqrt{L_e C_2}}$

$R_{ei} = \sqrt{\dfrac{L_e}{C_2}}\Big/M^2$

Figure 7-24 Buck converter with input filter and damping on input and output filters.

$$L_2 = \frac{V_o t_o}{\Delta I_o} = 406 \ \mu H$$

Then, if the output ripple (V_R) is to be 50 mV,

$$C_2 = \frac{I_o \tau}{4 V_R} = 420 \ \mu F$$

then

$$f_{OF} = \frac{1}{2\pi \sqrt{L_e C_2}} = 385 \ Hz \ll f_s$$

If the internal resistance of L_2 is 1 Ω, then

$$Z_o \ll \frac{R_e}{M^2} = 11.9 \ \Omega$$

to assure proper worst-case operation.

A target $Z_o = 2.5 \ \Omega$ is selected, even though the inductor and capacitor values will be large. Selecting $C_1 = 1000 \ \mu F$,

$$L_1 = 6.25 C_1 = 6.25 \ \ mH$$

Checking for $f_o < f_{OF}$,

$$f_o = \frac{1}{2\pi \sqrt{L_1 C_1}} = 63 \ Hz$$

The input filter must now be damped. Setting $|Z_o|_{mm}/R_o = 1$,

$$1 = \sqrt{\frac{2(2 + n)}{n^2}}$$

and solving by the quadratic equation, using the positive n,

$$n = 3.24$$

Thus, from the optimum Q equation,

$$Q_{o(opt)} = 0.687$$

it follows that

$$R_D = QR_o = 1.71 \quad \Omega$$

and

$$C_D = nC_1 = 3240 \quad \mu F$$

If $|Z_o|_{mm}/R_o > 1$, smaller values of n can be obtained to reduce C_D. For example, if $|Z_o|_{mm}/R_o = 2$, then $n = 1.28$, $Q_{opt} = 1.2$, $R_D = 3 \ \Omega$, and $C_D = 1280 \ \mu F$ for the previous example. The damping would not be as good since $Z_{o(max)} = 5 \ \Omega$, which is nearly half the limiting value of 11.9 Ω.

The output filter can be damped in a similar fashion, yielding for this example $R'_D = 0.675 \ \Omega$ and $C'_D = 1360 \ \mu F$ for $|Z_o|_{mm}/R_o = 1$.

For a $\bar{C}$uk, the damping network across C_1 in Figure 7-5 is altered as follows:

$$R_{D(\bar{C}uk)} = \frac{R_D}{D^2}$$

$$C_{D(\bar{C}uk)} = C_D D^2$$

In the dc isolated version of Figure 7-9a, the input damping goes across C_1, and

$$R_{D(\bar{C}uk)} = \frac{R_D}{n^2 D^2}$$

$$C_{D(\bar{C}uk)} = C_D n^2 D^2$$

As with any system, successful troubleshooting of switching power supplies requires a thorough knowledge of the system, consulting the available literature, using common sense and logical methodology, and having tenacity.

7.9 REFERENCES

Bloom, S., Unusual DC–DC Power Conversion Systems. In *Progress in Switching Power Supply Technology*. Dallas, TX: 1980 Midcon Professional Program, 1980.

Cattermole, P. A., Optimizing Flyback Transformer Design. *Power Conversion International*, February 1981, 74–79.

$\bar{C}$uk, S., Basics of Switched Mode Power Conversion: Topologies, Magnetics, and Control. In $\bar{C}$uk, S., and Middlebrook, R. D., (eds.), *Advances in Switched-Mode Power Conversion*, Vol. II. Pasadena, CA, Telsaco, 1981.

$\bar{C}$uk, S., Switching DC-to-DC Converter With Zero Input on Output Ripple. In $\bar{C}$uk, S., and

Middlebrook, R. D., (eds.), *Advances in Switched-Mode Power Conversion*, Volume II. Pasadena, CA: Telsaco, 1981.

Ćuk, S., and Middlebrook, R. D., A New Optimum Topology Switching DC-to-DC Converter. In Ćuk, S., and Middlebrook, R. D., (eds.), *Advances in Switched-Mode Power Conversion*, Volume II. Pasadena, CA: Telsaco, 1981.

Dendinger, S., Power Supply Circuits Head for Simplicity by Integration. In *Silicon General Linear Integrated Circuits 1980 Product Catalog*. Garden Grove, CA: Silicon General, 1980.

Middlebrook, R. D., Design Techniques for Preventing Input-Filter Oscillations in Switched-Mode Regulators. In Ćuk, S., and Middlebrook, R. D., (eds.), *Advances in Switched-Mode Power Conversion*, Vol. I. Pasadena, CA: Telsaco, 1981.

Middlebrook, R. D., Input Filter Considerations in Design and Application of Switching Regulators. In Ćuk, S., and Middlebrook, R. D., (eds.), *Advances in Switched-Mode Power Conversion*, Volume I. Pasadena, CA: Telsaco, 1981.

Middlebrook, R. D., Power Electronics: Topologies, Modelling and Measurement. In Ćuk, S., and Middlebrook, R. D., (eds.), *Advances in Switched-Mode Power Conversion*, Volume I. Pasadena, CA: Telsaco, 1981.

Middlebrook, R. D., and Ćuk, S., A General Unified Approach to Modelling Switching DC-to-DC Converters in Discontinuous Converter Mode. In Ćuk, S., and Middlebrook, R. D., (eds.), *Advances in Switched-Mode Power Conversion*, Vol. I. Pasadena, CA: Telsaco, 1981.

Middlebrook, R. D., and Ćuk, S., Isolation and Multiple Output Extensions of a New Optimum Topology Switching DC-to-DC Converter. In Ćuk, S., and Middlebrook, R. D., (eds.), *Advances in Switched-Mode Power Conversion*, Volume II. Pasadena, CA: Telsaco, 1981.

Neuman, W., and Mitchell, W., EMI Specifications and How Switching Power Supplies Can Meet Them. In *Progress in Switching Power Supplies*. Dallas, TX: 1980 Midcon Professional Program, 1980.

Pressman, A., *Switching and Linear Power Supply, Power Converter Design*. Rochelle Park, NJ: Hayden Book Company, 1977.

Van der Poel, J. M., Pick the Right DC/DC Converter for Your Switch-Mode Power Supply. *Electronic Design*, June 1978, 104–108.

CHAPTER 8

Stereo System Troubleshooting

Robert B. Thorne
Texscan Corporation
Benton Harbor, Michigan

8.1 THE GENERAL APPROACH

A complete stereo system may have a large number of modules,[1] each with its own chassis and enclosure. You may find it difficult to troubleshoot a complex system that is so diverse and "scattered" unless you have the specialized approaches and knowledge needed for stereo systems. This chapter presents the approaches and knowledge in three packages:

1. The generalized system discussion, presenting the system's modules and telling how they interact. It also gives important information about how you begin to find the actual module that is causing the trouble.
2. A general approach to troubleshooting at the modular level, including where to start and what to look for.
3. Specific discussions of the various system modules, for troubleshooting a module (or two) after you have identified a likely source of the trouble.

8.2 THE STEREO SYSTEM

Refer to Figure 8-1 for a modular-level diagram of a generalized stereo system. This diagram contains many modules that you might encounter in a typical audiophile's system. Usually, the modules may be considered as four groups:

---

[1] *Module* in this chapter indicates any component of a stereo system that is frequently available as a separate functional unit: amplifier, preamplifier, equalizer, and the antenna system are a few typical modules.

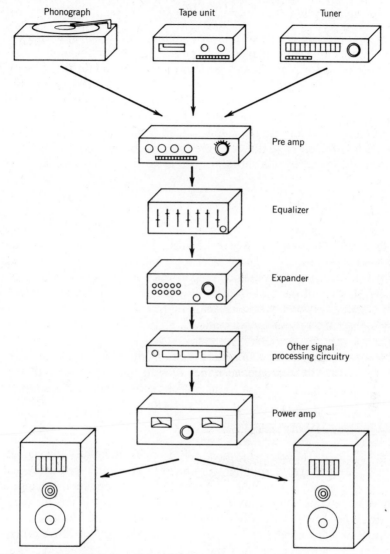

Figure 8-1 Modular diagram of a stereo system.

sources, processors, amplifiers, and audio transducers, but there are two additional groups that also need to be considered. Many modules are likely to have a power supply, which we will include as a fifth group. A sixth group is *interconnections* between the modules.

The following paragraphs deal with six basic families in the generalized audio system, discussing the function of each within the system and the problems you may trace back to each. The next section of the chapter deals with the modular approach to isolating system problems.

8.2.1 Power Supplies

Nearly every module has its own power supply, and these supplies are critical to acceptable audio performance. The power supply's obligation is to supply "clean" dc (without noise spikes or hum) and to maintain the dc within acceptable limits for the components in that module, regardless of the variations in load or line voltage. If the dc is not "clean," you will often notice hum or noise in the audio. If the dc is not within the acceptable limits of the components, it will probably cause the module to miss meeting one or more of its performance specifications. If you have not read the information in the preceding chapters about power supply troubleshooting, do so before going too far with your troubleshooting.

8.2.2 Signal Sources

A source module is anything that provides a signal to be processed, amplified, and transduced into audio. This gives the source two important obligations that you will want to consider. The first is for a high-quality signal. If the signal from the source has poor frequency response, clipping, phase shifting, or other problems, the rest of the system cannot be expected to function properly. The second obligation is that the signal must be free from noise. If the source has a noisy signal, the rest of the system will amplify that noise right along with the signal.

Problems that are due to a source are typically *noisy* or inaccurate signals and are fairly easy to locate because they tend to occur only when that specific source is being amplified. Refer to the various source modular descriptions for specifics.

8.2.3 Processors

There are many processors; they generally switch and modify the signal. They may select the source (preamplifiers, circuit switches), alter frequency response (phono RIAA curve preamplifier, equalizers, noise reducers), change levels (expanders), provide some amplification, provide output signals for recording, or separate the signal into output bands for biamping (electronic crossovers). Their function is to handle the signals from the sources (or other processors) cleanly without introducing noise or inaccuracies.

Signal switching processors usually cause a total or partial loss of the signal, or periodic bursts of noise when they are used. Signal-modifying processors may cause a loss of signal or degradation of signal through incorrect processing. Most processors can be checked by removing them from the system to see if the problem leaves with them. However, the preamplifier must often be checked with test equipment to evaluate the input signal against the output signal.

8.2.4 Amplifiers

Most systems have only one stereo amplifier, and it is often combined with the preamplifier into an *integrated amplifier*. However, it is frequently possible to separate the two, at the *preamp out* and *amp in* plugs at the back of the cabinet. The function of the amplifier is to amplify the signal it receives (noise and all) and drive the output transducer.

An amplifier's problems may include clipping, loss of output, thermal cutout, loss of volume (one stage of amplification), or poor frequency response. The amplifier is probably the culprit if the problem does not vanish as you change sources or remove processors. One last problem that belongs to amplifiers is the catastrophic damage that is easily traced because of the smoke and sizzling; amplifiers handle power—if it gets out of control, you know right away. Because of this, refer to the general hints at the beginning of Section 8.3 (Module Identification Hints) before you turn on a system suspected of having power amplifier problems.

8.2.5 Transducers

The transducer changes the electrical signal to the sound you hear. Only one transducer, you may think. The speaker system, right? Generally right, but don't think that speaker systems are simple. There are standard permanent magnet speakers, crystal-type tweeters, electrostatic speakers, and so on. But they all serve the same function and have the same obligations. They have to accept the signal and handle the power that the amplifier sends to them, and they have to show the amplifier the impedance it needs to see. And if the speaker system is a two-, three-, or four-way speaker, it may incorporate crossovers (bandpass filters) to channel the different frequencies to the speaker designed to handle that frequency band.

Speakers may distort the sound, add noise of their own, stop working altogether, or cause amplifier problems by presenting too little impedance. A convenient way to check a single speaker is by switching the amplifier output left and right connections to see if the problem "follows" the speaker. If it does, there is a speaker malfunction.

8.2.6 Interconnects

A problem inherent to modular systems is that they require electrical interconnections. These usually depend on wires and physical connection points. The only function of the interconnects is that they carry the signal (including ground) from point A to point B, but sometimes they can't seem to even get that right.

Connection points may corrode or oxidize (unless they are gold-plated) and require polishing. They can also loosen, producing signals that come and go or periodic bursts of noise as a result of any movement. Unshielded wires may pick up noise and hum. Two wires that are side by side will have additional capacitance that can present an incorrect impedance match to the modules they serve, especially affecting high frequencies in high-impedance situations. Contact points may even pick up radio signals. Signal hum and noise problems may be affected by the location of the wires (the closer to power supplies and power cords, the worse it becomes). Signal losses and capacitance are increased by increased wire length. Even the wires inside the different cabinets can pick up noise from nearby signals.

8.3 MODULE IDENTIFICATION HINTS

If you have the luxury of starting your troubleshooting with the stereo still set up in its normal configuration, you may have a definite advantage. The user's opinion about what is wrong could be misleading. (For example, if the user hears a hum while playing the phonograph, he or she could blame the turntable. But the hum could be from anywhere; the extra amplification for the phono input makes the sound loudest when it is on.) Now you can hear and decide what is wrong.

8.3.1 Preliminary Discussion

Before you turn the system on, ask some questions:

1. What did the unit *do*? (Don't ask what's wrong, instead ask what the symptom is.) What was the sound like?
2. When did the system symptom happen, randomly or all the time? For every source (AM, FM, phono, etc.)? At all volumes? Was there any powerful appliance on at the time?
3. Has it gotten worse slowly or all of a sudden?
4. Does the symptom change as the system warms up?
5. Has the user moved things, or changed or added components to the system lately?
6. Have there been any strange sounds or smells from it lately? Are the panel lights brighter or dimmer than usual?

If you hear any symptoms to indicate that the power amplifier is the culprit (no sound, universal distortion, snapping and crackling sounds, popping, or blinking of the unit's lights), *don't* turn the system on. The same is true for indications of power supply problems in any of the components. Those two problems are most likely to cause catastrophies, smoke—sparks—damage, if you turn a unit on.

If the owner indicates that the problem was getting worse every time the system was turned on, this is another possible reason for *not* turning on the system. In such a case, each turn-on may have presented a power surge that caused damage. Look for problems connected to power supplies, power switching, and power handling. Check those big capacitors around the power supply and the wiring to the power switch.

8.3.2 System Investigation

If you can feel secure that there is no "mortal" power problem, then you will want to turn the system on and hear, or view, the symptoms yourself. Remember that you are just collecting information. Avoid making a judgment until you have had time to collect as much evidence as possible.

Remember the following few considerations for finding the faulty module or modules:

1. Check for hum by removing various blocks from the system. If necessary, switch them out, turn them off, unplug them, or disconnect all the cables.
2. If the problem seems to be in a number of places, check the system processors that are used in all modes of operation. Then move down the line towards the end (speaker).
3. Don't forget cables and connectors! They can pick up the hum, and their connections can cause snapping, crackling, and frying noises.
4. Don't be afraid to wiggle connections. You may find any mechanical problems right off.
5. If there is a new module in the system, suspect the connection of the module as much as the module itself. (Did the user connect it right?)

8.3.3 Review the Module

When you decide that a specific module needs work, consider these points. Begin with a thorough visual inspection. Look for anything out of the ordinary:

1. Signs of heat.
2. A clean area when the rest has dust will show that someone has tried to fix something in that area before.
3. A misplaced component where someone has previously fixed or modified the circuit.
4. Foils that have heat damage.
5. A component that doesn't fit (different size, age, manufacturer, etc.).
6. Mechanical considerations (cracked circuit boards, screws touching where they shouldn't, plugs that are loose or have oxidized, wires that show heat damage or are cracked from old age).

In the case of pops and sizzles, begin by suspecting capacitors (especially tantalums) and transistors (unless the noise obviously comes from moving a control).

For any problem that is in only one channel of the system, remember that you have a complete identical channel for comparison.

8.4 TROUBLESHOOTING THE MODULE

For the remainder of this chapter, we shall look at each stereo component module one at a time. First, there is a brief discussion of the basic function of each module. Then, we discuss a simplified circuit block diagram. Finally, we include some common symptoms and troubleshooting tips. The discussion of how the module operates will tell a capable technician what to expect and where to look while working through the device.

8.4.1 FM Tuners

The basic function of an FM tuner is to receive a frequency-modulated signal, separate the audio information, and present it to the system for use. First, make sure you are receiving a good signal for the tuner to work with. An outdoor antenna system is more likely to degrade with age than is a well-protected tuner. Refer to Figures 8-2 and 8-3 for the block diagram and frequency spectrum of an FM tuner.

Signal   The signal is frequency-modulated, with the center frequency at one of the assigned frequencies between 88.1 and 107.9 MHz. The maximum allowable modulation (deviation from the center frequency) is ± 75 kHz, and an additional safety band of 25 kHz is provided above and below to minimize the interference between adjacent stations. This spaces the stations at 200-kHz intervals along the FM band.

RF Amplifier   The radio-frequency band (88 to 108 MHz) signal from the antenna is amplified by an rf amplifier (with a passband of 200 kHz) tuned to the frequency you wish to receive. The amplification strengthens the signal *before* any noise is added by the subsequent processing steps. A low-noise front end can help the noise figure of the tuner considerably. Problems due to the rf amplifier can include noise and multiple (''ghost'') stations due to overdriving the transistors (often FETs are used here to reduce the likelihood of overdriving). This block must be able to handle frequencies up to 108 MHz.

Oscillator   The oscillator provides a signal to heterodyne, that is, mix with or beat against, the rf. Its frequency is variable and is kept at exactly 10.7 MHz above the station being tuned by the rf section. Oscillators may use mechanically variable capacitors, voltage-variable capacitive diodes, phase-locked loops, or other means of tuning. Automatic frequency control (AFC) voltages from

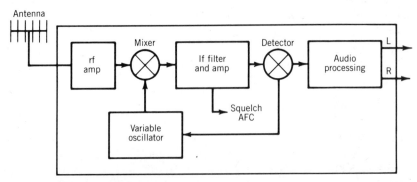

Figure 8-2   FM block diagram.

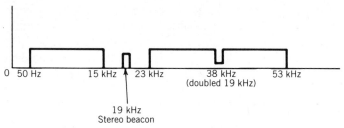

Figure 8-3   FM frequency spectrum.

the output of the intermediate frequency (IF) may be fed back to the oscillator to adjust its frequency as needed for the strongest output to the IF.

Mixer   The rf and the oscillator frequency are mixed to provide the 10.7 MHz for the IF section to process. This is why they must move together when you tune a new station. If part of the 88- to 108-MHz bandwidth is distorted or attenuated, the oscillator and rf amplifier are apparently not shifting together properly to produce the required 10.7 MHz IF for all tuned frequencies.

Filters and IF Section   Other frequencies reach the IF through the mixer besides the desired 10.7 MHz. The desired signal is selected from the rest by the frequency response of the IF and possibly by filters at the IF input. Not only does the IF select the desired signal, it also eliminates any amplitude modulation of the signal by limiting (removing the rf voltage peaks so that each cycle has the same amplitude). The input FM sinewave is amplified to produce a chopped-off frequency-modulated squarewave. Often, the loss of output will indicate that one of the amplification stages in the IF has failed. Check the input and output at each stage of amplification.

While many consider alignment to be the primary fault in the IF, you should not find it a frequent problem. Modern units are rugged, and they do not suffer from the heat that used to plague tube devices. You can correct poor IF alignment (indicated by poor sensitivity or distortion of the audio) as follows: Inject a small unmodulated 10.7-MHz signal at the first stage of the IF and adjust for the maximum voltage output at the end of the IF strip. Always begin adjusting at the first stage and work your way through to the end. Remember to reduce the input as you progress through to avoid limiting (at which time you can no longer tell if your adjustments were improving the alignment).

Detector   The detector separates the information from the modulated 10.7-MHz carrier. The amplitude of the audio signal is determined by how much the IF frequency varies from the 10.7-MHz center frequency; the **frequency** of the audio signal is determined by how rapidly the IF frequency deviates above and below the 10.7 MHz. Figure 8-3 shows the remaining signals: L + R (50 to 15,000 Hz), 19-kHz stereo indicator; and L − R on a 38-kHz carrier (covering 25,000 to 53,000 Hz). If the detector is faulty, it will attenuate or block the signal it receives.

Audio Processing   The L + R signal represents the normal monophonic signal for nonstereo receivers and is available here in the tuner's circuitry. A stereo receiver can use a 38-kHz doubler oscillator (doubling the 19-kHz stereo indicator) to mix and detect the L − R signal (centered around 38 kHz) down to 50 to 15,000 Hz. With both signals returned to the audio range, the tuner can then separate the L and R signals. Adding L + R and L − R gives 2L, and subtracting them gives 2R (L + R + L − R = 2L and L + R − L + R = 2R).

Noise associated with FM broadcasting is generally above 2000 Hz. To minimize the effects of this noise, the broadcast signal has the higher-frequency audio information boosted in strength. This is the 75-μsec *preemphasis* (the 75 μsec refers to the RC time constant corresponding to the 2000 Hz where this takes place). Here, in the audio-processing area, the higher frequencies are *rolled off,* or attenuated, to compensate for the broadcasted preemphasis; while reducing the highs above 2000 Hz back to normal, it serves the desired function of also reducing the noise.

Other Circuitry   You may encounter many of the following circuits in a typical FM tuner:

AGC. Automatic Gain Control comes from the IF to the rf to keep strong stations from overloading and to maintain a constant signal level to the IF while tuning from strong to weak stations.

Squelch. Many of the finer tuners have a circuit that detects noise level in the signal and turns the output off if the noise is too loud. One way of doing this is to look for a frequency component that should not be there. (The Heath AJ-1510 digital tuner samples the IF for a 100-kHz signal and shuts down if a preset level is exceeded.)

AFC. Automatic Frequency Control comes from the detector to the oscillator, to adjust for any shift in either the oscillator or the received signal. As the quality of electronic circuitry has improved, the need for AFC has decreased, because the circuits do not shift as much.

Synthesized Tuners   A development that has recently had an effect on audio, particularly FM tuners, is digital electronics. Other portions of this book deal directly with digital circuitry.

8.4.2 Antenna System

The antenna, antenna amplifier (if used), and down-lead coaxial cable are responsible for providing a strong, clean signal to the FM tuner. All these devices however may lose their ability to perform. One way to check the antenna pack credibility is to connect it to another FM receiver—an inexpensive radio should suffice.

Antenna   The primary function of the antenna is to collect FM signals. Bent or broken elements affect its ability to collect a signal and to resonate at the right frequency. Proper aiming is critical on higher-gain antennas *because* they reject signals that are not within a few degrees of their aim. (This can help your tuner ignore multipath signals that bounce off nearby buildings; see Figure 8-4.)

Another antenna consideration is the connection to the antenna amplifier or the down-lead. Weather and wind can damage this connection and eventually break the wire. Check and repair these connections.

Antenna Amplifier   The antenna amplifier gives the advantage of amplifying the signal before any additional noise is introduced during the trip to the tuner, because it is close to the antenna. As with the antenna, verify that all connections are good. Then, check to see that the antenna amplifier has power, usually provided from below on the down-lead. The amplifier itself is a simple broadband amplifier. You may be able to verify its performance using a small FM radio tuned to a weak station. Connect the down-lead to a small FM radio at the input and then the output sides of the amplifier. A good signal at the amplifier output means that any trouble must be occurring farther on along the down-lead. *Note:* Don't forget that some TV antennas have FM traps that intentionally reject FM; the user should not be using this type, of course.

Down-Lead   The down-lead always introduces some noise and signal losses. However, cracks in the insulation and sharp bends (which may change the

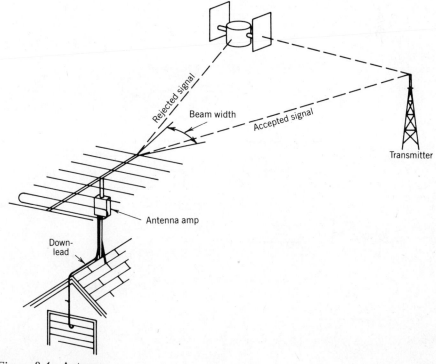

Figure 8-4   Antenna.

cable impedance) in the down-lead will make the situation much worse. Also, keep the down-lead away from electrical lines and other metal objects. Finally, remember that poor impedance matching (300-Ω cable to a 75-Ω connection, for example) can cause the loss of over two-thirds of your signal. Be sure the cable you use is of the same impedance as your amplifier and stereo system input. Impedance matching devices can be purchased if your antenna and FM tuner have different impedances. Normally, the impedance matching device is installed indoors between the down-lead and the tuner. The down-lead and amplifier impedances should be equal.

8.5 TAPE RECORDERS

When you consider the tape system, consider its component parts as shown in the block diagram of Figure 8-5. Problems can occur in any of the blocks.

Problems Many of the typical tape recording problems are listed below along with the areas of this chapter that will help you find their cause:

1. Noise—see "Tape and Heads."
2. Distortion—see "Tape and Heads" and "Bias."
3. Poor frequency response—see "Tape and Heads" and "Equalization."
4. Transport problems and flutter/wow—see "Transport System."
5. Loss of sound—see "Amplifier."

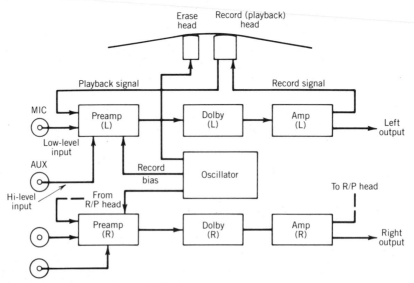

Figure 8-5 Tape recorder block diagram.

Tape and Heads   Recording tape is a thin film of plastic with a magnetic substance bonded to it (originally iron oxide; but now, others such as chromium dioxide are used, each with different magnetic properties). If the machine's bias is not correctly set for the tape in use, it will not get the proper frequency response.

Tapes. Tapes can cause problems, too. Dropouts—momentary losses of audio—occur when tapes lose some of their magnetic particles in one spot. Correct this by using higher-quality tape. Particles from the tape may build up around the capstan (the mechanical structure that feeds the tape past the head) and the record/playback head(s). At the capstan, this may cause uneven movement of the tape. At the heads, the particles can prevent the tape from becoming magnetized while recording, or inducing current in the heads during playback. Correct these problems by cleaning the head and capstan and any other dirty parts of the drive mechanism. Press the "play" button with the machine off to get the heads out where they are easy to see, and use a commercially available head-cleaning solution on a cotton swab.

Tapes can get wound around the tape transport mechanism if they are very thin; avoid 0.5-mil tape in all but the most gentle transport mechanisms (note that dirty transport mechanisms also contribute to this problem). Tape heads will wear due to the abrasiveness of the tape rubbing across them; cheap tapes may be much more abrasive than those whose quality is carefully controlled. Print-through (usually heard as *pre-echo*) is commonly caused when one layer of tape on the cassette magnetizes the layer next to it. This occurs primarily in very thin tapes, because the magnetized layers are much closer together. Remember the rule of squares: doubling the tape thickness from 0.5 mil to 1 mil will mean one-fourth as strong a print-through signal; use 1- or 1.5-mil tape to minimize print-through.

Heads. The placement and positioning of a tape machine's record-and-playback head can make tapes recorded on other machines sound dull and lifeless. Heads are used to record (magnetize) the tape and to play back (sense the magnetism) on the tapes. Figure 8-6 shows where stereo information is placed on a cassette tape. Note that the vertical placement of the head gaps is critical; if the head is not in the right place during playback, it will not receive all the information on that track. Instead, it may receive noise or even audio information from an adjacent track.

Figure 8-6*b* shows what would happen if head placement (azimuth) were incorrect. This causes the loss of high-frequency clarity. *Note:* A tape recorded with crooked heads will sound quite good when played back with the same crooked heads; however, that tape will lack highs and sound unclear on any normally adjusted recorder. Tapes traded between recorders can have the same problem. Poor head alignment is an infrequent problem in inexpensive tape recorders because their heads are permanently fixed in place. If you have adjustable heads and are certain they need correction, use a high-quality commercial test tape, or a commercial tape with full orchestral music. The amplitude

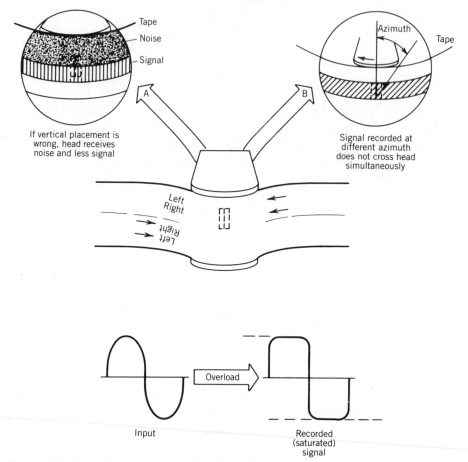

Figure 8-6 Tape player head considerations. (*a*) Tape head. (*b*) Azimuth. (*c*) Saturation.

of the playback signal will be maximum when the vertical placement is correct; the upper frequencies will be strongest when the azimuth is correct.

Erasure. A major long-term problem for tapes is the erasure of the recorded signal with use (primarily the high frequencies, which are the weakest). This is because the record–playback head gets magnetized gradually during record and playback sessions and needs to be demagnetized periodically. A commercially available "demagnetizer" will use the hysteresis effect to reduce the residual magnetism to zero in a few seconds as you slowly move it away from the head. Estimates of how often this should be done vary widely; every 10 to 15 hours of use may be a good, safe rule. Another good idea is to do both, clean the heads and demagnetize, whenever you do either.

Figure 8-6*c* shows how tape can be overmagnetized by trying to record too strong a signal. The magnetic properties are not linear, so the tape can be overmagnetized causing the signal peaks to be outside the small, roughly linear portion of the tape characteristics. The recorder meters are supposed to prevent this by showing a warning about 3 dB before reaching this overload. If the audio goes into saturation the sound will distort. If you find that a 0-dB meter reading produces distortion, you have the option of recalibrating the meter for the tape being used (using the "cut and try" method) or recording at a level that doesn't reach 0 dB.

Tape Bias   Because the magnetic properties of recording tape are inherently nonlinear, a nonlinear (distorted) recording results if only the audio signal is applied to the head. However, this nonlinearity can be overcome by applying a second, very strong high-frequency signal along with the audio signal. The signal is called the *bias signal* and forces the magnetic particles into a more linear response. Because the bias frequency is at least four times the audio limit (typically around 100 kHz), it is never recorded on the tape.

Equalization   Tape recordings have some frequencies boosted during recording to cover tape noise that would otherwise be offensive. This gives a better signal-to-noise ratio. To prevent an unnatural sound during playback, these frequencies are attenuated, suppressing the noise that was there and returning the natural sound.

Transport System   Like a record player, the tape transport is basically a mechanical system driven by electric motors. Improper action can be caused by malfunctions in the mechanical devices, and it will require patience and keen observation to locate the problems. You should first check to see that certain parts of the system are clean. The capstan and pressure wheel must be free of tape particle buildup or the tape can slip, jerk, or rip, or even wind the tape around the mechanism.

Amplifiers   The electronics for the tape machines consist primarily of amplifiers that boost the signals into and out of the heads, provide the specified equalization, and create the bias signal for the record and the erase head. As with other stereo devices, you have the advantage of having two complete circuits, and you can compare the left channel to the right when looking for abnormalities.

8.6 TURNTABLE AND RIAA PREAMPLIFIER

A turntable assembly (see the block diagram, Figure 8-7) can be difficult to repair because it is mechanically complex.

Problems   The primary problems with record players are mechanical malfunctions, audio distortion, and low-frequency noise (not noise from the record). For mechanical problems, read the changer mechanism section. For

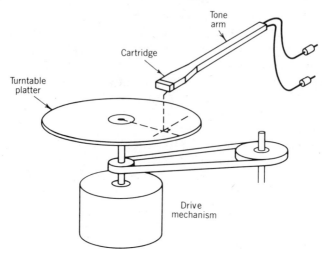

Figure 8-7   Record player block diagram.

distortion, pay special attention to the following sections on the platter, tone arm, and cartridge; and also check to see that the cartridge is providing the correct input signal level to the preamplifier. For low-frequency noise (60-Hz hum or acoustic feedback), refer to the changer mechanism and the preamplifier sections.

Platter and Driver   The platter and its supporting equipment can cause several problems. If the platter is not level, the tone arm may jump or distort. If the platter system does not provide adequate isolation from vibration, the tone arm may jump when anyone walks across the room, or may feed back low-frequency noise from the speakers or other vibrating devices, such as the refrigerator running in the next room. Check to see that isolation springs around the platter system are still in place and are not locked (manufacturers often do this during shipping). Similarly, check the phonograph's feet (when applicable) to see that they are resilient enough to isolate the unit.

Another fault associated with the platter system is static electricity, since records can get a static charge, especially in a dry environment. When charged, they attract dirt and can also pull on the tone arm enough to cause distortion. Static removers, sprays, and wipes are all available, but the simplest and least involved cure is to use a special conductive mat on the platter. It allows static charges to drain off (this is good reason to have the turntable properly grounded into the system).

A number of drive motors are available to turn the platter at the desired speed. Induction and synchronous motors turn much faster than the platter and use either a belt system or capstan-and-idler wheel to transfer the drive power. Different speeds, and allowances for 50-Hz supply voltage, are usually accomplished mechanically by changing the size of a pulley or drive wheel. Check these systems for mechanical problems, then troubleshoot the motor.

Servo-controlled motors are now popular for direct platter drive, and they claim the advantage of a low-speed motor to produce less noise in the audio spectrum. Check these systems for the speed sensor that tells the system how fast the platter is turning. They are usually magnetic, but they can also be optical, and they are normally located at the edge of the platter or underneath. This should feed back to the motor speed control for comparison with the desired speed to modify the motor drive speed.

Changer Mechanism   If your module is a changer or a semiautomatic turntable, there will be some mechanisms to lift, position, and set the tone arm on the record. Mechanisms may also sense the size of the record and drop one record on top of another. A keen eye and patience are needed for the successful repair of a malfunctioning changer. Often, you can cycle the changer a few times as you watch from underneath with the bottom panel removed (making sure you have good light) to see how the mechanism operates. Avoid doing anything until you are sure of exactly what is going on. If the turntable is more than a couple of years old, its original lubricant may have dried out. Scrape out the old lubricant and replace it with lithium grease or something similar.

Tone Arm   A tone arm may be at fault if it doesn't (1) hold the cartridge tangent to the groove and at the correct tracking angle; (2) move vertically and horizontally with ease, to keep the stylus in the groove; (3) avoid skating in toward the center of the record or out away from it; (4) provide just enough downward force on the stylus to keep it in continuous contact with the sides of the groove. Record changer and cartridge manufacturers provide ideal tracking weights with their cartridges.

The two items listed under fault 1 above cannot normally be heard, even by a critical listener. Tangency is a function of the mechanics of the arm and can usually be adjusted by changing the length of the arm at the mounting gimbal. Check the manufacturer's literature for the recommended length. Stylus angle is due to the arm's construction and the mounting of the cartridge. Usually, the cartridge is mounted parallel to the disk, and the stylus sets itself at the correct angle. Again, check the manufacturer's literature. Remember that adding a stack of records on a changer will change the angle between the tone arm and the record, so set the arm with the recommended number of records on the platter.

The mechanics of the arm determine its movement, but dirt in the arm's mounting gimbal or a damping adjustment may limit the ease with which it moves.

Fault 3 is prevented by antiskating devices in some players and is ignored in others. There may be a spring or some other device to press slightly out on the arm while the record is playing. Check the calibration on the antiskating controls.

Fault 4, tracking force, should be easily adjustable with a calibrated spring or weight on the tone arm. Remember also to check for dirt or dust on the tone arm gimbal. If the stylus skips around at the slightest provocation, the tracking may be too light. If the stylus bends appreciably when the arm sets down, the tracking is too heavy.

Note that most tone arm tracking problems are frequently caused by friction from dirt in the gimbal bearings, which will often cause the needle to jump back, playing the same groove over and over. A mechanical misadjustment in the arm or the changer mechanism can also cause tracking problems. An unlevel platter or inadequate isolation from sources of vibration are frequent causes for apparent faults in the record player.

Pickup Cartridge   Generally, there is not much to go wrong or to fix on a cartridge—either it works or it doesn't. One potential problem area is the tip of the stylus. It must be kept very clean. A deposit of dirt, hair, or anything sticky can change the way it behaves and cause serious distortion. Check also to see that the stylus is not bent and that it is fully inserted in the cartridge. This area is very light and flimsy in comparison with your finger; treat it with care and hold the cartridge only by the case. Another reason for not touching the needle with your finger is that oils on your finger can damage it. Preventative care is actually the most important factor in the performance of the cartridge. Eventually, the stylus will wear; but properly cared for, it can last many years.

A second area of consideration is the connection between the wires in the tonearm and the cartridge (see Figure 8-8). Make sure the clips of the wires (if used) are firmly on their posts and that there is no corrosion to hinder the electrical connection. If one of the two stereo channels is missing or intermittent, check those connections first. Reverse the connections, at the cartridge and again at the preamplifier. If the silent channel changes speakers (left to right) in correspondence with either change, the problem is located before the area you changed.

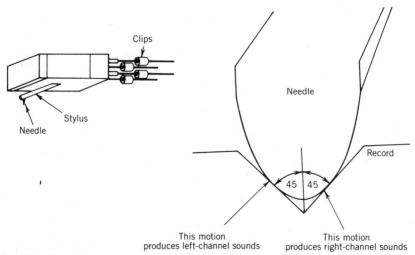

Figure 8-8   Cartridge.

As a final consideration, if the problem has been there as long as the cartridge, check the cartridge compatibility with the rest of the system in terms of voltage and impedance. A crystal cartridge is capable of an ac output in the range of 1 V or more, so it should feed a high-level input to the preamplifier. A moving-magnet cartridge is the most common and generates voltages of a few millivolts, equal to the typical phono input requirement of preamplifiers. Finally, there is the moving-coil cartridge. It is highly praised and costly, but has an output that is so small that it must have its own preamplifier, before the phono input to the player preamplifier. Make sure the system is equipped to take the type of cartridge it is using. One consideration gaining importance as frequency responses go up and signal voltages go down is that of capacitance; wiring capacitance in long leads to the preamplifier, and capacitance in the preamplifier itself, can degrade the high-frequency performance of a system.

RIAA Preamplifier   First, make sure that the signal is getting from the cartridge to the electronics and that it is not picking up any hum. This means that you must check the wiring from the cartridge to the preamplifier. Motors, transformers, and power cords may cause 60-Hz hum in the connecting wires if they are close enough. Remember too that hum can be present if the player is not properly integrated (grounded) into the system. Because of the relatively low-level signals in this part of the system, many manufacturers provide a grounding wire for connection to the amplifier chassis. Finally, check the runs and connections of the shielded cable between the player and the preamplifier.

The RIAA (Radio Industry Association of America) preamplifier itself is unique to record players. It works like other noise reduction systems (Dolby and FM preemphasis). Since inherent record noise is in the higher frequencies, they are boosted during recording to overcome the noise, and are then reduced during playback (also reducing the noise) to restore the proper balance with other frequencies. The specific curve for this reduction is the RIAA curve. If the sound from recordings is unnaturally bright, the frequency response of the preamplifier may have drifted.

8.7 EQUALIZERS

An equalizer separates audio information into different frequency bands and controls the relative strength of each band, based upon the way the user sets the controls. The more sophisticated equalizers allow the user to select the width of the selected band (usually between $\frac{1}{3}$ octave[1] and 1 octave) and even the center frequency of the band. This is done by changing the values of components in the filter circuitry, as shown in Figure 8-9.

The novice may purchase an equalizer and soon decide that it is not working right. If this is the case, suspect the user first, not the equalizer. If the user has specific complaints about noise, distortion, or controls, there may

---

[1] An *octave* is a doubling or halving of the frequency.

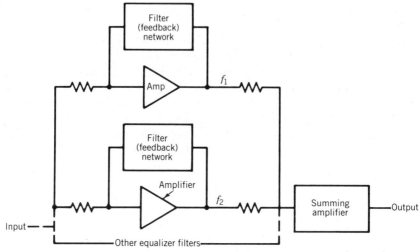

Figure 8-9 Equalizer block diagram.

be a real problem; but if the complaint is about overall sound, first consider the following section.

Overall Sound   Before doing any real service, use a record or FM interstation noise to check the system's sound with the equalizer set flat, both in and out of the system. There should be no audible change. Then, move each control separately and check by ear to verify that it seems to be affecting the sound in the general manner anticipated. (Below 250 Hz, you may need a full orchestral record; above 250 Hz, you may use interstation noise very effectively.) If these quick tests indicate that the equalizer is operating correctly, then the problem may be that the user doesn't fully understand the use and capabilities of the equalizer.

Equalizer Use and Setup   All an equalizer can do is increase or decrease the amplitude of various frequency bands as they are processed through it. Consequently, a speaker that previously would not produce a nice, crisp 50-Hz tone still will not. All the equalizer will do is make uncrisp sound louder (and probably more distorted since the speaker will be driven harder). The equalizer also will not put high frequencies back in FM broadcasts, which are cut off above 15 kHz.

The following are the things that an equalizer can do, if properly set: It may help compensate for a mild dip in a sound system's response curve (even at the high or low end), but it will not improve the quality of that signal. The equalizer may also help compensate for dips or peaks caused by the acoustics of the listening environment. (These however must first be discovered before they can be treated effectively; see section on Acoustic Analysis.) Finally, an equalizer may be used to attenuate a specific frequency band of noise (and, unfortunately, any signal in that frequency band), such as the noise in an old radio recording.

The typical adjustment of an equalizer to give a good perceived sound will give the approximate response curve shown in Figure 8-10. Note that this curve is not "flat" at the ends. A flat curve will be tiring and harsh to the listener. Unless you have a calibrated ear, you will not be able to accomplish much in a short time without measurement equipment, such as that discussed in the Acoustic Analysis discussion. Probably the best rule to remember when working without an analyzer is to avoid anything too drastic in the setting of the controls. Most controls give a graphic display of what they are doing; don't allow them to look like mountain peaks—try to approximate a gentle, basically flat curve. Note that regular box-shaped rooms may have resonances based upon the room's dimensions; a control may end up being out of line to correct for the resonances.

Whenever you try to adjust the controls by ear, remember to make all decisions based upon how things sound *at the specific spot where you listen most*. You may use interstation FM noise, with the system in monaural to help make the speakers match. This is necessary because the position of walls and other objects will affect how each speaker sounds.

Circuitry A typical equalizer circuit is represented in Figure 8-11. Each frequency band is selected by its own filters. In most high-quality devices, the filter portion has active components and amplifies the signal as part of its operation. Some use passive filters and an amplifying adder at the end. If the Q (the shape of the bandwidth curve) and the center frequency are variable, you will find additional controls. Resistors in parallel with the resonating circuitry, or to ground, can lower the Q to broaden the bandpass. Variable RC circuits in the filter feedback are used to select the center frequency.

8.8 NOISE REDUCTION DEVICES

Several devices have been designed to improve the audio signal-to-noise ratio; most work on a single principle. That principle is to boost the signal strength

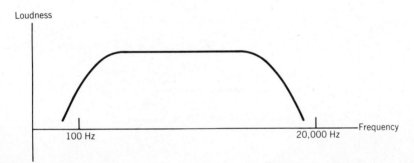

Figure 8-10 Frequency curve for "perceived" good sound.

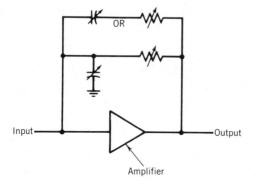

Figure 8-11   Typical equalizer circuit.

in the frequency where the noise is worst. This makes that part of the signal easy to hear but too loud to balance with the rest. So the playback system has a built-in noise reduction circuit that attenuates the boosted frequency (and at the same time attenuates the noise), returning a normal frequency response curve with the improved signal-to-noise ratio (see Figure 8-12).

The curves for augmenting/decreasing the signal are generally controlled by two things. The first is the frequency spectrum for the noise; you need to select the spectrum with the offending noise. The second is the electronics; you want the electronics to be simple and the adjustment to be noncritical. Consequently, the circuits are frequently single-pole RC filters. They seldom fail.

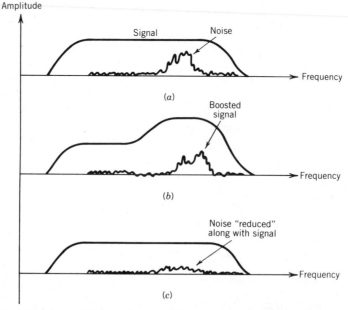

Figure 8-12   Noise reduction process. (*a*) Signal + noise curve. (*b*) With signal boosted. (*c*) Final (reduced noise) curve.

Examples   Phonograph equipment uses the RIAA curve, and FM broadcasting uses the 75-μsec preemphasis. Tape recorders use Dolby or similar systems, but with an added touch of sophistication. Loud signals that are within the designated noise frequency are not emphasized—they are already loud enough to have a good signal-to-noise ratio, and increasing them further could be more than the tape could handle. However, a level detector senses when the signal in that frequency band is not loud enough and begins to add some gain during the recording. The variable gain is replaced by an equal attenuation during the playback process.

Other Noise Reduction   More sophisticated approaches are also possible. Digital signals can be recorded along with the audio signal at very low or very high frequencies, out of the audible spectrum, or they can be recorded on a totally different track on tape. The digital information is used to control the gain of the playback system. With such a system, the recording could be made with the level set for the best signal-to-noise ratio and distortion. The recording would have little or no difference in volume from softest passages to loudest, but the volume would change during the playback based on the digital information.

In systems where you suspect the noise reduction circuitry is malfunctioning, look for the filtering operation (for simple phono and FM devices). In the more sophisticated circuitry, you must identify the level-sensing circuitry as well. To troubleshoot the more complex noise reduction circuitry, you will need at least a VOM (an oscilloscope is better) and possibly a signal generator. After you determine that the level-sensing circuitry has the proper output (if it does), troubleshoot the rest of the circuitry with the techniques discussed in Chapter 5.

8.9 EXPANDER/COMPANDERS

Live performances often have a dynamic range of more than 85 decibels. Tapes and records cannot record this range without distortion at the loud end and a very poor signal-to-noise ratio at the soft end. Consequently, electronic circuitry is used to compress (compand) the range during recording; in this way, it is possible to use electronics to restore at least part of that range during playback for more realistic sound.

A basic expander is shown in Figure 8-13. This device senses the level of the input signal. It reacts by increasing the expander's gain for larger inputs and decreasing the gain for smaller inputs. The filter circuit isolates some representative portion of the audio spectrum (700 Hz to 7 kHz for the Heath AD-1706) to be sensed. Then, the rectifier and detector circuits convert the filtered audio into a variable-dc voltage that changes in proportion to the input (ac audio) level. This variable dc is used to control a voltage-controlled, or current-controlled, amplifier, which processes the audio signal. Note that there is a capacitor which determines how fast the gain can be allowed to change. It is chosen so the change is least noticeable to a listener. If it is too slow,

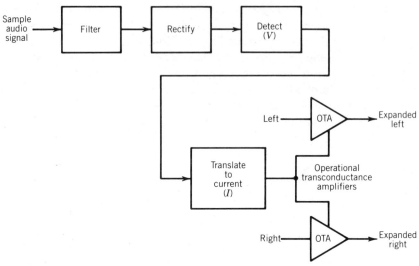

Figure 8-13 Expander block diagram.

music peaks are lost; if it is too fast, noises such as a record pop will cause the sound level to jump briefly. This is inherent to this type of expander, and the user is well advised to defeat the expander if playing an old record full of pops and clicks.

You can service these devices by generating and tracing a control signal through the control portion of the circuitry all the way to the variable amplifier. Look for bad diodes and capacitors that could alter the transmission of the signal through the control path. If the control voltage is present and reacts proportionally to the input, the trouble must be at the variable amplifier. It is usually an IC, but check input and output pins before replacing it.

Note: Simple expanders such as this have this disadvantage: a single loud note in the recording can increase the gain for the entire spectrum, making *all* notes louder. The obvious way to overcome this is by separating the audio spectrum into more and more separate bands and treating each band with separate expander circuits. The complexity of the circuitry and of the coordination between bands increases very rapidly in such a situation.

8.10 AMPLIFIERS

8.10.1 Preamplifiers

Essentially, the preamplifier section is a combination input switcher, equalizer, phonograph RIAA curve preamplifier, and the loudness compensation. Often, there are potentiometers at the inputs to make the different input levels match at the output. Check by tracing the inputs through the circuitry, and treat the amplifiers and tone controls as separate units.

In general, a stereo amplifier contains a preamplifier and a power amplifier. They are usually integrated units, although separate preamplifiers and power amplifiers are used. As a stereo system becomes more complex and expensive, the probability increases that separate power and preamplifier units are used.

8.10.2 Power Amplifiers

Troubleshooting Warning   Remember the warning from the beginning of this chapter: If the power amplifier is suspect, avoid running power through it. A power amplifier has the greatest potential of damaging itself and the speakers, since it generally handles more power than the other blocks. Section 8.2 deals with some of the typical symptoms of amplifier trouble. If it sounds like you have such a problem, unplug the amplifier and open it for a good visual check. Follow this with some resistance checks, and refer to the following for information about amplifier circuits.

Power Amplifier Versus Preamplifier   This section deals only with the power amplifier, since the preamplifier is a processor and is handled elsewhere. Many integrated (power- plus pre-) amplifiers can be readily separated into different sections to meet this definition. The back panel will have a set of jacks for "preamp out" and "power amp in." They provide this separation so that tone controls can be inserted between the preamplifier and the power amplifier. The point between the preamplifier and the power amplifier is where you may check a system to help localize a problem.

Standard Amplifier Circuitry   Despite its sophistication, a power amplifier is still relatively straightforward. You may use an oscilloscope to trace the signal through the amplifier circuitry and, because there are two identical sets of circuitry, you always have something to compare against.

Remember that amplifier problems can also come from the power supply and the speakers. The amplifier and the power supply are both suspect if both channels of amplification seem to be affected. The speakers can cause problems if they present an impedance that is too low for the amplifier to handle, causing the amplifier's protective devices to shut down the amplifier. Occasionally, the same problem may cause the speaker's fuses to blow.

8.11 SPEAKER AND HEADPHONE SYSTEMS

Each speaker or headphone system is composed of one or more transducers in an enclosure. The transducers translate an electric signal into sound. If this sounds simple, it is so in theory only.

A basic three-way speaker system, with built-in crossovers, will be something similar to Figure 8-14. A low-pass filter passes the lowest frequencies to the bass speaker, a high-pass filter passes the high frequencies to the tweeter, and the remaining (middle) frequencies go on to the midrange speaker. There

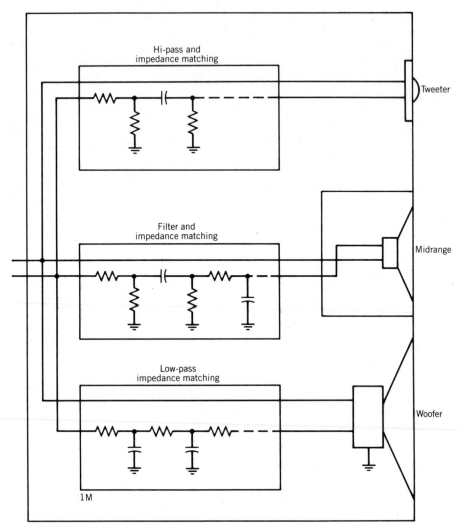

Figure 8-14　Speaker block diagram.

may be some resistors or other circuitry to present a fairly constant impedance to the amplifier over the entire frequency range. The frequency response of the transducer also contributes to the filtering (roll-off) action. The shift from one transducer to the next is gradual to smooth over any difference in sound quality; this means that simple one- or two-pole filters are usually employed. However, other minor variations on the basic speaker include more (or fewer) transducers and multiple amplifiers with prefiltering (see Figure 8-15).

A Note About Servicing Speakers　Remember that you are working in a different world with speakers. In this world, mechanical issues must be considered; the tolerances of fit and motion are as important as electrical tolerances. You must be a craftsman as well as an electronic wizard.

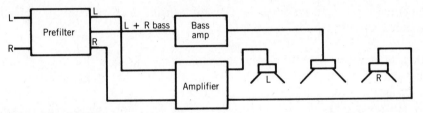

Figure 8-15   Biamping.

The most common way into a speaker enclosure is not by the back, but by removing the grille and then unscrewing and lifting out the largest of the speakers. You can verify this by examining the enclosure's back panel for any other way to get into it. The grilles are usually held in place with nails, screws, or patches of velcro fabric.

Note also that speaker cabinetry is usually kept air tight. If you create an air leak, two things could happen. Air could whistle through the leak during low notes, and the frequency response of the system could be altered. Remember to reseal the enclosure when you finish.

Missing Sounds   Frequently, a speaker may lose the sound from one or more of its transducers. A total loss of sound probably means that there is an open circuit, but not necessarily in the speaker. First make sure the amplifier's signal is getting to the speaker terminals. A normal listening level usually requires an ac voltage between 100 mV and 5 V at the speaker terminals (speakers vary widely in their efficiency). If there is voltage there, you should check each transducer. If the voltage is not present at the transducer, something is wrong with the crossover (filter). If the voltage does reach the transducer but no sound is made, the problem must be within the transducer. The impedance will be near zero for dynamic transducers tested with an ohmmeter. At frequencies in the speaker's range, the impedance should measure near its rated impedance. Refer to Figure 8-16 for a setup to verify the impedance. The two impedances will be proportional to the voltage drops across them. A low impedance at a specific frequency could indicate a crossover problem or a design flaw and could cause problems with the amplifier.

If the ohmmeter test shows that the speaker impedance is infinite, the coil is probably open and you cannot repair it. Check the back of the transducer to make sure that the wires from the terminals to the coil have not been broken;

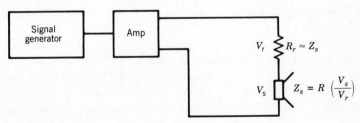

Figure 8-16   Resistor speaker impedance measurement.

you could repair that quite easily. Crystal transducers which are often used as tweeters can sometimes be repaired.

Unusual Sounds From Speaker   Rubbing in the transducer or damaged speaker cones can cause muffled or buzzing sounds. A small possibility exists that you may repair this without special equipment.

Buzzing and scraping noises can come from the speaker if the coil is rubbing the housing. The mechanical tolerances around the coil are very fine, and even a slight bend in the transducer's frame can pull the coil to one side. You may be able to identify a rubbing coil by gently moving the diaphragm in and out with your fingers and listening closely. If you hear or feel it rubbing, check to see that the transducer's frame is not bent in mounting to the enclosure. Try loosening the mounting hardware to remove or change any pressures on the frame. If you cannot correct the rubbing, you will probably have to replace the transducer.

A similar noise may come from the diaphragm itself. Older speakers in particular may have diaphragms that wrinkle, tear, or rub somewhere around the mounting edges. Sometimes, a little rubber glue at a tear or wrinkle may improve the diaphragm's sound, but this will probably be temporary, especially if the speakers are played at high volume.

A harder, clanking noise can result at high volume if the signal drives the coil beyond its limits and into the magnet structure. This will also produce considerable distortion since the diaphragm is not designed to allow such long excursions. The only cure for this is to decrease the bass drive (it is the bass that causes these long excursions) or replace the speaker system with one that can handle the bass power.

Exotic Speakers   Special speakers may have their own equalizing amplifier (with a response curve designed to complement the speaker transducers) or other special drive systems, and some may have special impedances. However, the principle of tracing the signal to the enclosure and then to the transducer's terminals will serve to locate most faults that can be readily treated.

8.12 ACOUSTIC ANALYSIS

Users who are seeking the ultimate sound often include many accessories in their systems and then use other equipment to maintain and adjust everything for peak operation. One popular approach is to "tune" the system to the room and furnishings in which it plays. This is done with a special sound generator and an analyzer.

The sound generator produces broadband noise, which is fed through the system and reproduced through the speakers of the system. The analyzer is then used to see how the output has changed from the input as a result of the system and the room acoustics. Testing must be done at specific listening positions, since every position in the room will produce different readings.

These tools are useful not only to check the whole system, but for such things as learning how changing the tape bias will affect the frequency response of a tape deck and setting the equalizers. Only now are generators and real-time analyzers becoming available at prices individuals can afford. By the time this book is published, Heath Company, Benton Harbor, should have its reasonably priced noise generator and real-time analyzer available. This is because of improved IC technology and broadened horizons of use for microprocessors.

CHAPTER 9

Analog Communications Systems

Harold B. Killen
University of Houston

9.1 INTRODUCTION

If two similar signals are used to modulate two separate carrier frequencies, two distinct frequency bands or spectrums are formed. If the two carriers are far enough apart, the two bands can be transmitted over the same medium with no interaction. This result is referred to as *frequency division multiplexing* (FDM). Bandpass filters can be used to separate the frequency spectrums at any desired point.

FDM/carrier systems are used to transmit large numbers of communications channels simultaneously over the same transmission media. Each frequency channel is allocated a unique part of the frequency spectrum, and the transmission media may consist of microwave radio links, coaxial cables, fiber optic cables, and so on. A typical microwave radio link can consist of up to 1800 channels on each rf carrier. Coaxial systems may carry up to 3600 channels, and some high-capacity systems (60 MHz coaxial) can carry up to 13,200 channels.

The actual transmission spectrum is quite complex. In addition to the communication channels, the spectrum also contains residual carriers, pilot tones, signaling, and test tones. These signals are used to monitor the workings of the system while it is currently carrying traffic.

The common objective of the various links mentioned above is to transmit information. Consequently, it is necessary to limit distortion in the link and to keep intermodulation noise to a minimum. These factors serve to reduce the capacity of the channel to transmit information. In this chapter, we discuss the theory and measurement techniques necessary to maintain an existing FDM communication system.

9.2 FREQUENCY MODULATION (FM)

In frequency modulation, the signal deviates in frequency by an amount proportional to the data signal amplitude and at a rate equal to the data signal frequency. In Figure 9-1, the sinusoidal carrier is frequency-modulated by the lower-frequency sinewave. Modulation by a complex waveform is essentially the same.

The ratio of the carrier frequency deviation Δf_c to the modulation frequency f_m is defined as the modulation index D:

$$D = \frac{\Delta f_c}{f_m} \tag{9.1}$$

In FM systems, the value of the modulation index has a number of important implications. For one thing, it is an indication of the attainable accuracy of the transmitted data. The output signal-to-noise ratio $(S/N)_o$ at the output of an FM demodulator is given by

$$\left(\frac{S}{N}\right)_o = \sqrt{3}\left(\frac{S}{N_r}\right)D^{3/2} \tag{9.2}$$

where N_r is the RMS noise voltage measured over $2\ \Delta f_c$ and S is the RMS signal. If the modulation index D is doubled, $(S/N)_o$ will increase by 2.83. If D is tripled, $(S/N)_o$ increases by a factor of 5.2.

The effect of increasing the modulation index is observed most readily in terms of the spectrum (Figure 9-2). Note that the frequency spectrum consists of a center frequency spectral line with sidebands on either side. The spacing between each sideband is exactly equal to the modulating tone f_m. The number

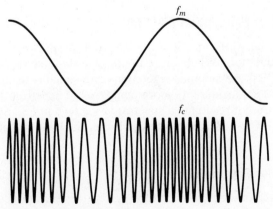

Figure 9-1   Frequency modulation waveform.

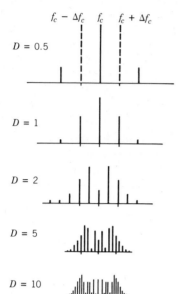

$f_c - \Delta f_c$ f_c $f_c + \Delta f_c$

$D = 0.5$

$D = 1$

$D = 2$

$D = 5$

$D = 10$

Figure 9-2   FM spectrum with constant Δf_c and different deviation ratios.

of sidebands produced in an FM spectrum is theoretically infinite. The amplitude and spacing of the most significant sidebands must be preserved to prevent distortion. As a rule of thumb, the bandwidth (*BW*) needed to accomplish this is given by

$$BW \approx 2(f_c + f_m) \quad \text{Hz} \tag{9.3}$$

We observed (Equation 9.2) that as D increases, the $(S/N)_o$ also increases. We see from Figure 9-2 that as D increases, the number of significant sidebands increases while the amplitude of the sidebands decreases. Consequently, this has the effect of spreading the FM signal across the transmission bandwidth more uniformly and thus more efficiently.

9.2.1 FM Carrier Modulation (FM/FM)

The spectrum illustrated in Figure 9-2 is an FM spectrum that results from sinusoidal modulation. When the outputs of a number of voltage-controlled oscillators (VCOs) are summed, the composite signal is used to modulate the transmitter, and a considerably more complex FM spectrum results (Figure 9-3). Each VCO output results in a spectrum as shown in Figure 9-2. The overall modulation index D is a composite of all the individual modices (modulation indices).

9.2.2 Subcarrier Preemphasis

The performance of a communications system depends directly on the signal-to-noise ratio (SNR). In an FM system, it can be shown that in the presence

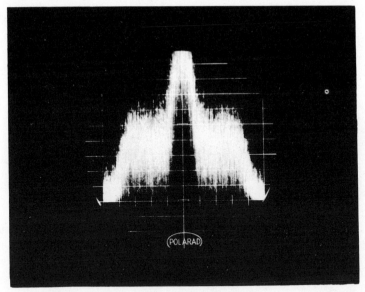

Figure 9-3   RF spectrum display with 32 channels (VCOs).

of a uniform noise spectrum (white noise), the interference caused by this noise increases at a parabolic rate as the subcarrier frequencies increase. A subcarrier is the signal, at a frequency lower than the final output frequency, upon which only a portion of the output information is modulated. In Figure 9-6, the 730-Hz, 560-Hz, and 400-Hz are subcarriers. Indeed, for proportional bandwidth channels—channels whose bandwidth is proportional to the assigned center frequency—interference increases as the $\frac{3}{2}$ power of the subcarrier frequencies.

At low subcarrier frequencies, this has little effect. At higher frequencies, this is not true. In fact, if the SNR for all the channels is to be the same, it is necessary to increase the rf power placed in the higher-frequency subcarriers. That is, these channels must be "preemphasized" at a 9 dB/octave rate to offset the $\frac{3}{2}$ power increase in noise. In practical systems, this preemphasis schedule is not exact due to nonlinearities in transmitters and receivers. Adjusting the share of rf power allocated to a given channel is a matter of changing the individual modulation index. Most manufacturers provide adjustable outputs on their VCOs for this purpose. Care must be exercised in making this adjustment in that, if a channel is emphasized too much, its spectrum may extend into the spectrum of an adjacent channel. Also, intermodulation distortion may arise if amplifiers, transmitters, or other circuits are overdriven.

9.2.3 FM Improvement Threshold

The improved signal-to-noise performance of FM over AM is one of the reasons it is preferred for rf transmission. A vector representation of an FM signal with noise is shown in Figure 9-4. The quantities a and b represent the signal and noise, respectively, at the input to the receiver. Both components a and b rotate

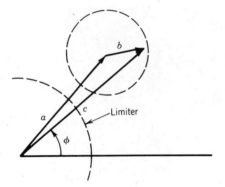

Figure 9-4   Vector representation of FM
signal: *a*, signal; *b*, noise; *c*, resultant.

at their frequency rates around the origin. The resultant, *c*, is the vector sum
of *a* and *b*. Note that *c* has a maximum and minimum amplitude of *a* + *b* and
a − *b*, respectively. The amplitude variations are removed by the limiter in
the receiver. The rate of change of the angle *φ* is the total effective frequency
modulation.

The effect of *b* on *c* does not become appreciable until the ratio *a/b*
approaches 1, when the noise and signal amplitudes are equal. When *b* becomes
greater than *a*, the noise becomes the dominant signal and the signal represents
the variation. The transfer represents the "improvement threshold." For *a* >
b, the output is practically noise free. For *b* > *a*, the output is practically all
noise.

9.3 FREQUENCY DIVISION MULTIPLEXING (FDM)

We indicated in the introduction to this chapter the basis for FDM systems.
One direct application of FDM systems is its use in telemetry systems. Telemetry is a marriage of the sciences of instrumentation and communications—the
technique of measuring something at one place and reading the measurement
at a remote location. An example of telemetry is the monitoring of signals in
aircraft tests. The signals of interest, such as wing stress, are often sent to a
ground station by FDM. The layout for a typical telemetry system is shown in
Figure 9-5.

Note that the different channels are combined by the multiplexer and that
the composite signal is used to modulate the rf transmitter. At the receiver,
the inverse operation takes place and the channels are separated by the de-
multiplexer. A typical arrangement for frequency division multiplexing is shown
in Figure 9-6. The VCO (voltage-controlled oscillator) output frequency will
shift as the channel input voltage varies. Thus, the VCO of each channel changes
the voltage amplitude information to frequency-modulated information.

The mixer in this system is actually a summing amplifier. The composite
output signal forms a frequency band ranging from 370 to 785 Hz. Note that
there is no overlapping of the channels, and unused "guard" bands are placed

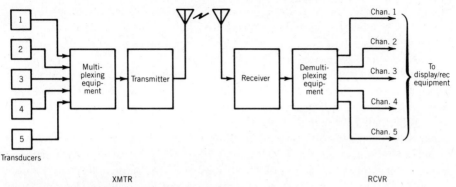

Figure 9-5   Block diagram of a typical telemetry system.

between the channels to ensure separation. At the receiving end, bandpass filters are used to separate the channels. The filter outputs are routed to discriminators where original signal is recovered.

The composite output from the summing amplifier is often used to FM modulate a radio transmitter. This system is known as an FM/FM system because two steps of frequency modulation are involved. If the transmitter is phase-modulated, the system is referred to as an FM/PM system.

The "peak deviation" of the VCO is generally a certain percentage of the center frequency. For example, if the center frequency is 400 Hz and the percent

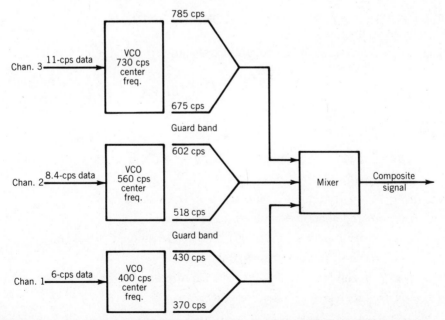

Figure 9-6   Typical arrangement for frequency division multiplexing. Channels are separated by unused "guard bands."

deviation is 7.5%, the peak deviation is ± 30 Hz. As discussed earlier for FM, the peak deviation is related to the data frequency by the modulation index. If the VCO described above is operated at a modulation index of 5, the maximum frequency deviation is 6 Hz.

In telemetry, two types of systems are often employed: (1) proportional bandwidth and (2) constant bandwidth. In proportional-bandwidth systems, channel bandwidth is proportional to the assigned center frequency. Thus, the widest channel is at the highest frequency. For example, IRIG standards (Inter-Range Instrumentation Group of the Range Commanders Conference) define 18 subcarrier channels for FM/FM and FM/PM systems. For proportional-bandwidth systems, the deviation of all channels is the same in terms of percent of center frequency ($\pm 7.5\%$). A system such as this allows assigning slowly varying data to the narrower channels.

In many cases, it is desirable to use constant-bandwidth channels. In this system, each channel has the same bandwidth regardless of center frequency. We see that the fractional deviation of each channel decreases with increasing frequency. Eventually, the deviation becomes too low in terms of oscillator stability. The solution to this is to operate the oscillators at relatively low frequencies and then perform frequency translation to move the frequency-modulated channel into the desired portion of the spectrum. For example, if a 16-KHz subcarrier is heterodyned (mixed) with a translation frequency of 112 KHz, we have sum and difference frequencies at 96 and 128 KHz. Filtering is then used to obtain the sideband of interest. Both sidebands contain all the information.

9.4 ADDITIONAL PRACTICAL CONSIDERATIONS

The basic expression for the required transmitter power in an FM telemeter system is given by

$$P_t = (4\pi R)^2 \left(\frac{S}{N_i}\right) \left(\frac{BLN_r}{G_t A_r}\right) \tag{9.4}$$

where

P_t = required transmitter power

R = path length of rf signal

S/N_i = required input S/N to reach FM threshold

B = predetection bandwidth of receiver

L = signal path loss through atmosphere, antenna loads, etc.

G_t = transmitting-antenna gain

A_r = receiving-antenna effective capture area

N_r = receiver thermal-noise power

The required S/N_i is on the order of 10 dB to be at the FM threshold signal-to-noise ratio.

9.5 FDM/CARRIER SYSTEMS TESTING

The design, installation, and maintenance of FDM/carrier systems requires that the capability exists to make several different types of measurements. A number of general-purpose instruments exist for use such as network analyzers, power meters, and signal generators. Many specific measurements however require dedicated instruments such as a selective level meter and a level generator. Typical measurements which are generally made are:

- Reference and line pilots
- Channel signal power
- Channel noise power
- Supervisory and test tones
- Frequency response
- Crosstalk
- Spectrum analysis

In the design of analog systems, it is customary to allocate the noise, phase, and amplitude tolerances of the transmission system on the basis of the overall accuracy required. The errors that occur in analog systems may be much more difficult to detect and correct than in a digital system. As a result, it is difficult to build an analog FM system and provide a transmission path such that it will have the same accuracy in all channels. Noise may be more difficult to control in the wider-spread channels. Also, a transmission path may not be phase coherent, with the result that there is a slight frequency shift in the data system. Unless this condition is corrected, the information being transmitted over the data system will be erroneously biased. This problem can be avoided by transmitting a reference tone and correcting the data in accordance with the frequency shift it experiences.

Wide variations of circuit noise can occur on a data communications system. On a per-channel basis, the amplitude and phase variations of the circuit can be expected to be relatively stable, particularly after the circuit is established and equalized. Under heavy-load conditions however, cross modulation in the circuit increases the circuit noise. In the more complex switched networks, where the circuit connecting two points may be made up of several switched links in tandem, the gain and phase responses will be variable. Obviously, maintenance and troubleshooting require that these parameters be measureable.

9.5.1 Microwave Radio Testing

Frequency division multiplexing of many carrier frequencies is the most common method of bundling individual communications channels together. Usually, the carrier modulation is either AM or FM. The most common type is single-sideband suppressed-carrier modulation. A single-voice channel needs only 3.1 KHz of bandwidth. The carrier is typically suppressed 23 dB below the sideband level. A properly equalized voice channel can handle a single data channel at rates of over 4800 bits/sec, or 12 or more teletypewriter channels. Standard groups of 12 channels, each 4 KHz wide, are combined into a 60- to 108-KHz bandwidth channel. Guard bands are included between each channel, and pilot tones are transmitted to regulate amplitude levels, synchronize individual carrier frequencies, and provide alarm and monitoring signals.

The next level in the organization of the signal structure combines five 12-channel groups that occupy the 312- to 552-KHz range. This arrangement is referred to as a supergroup. Ten such supergroups, which contain 600 channels, are then often combined to cover 564 to 3084 KHz. Additional combinations can result in a 1632-channel, 4 GHz TD$_2$ microwave link or an L3 Bell coaxial cable system with 1800 voice channels. A typical channel grouping is shown in Figure 9-7.

A spectrum analyzer is recommended for monitoring the 1800 channels with their suppressed carriers, pilot tones, and intermodulation products. This represents a spectrum on the order of 10,000 signals. Using a modern analyzer, all 1800 channels can be displayed with a single sweep. Any portion of the total spectrum can be expanded for further study. See Chapter 3 for more information on testing communications systems using the spectrum analyzer.

As mentioned previously, more detailed level and noise measurements of FDM systems require the use of a selective level meter. The Cushman CE-21A selective level meter provides fast digital synthesizer tuning to 9.1 MHz

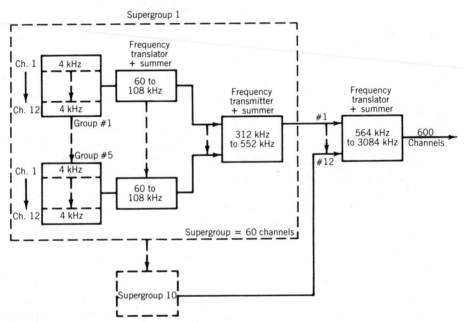

Figure 9-7   Typical channel grouping to form a multichannel communications system.

in 1-KHz steps. A resolution of 25 Hz is attainable with vernier tuning, and direct digital readout of frequency and level is available.

The third specialized instrument required is a tracking signal generator. This instrument, in combination with the other two, provides a combination that can be used to spot spurious signals, measure excessive noise and intermodulation products, and analyze other failures that occur in FDM systems.

A simple test used to maintain and check a microwave link's overall performance and to identify degraded channels is the noise power ratio (NPR) test. To perform this test, a noise loader, such as the AT-9003 manufactured by W and G Instruments, Inc., can be used to provide the signal source. Instruments such as this contain plug-in limit and bandstop filters which allow the instrument to be tailored to almost any requirement with a baseband up to 12.5 MHz. A selective level load meter or a specialized high-quality noise receiver is necessary at the receiving end.

The NPR is determined by the introduction of band-limited white noise at a standard reference level. For example, in an 1800-channel system, 0 dBm of white noise is typically used for a 9-MHz band. The noise is measured in a narrow-frequency slot one channel (4 KHz) wide. Next, a bandstop filter is inserted at the system's input to keep out frequencies within this slot. The residual noise power in this slot at the output now includes only thermal noise, noise pickup within the channel, and intermodulation products. The dB difference between the reading with and without the bandstop filters is the noise power ratio. An NPR of 50 dB or greater is desired for an 1800-channel system.

Two distinct types of measurements are required on microwave systems—qualitative and diagnostic. Test signals that simulate normal traffic are used to make qualitative measurements. These measurements are intended to ensure the operator that the microwave link is performing satisfactorily. These measurements are normally made end to end (baseband to baseband) and yield an indication of overall system performance. Diagnostic information is missing if a discrepancy is noted. Table 9-1 lists qualitative measurements for both video and FDM traffic, together with the capabilities of the HP 3724A/25A/26A baseband analyzer.

Qualitative measurements may be made on individual sections of a mi-

Table 9-1   **Qualitative tests to verify radio-system performance**

Test	FDM	Video	3724A/25A/26A
1. Insertion gain	•	•	•
2. Frequency response	•	•	•
3. Envelope delay distortion		•	
4. Spurious interference tones	•	•	•
5. Thermal nose	•	•	•
6. White noise loading	•		•
7. Video waveform tests		•	
8. Video system program channel (subcarrier) tests		•	

crowave link on a hop-by-hop basis. This may require the use of a modem, since baseband signals may not be available in nondemodulating repeaters. If these tests show that the distortion in a link is too high, diagnostic tests are then required to locate the problem.

A set of diagnostic tests are itemized in Table 9-2 along with the capabilities of Hewlett-Packard link analyzers. The main contributors to distortion in FM microwave radio links are the modulators, demodulators, and rf and IF amplifiers. Distortion parameters of these circuits are measurable in terms of nonlinearity, amplitude variations, and group delay variations. The test equipment must interface with the link at baseband, IF and rf. Maintenance of microwave link equipment involves keeping the distortion parameters at a minimal level.

9.5.2 Data Communications Analysis and Maintenance

Communications carriers are now called upon to guarantee the quality and reliability of their transmissions. This has resulted in special-purpose equipment to perform measurements. Some of these, such as the selective level meter and the frequency synthesizer, were discussed in the previous section. For detailed measurements and fault analysis of individual channels, Hewlett-Packard's 4944A transmission impairment measuring set measures all the parameters tariffed by the FCC for high-performance data transmission over conditioned voice grade circuits. Specifically, this instrument measures:

- Loss
- Message circuit noise
- Attenuation distortion
- Envelope delay
- Noise with tone
- Signal-to-noise ratio

Table 9-2   **Diagnostic tests to maintain radio-system performance**

Measurement	BB	IF	RF
1. Module power levels, gains and losses	•	•	•
2. Modem center frequencies		•	•
3. TX and RX local oscillator frequencies			•
4. Transmitter rf output frequency			•
5. Spurious tones	•	•	•
6. FM mod + demod deviation sensitivity	•	•	•
7. FM mod + demod linearity	•	•	•
8. Return loss	•	•	•
9. Amplitude flatness	•	•	•
10. Group delay		•	•
11. Differential gain and phase		•	•

- Intermodulation distortion
- Impulse noise

A large number of the channel tests relate primarily to data. This occurs because data traffic requires higher-quality channels than voice. These tests include measurement of transient effects, envelope delay, and phase jitter. These parameters ordinarily do not impair voice communications. The transient effects may be broken into four categories: dropouts, gain hits, phase hits, and impulse noise. The sporadic nature and difficulty in separating these events into separate categories make obtaining repeatable results difficult.

The HP 4944A measures all the necessary parameters to completely describe the ability of a voice band channel to carry medium- and high-speed data. This instrument can simultaneously record hit, dropout, and impulse noise events, distinguishing one from the other. A dropout is often defined as a loss of signal carrier level of at least 12 dB and lasting for at least 10 msec. Gain hits (rapid changes in channel gain of from 2 to 8 dB) are difficult to distinguish from impulse noise. Impulse noise generally consists of a burst of narrow spikes. Thus, a measuring system that responds only to longer level changes (4 msec) is needed to help discriminate between the two effects. The 4-msec delay is faster than the automatic-gain-control (AGC) response time of a typical modem but longer than most noise spikes.

Phase hits are rapid phase changes in the channel. A phase mismatch with a 4-msec delayed signal indicates a phase hit. The time constant of the instrument should correspond to the time constant of a typical phase-sensitive modem.

Envelope delay distortion represents yet another important measurement for data communications. Envelope delay distortion is often confused with phase delay distortion. Phase delay is the phase shift of the channel divided by frequency (ϕ/ω). Envelope delay is the rate of change of phase delay with respect to frequency ($d\phi/d\omega$). Of the two delays, envelope delay is the most frequent problem in data communications.

The HP 4944A measures envelope distortion by a method that is compatible with Bell methods. A 50% amplitude-modulated signal using an $83\frac{1}{3}$-Hz modulating signal is transmitted on a carrier frequency selected in the range of 300 to 3904 Hz. At the receiving end of the channel, the signal is demodulated, and the recovered modulation is sent back on a midband carrier—approximately 1800 Hz. The phase of the returned modulation envelope is compared with the original to determine the phase difference. The midband carrier is used as a reference because, in a typical channel, the envelope delay is usually flat at the frequency and is also a minimum.

The trend in the maintenance and repair of FDM/carrier systems is toward the use of multifunction instruments. Among these are instruments for digital signal analysis and automated network testing. Minicomputers have made centralized testing, diagnosis, and network management practical. Where formerly test engineers and technicians had to improvise test systems, specialized instruments—some of which are discussed in this chapter—now provide the precise measurements needed. A knowledge of how FDM systems work and the availability of test equipment provides the basis for maintenance of modern communications carriers.

9.6 REFERENCES

EMR Telemeter, **No. 1,** September 1965.

EMR Telemeter, **No. 4,** September 1966.

Grossman, M., Test Equipment Proving Its Worth in Ensuring Transmission Quality. *Electronic Design,* **13,** June 21, 1975, 60–66.

Hewlett-Packard Electronic Instruments and Systems Data Book, 1982 edition.

Hewlett-Packard Technical Data Specification, Selective Level Measuring Set—50 Hz to 32 MHz—Model 3746A. June 1982.

Jameson, B. W., Analyzing Data-Comm Channels Requires Special Equipment Doing Special Measurements. *Electronic Design,* **21,** Oct. 16, 1979, 144–148.

James, R. T., Data Transmission—The Art of Moving Information. *IEEE Spectrum,* 1965, 66–83.

CHAPTER 10

Electro-optic Links

Harold B. Killen
University of Houston

10.1 INTRODUCTION

Optical communications using a single or multimode fiber waveguide has evolved into widespread usage in telecommunications. The fiber optics consists of a silicate glass or similar material fiber having a diameter in the range of 10 to 400 μm encased in a cladding with a slightly smaller dielectric constant. This results in a very-low-loss cable: as low as 0.5 dB/km for a high-silica fiber. A major advantage of fiber is its noninductive nature, which results in it being immune to electromagnetic interference.

The performance and operation of a fiber optics communications system can be grouped into three main categories: (1) the light source and associated drive circuitry, (2) optical fiber and mechanical cable construction, and (3) photodetector and receiver circuitry. An understanding of these components is necessary for the maintenance and repair of a fiber optic link.

The two parameters that define the performance limitations of a fiber optics system and, consequently, repeater spacing are bandwidth (dispersion) and optical power. Since each repeater requires a back-to-back photodetector/transmitter arrangement, repeater spacing is a critical feature of a fiber optics system. Dispersion and attenuation are set by the fiber design. Failures in connectors or damaged fibers can result in changes that cause insufficient power to be present at repeaters. Other factors that must be considered are modulation rates, optical radiance, detector responsivity, rise time, noise factors, and others.

The expected dispersion varies with the fiber type. The types of fiber to be expected are discussed in the following section.

10.2 FIBER TYPES

There are three basic fiber types: multimode step index, multimode graded index, and single mode. It is worthwhile to discuss briefly the characteristics of each fiber type. Light itself is an electromagnetic wave. Propagation of light along the fiber dictates that the connector and the fiber be properly aligned. The refractive index of a fiber varies as a function of radial distance from the center of the fiber. The "index of refraction" is defined for both glass and plastic as

$$n = \frac{c}{v} \tag{10.1}$$

where c is the speed of light in a vacuum (3×10^8 m/sec) and v is its speed in the fiber. The "index profile" of an optical fiber refers to how its refractive index varies as a function of radial distance from the center of the fiber.

10.2.1 Step Index Fiber

The index profile for a step index fiber is shown in Figure 10-1. Note that the refractive index undergoes an abrupt change in value at a radius r_c. We see that the cross section of the fiber divides into two regions: a circular central core and the surrounding annular cladding. Within the fiber, optical-energy propagation takes place through total internal reflection at the core–cladding interface. The refractive index of the core is greater than that of the cladding. In fact, air with an index of 1.0 could serve as the cladding. The cladding

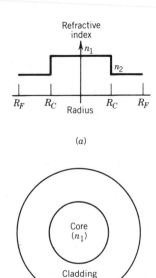

Figure 10-1   The index profile takes a downward step at the core radius (*a*) in a step index optical fiber. This step divides the fiber's cross section (*b*) into a central core with index n_1 and a surrounding cladding with index n_2.

however serves to support the core. Typical values for the refractive indices are $n_1 = 1.48$ and $n_2 = 1.46$.

The bandwidth of the fiber depends on the propagation characteristics of the fiber. A rigorous analysis of optical propagation involves a solution of Maxwell's equations with the associated boundary conditions. This analysis shows that the propagating energy is distributed among a discrete set of superimposed field solutions called *modes*. Differences in the propagation characteristics of these modes give rise to *modal dispersion*. This is one factor that limits fiber bandwidths. A geometrical ray approach is often used to illustrate modal dispersion (Figure 10-2).

From Snell's law, the minimum angle that will support total internal reflection is

$$\sin \theta_{min} = \frac{n_2}{n_1}$$

or

$$\theta_{min} = 80.6° \tag{10.2}$$

Rays that strike the core–cladding interface at angles less than 80.6° will be lost in the cladding. At this angle for Figure 10-2, the total path length is 1014 m. If this is compared with a ray that propagates down the central axis of the fiber, there is a difference of 14 m. The speed of travel is given by

$$v = \frac{c}{n_1} = \frac{(3 \times 10^8 \text{ m/sec})}{1148}$$
$$= 2.03 \times 10^8 \text{ m/sec} \tag{10.3}$$

If both rays start out at the same instant, the bouncing ray reaches the end of the fiber 69 nsec after the axial ray. The temporal delay (dispersion) in the arrival time produces bit-smearing or intersymbol interference in a pulsed-data system and delay distortion in an analog-modulated system. If we regard the reciprocal of the relative time delay as an order-of-magnitude estimate of the modal dispersion-limited bandwidth, we have a figure of 14.5 MHz.

10.2.2 Graded-Index Fiber

In a graded-index fiber, the refractive index decreases continuously with radial distance from the center of the fiber. Light propagation occurs as a continual

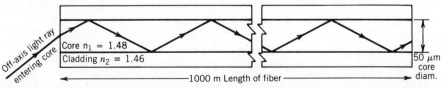

Figure 10-2   Reflecting back into the core each time it strikes the core–cladding interface, an off-axis light ray follows a zig-zag path 1014 m long. Compared with an axial ray, this extra 14 m produces an arrival time difference of 69 nsec.

bending of the ray toward the fiber's optical axis (Figure 10-3). The light travels faster in the lower index regions (outer extremities) of the fiber, resulting in reduced differences in arrival time and less dispersion. Minimum dispersion occurs if the index profile is nearly parabolic.

10.2.3 Single-Mode Fiber

Field analysis yields a parameter referred to as the normalized frequency ν, which is defined as

$$\nu = \frac{\pi d}{\lambda} \sqrt{n_1^2 - n_2^2} \tag{10.4}$$

where d is the fiber core diameter and λ is the optical wavelength of the source.

The number of propagating modes, N, can be estimated by

$$N = \frac{\nu^2}{2} \tag{10.5}$$

Equation 10.3 suggests a number of ways to reduce the number of modes, N. Studies indicate that if $\nu < 2.405$, only a single mode will propagate. For the fiber in Figure 10-2, a source wavelength of 0.82 μm results in a maximum core diameter of 2.6 μm. A fiber such as this offers the ultimate in bandwidth. However, projecting light into such a small core is extremely difficult. In addition, alignment problems occur in making a fiber-to-fiber splice.

10.3 BANDWIDTH MEASUREMENTS

The bandwidth of a fiber optics system can be determined by measuring the impulse response of the system. The connection between frequency response

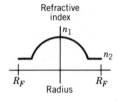

(a)

(b)

Figure 10-3   The graded-index profile tapers off parabolically with radius from its central-axis value n_1 to a lower value n_2 at the fiber radius r_f (a). The cross section (b) shows that light travels slower near the shaded center region than it does away from the center, resulting in more consistent arrival time and less dispersion.

and impulse response is rooted in linear systems theory. With this technique, narrow light pulses are projected into the fiber (Figure 10-4).

The output of the photodetector is a spread current pulse. Pulse spreading results from modal dispersion, material dispersion (wavelength-dependent fiber), detector rise time, capacitance, and so on. A spectrum analyzer indicates system frequency response. Faulty components that may limit bandwidth can be found by substitution. The bandwidth of the optical fiber is generally specified by the manufacturer as a bandwidth length product, with units of megahertz, kilometers, or a pulse dispersion denoted in nanoseconds per kilometer (see Figure 10-5). The spectrum analyzer displays the power spectrum of the output.

10.4 SYSTEM LOSSES

The end-to-end losses of a fiber optics system arise from a combination of factors. In a given system, sufficient power must be present at the photodetector to ensure an acceptable signal-to-noise ratio (SNR) or bit error rate (BER).

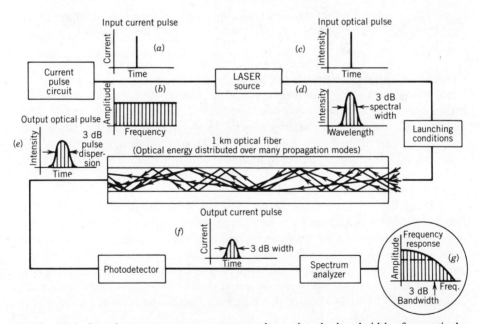

Figure 10-4 Impulse response measurements determine the bandwidth of an optical fiber. Narrow time domain electrical pulses (*a*) with a flat spectrum in the frequency domain (*b*) are converted to light pulses (*c*) with a wavelength distribution characteristic of the LASER source (*d*). Depending on the input conditions, light energy propagates through many modes of the fiber under test. Because of modal and material dispersion, the output light pulse spreads with time (*e*) and is photodetected as a widened current pulse (*f*). A frequency domain analysis then yields the fiber's 3-dB bandwidth (*g*).

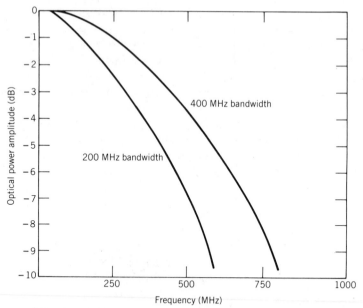

Figure 10-5 A fiber's 3-dB frequency is identified by taking the Fourier transform of the output signal. Specifications are usually quoted for 1-km fiber lengths.

Because of various optical losses between the transmitter and receiver, only a fraction of the source's total radiant power reaches the photodetector. These losses can be divided into input-coupling losses, connector/splice losses, fiber attenuation, and output coupling losses. In a given system, the losses are already accounted for in the loss budget of the original design. If a system is defective with insufficient power at the photodetector, the troubleshooting procedure involves determining which of the above parameters is at fault. The light source should also be checked.

A defective source could result in a complete loss of power at the photodetector. The input-coupling losses occur at the source–fiber interface. Usually, a short length of fiber (called a pigtail) is attached to the source's emitting area in single-fiber communications. Any mismatch that exists between this emitting area and the pigtail's core area results in the first input-coupling loss factor. Additional loss will occur if the core area is smaller than the source emitting area. A rough value of this fractional loss is the ratio of the core area to the emitting area.

Another input-coupling loss factor relates to the light-gathering ability of the fiber itself. The acceptance cone half-angle for a fiber is shown in Figure 10-6. The *numerical aperture* (NA) is the important parameter. It defines the half-angle θ of the acceptance cone of the fiber. Light injected at angles that lie within this cone will propagate down the fiber. Light entering at steeper angles will propagate into the cladding and be lost. Mathematically, the numerical aperture is defined as

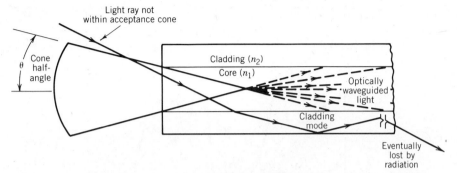

Figure 10-6   A fiber's acceptance cone half-angle derives from the fiber's numerical aperture. Light projected into the acceptance cone undergoes waveguiding in the core, while rays outside the cone reflect into the cladding and are eventually lost.

$$NA = \sin \theta \qquad (10.6)$$

For step index fibers, it is also equal to

$$NA = \sqrt{n_1^2 - n_2^2} \qquad (10.7)$$

A numerical aperture of 0.25 results in an acceptance cone half-angle θ of

$$\theta = \sin^{-1}(0.25) = 14.5°$$

To evaluate the input-coupling loss of the numerical aperture, it is necessary to have a knowledge of the source's emission profile. This will be discussed later. A small input-coupling loss comes from light reflection at the input end of the fiber. It is usually on the order of 0.2 dB. Losses at the receiving end of the fiber are generally not as severe as those at the transmitting end. Generally, the total output losses should not exceed 1 dB.

Losses between the system input and output occur as a result of fiber attenuation losses as well as losses arising from splices and connectors. Attenuation in the fiber results from scattering, the deflection of light by impurities in the fiber, and absorption. The attenuation varies with the wavelength (Figure 10-7). The wavelength of the source must fall in an area where the attenuation is low.

10.5 SOURCES AND RECEIVERS

Next to fiber attenuation, input coupling contributes most to system losses. We previously identified these as:

Unintercepted-illumination (UI) loss resulting from an area mismatch between the source's illumination spot (in the plane of the fiber end) and the fiber core area.

Numerical-aperture (NA) loss.

Reflection (R) loss.

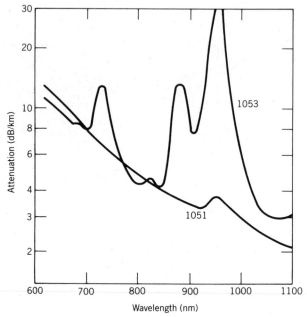

Figure 10-7 Fiber attenuation varies dramatically with wavelength, so good overall system performance requires a good match between source and fiber characteristics.

If the emitting area of the source is larger than the fiber core area, all the light cannot be coupled into the fiber. If the source is smaller than the core, problems may still exist (Figure 10-8). Any separation between the elements allows emitted light to miss the core and be lost. An approximation to this loss is given by

$$\text{UI loss} = 10 \log \left(\frac{A_c}{A_s} \right) \tag{10.8}$$

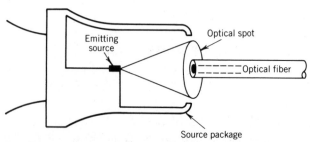

Figure 10-8 Unintercepted illumination loss can be a problem when the light-emitting surface is separated from the end of the fiber core. You can minimize this loss however by using an uncapped source diode or one with a "pigtail" already installed by the manufacturer.

where A_c is the fiber core area and A_s is the area of the source's projected optical spot in the plane of the fiber end.

The magnitude of the loss depends on the source's angular emission profile and the fiber end. All sources have rapidly divergent beams. Consequently, separation between the emitting surface and the fiber end must not be greater than about two to four times the core diameter (on the order of 50 μm).

10.5.1 Launch Profile

The launch profile is important when considered in terms of the numerical-aperture loss. A considerable loss of light results because of the conflict between the relatively small acceptance cone angle of the fiber (10° to 14°) and the broad divergence of both LED and injection laser diode (ILD) emission beams (Figure 10-9).

The NA loss can be estimated using the source beam profile. Typically, manufacturers plot the source beam pattern on a polar diagram (Figure 10-10). The emission power profile (P) of a "Lambertian source" is given by

$$P = P_0 \cos \phi \qquad (10.9)$$

where P_0 is the radiant intensity along the line $\phi = 0$. Sources with narrower beams can be described mathematically by

$$P = P_0(\cos \phi)^m \qquad (10.10)$$

A set of curves for various m values is shown in Figure 10-10b.

The amount of power coupled into the fiber (P_c) is given by

$$P_c = P_t[1 - (\cos \theta)^{m+1}] \qquad (10.11)$$

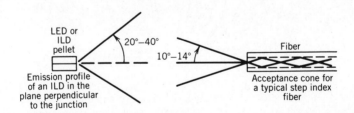

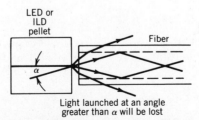

Figure 10-9 Lots of light is not necessarily a blessing at the source–fiber interface. Both LEDs and ILDs have broadly divergent emission beams, and all light radiated outside of the fiber's acceptance cone angle contributes to numerical-aperture loss.

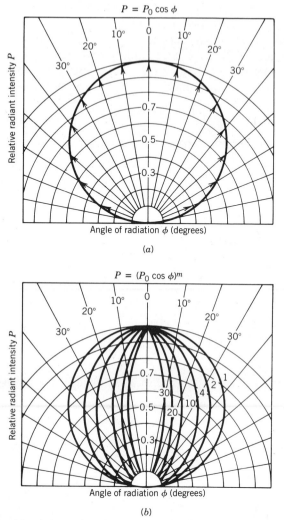

Figure 10-10   A wide variety of emission beam patterns appears in sources available today. While uniform surface emitters typically have Lambertian-type profiles (*a*), some sources (*b*) exhibit much narrower-beam profiles.

where P_t is the total source power and θ is the fiber acceptance cone half-angle. Note that for a Lambertian source when $m = 1$ and $NA = \sin \theta$, Equation 10.10 reduces to

$$P_c = P_t (NA)^2 \qquad (10.12)$$

The NA loss, in dB, can be obtained from the following expression:

$$NA \text{ loss} = 10 \log \left(\frac{P_c}{P_t} \right) \quad dB \qquad (10.13)$$

10.5.2 Reflection Loss

The reflection loss is almost negligible when compared to the NA loss. However, it is important in fiber splices. At the air–cone interface, a change in the index of refraction occurs. The reflection coefficient ρ, which gives the fraction of incident power to power reflected from the core, is given by

$$\rho = \left(\frac{n_1 - 1}{n_1 + 1}\right)^2 \tag{10.14}$$

In decibels, the reflection loss (R) is

$$R \text{ loss} = 10 \log (1 - \rho) \quad dB \tag{10.15}$$

A core index of 1.5 yields 4% reflection. This is equivalent to about a 0.2-dB loss.

10.5.3 Source Requirements

In a given fiber optic system, link budget calculations are used to establish repeater spacings. Generally, the source should be as intense as the state of the art allows. The emitted beam pattern must be nearly collimated to fit into the fiber's acceptance cone. In addition, it must be nearly monochromatic to avoid material dispersion. Digital systems require sources with fast rise and fall times. Analog systems require the output optical power to be linear in relation to the drive current or voltage over a wide dynamic range. LEDs are best suited to analog systems.

The power output for surface-emitting devices is usually specified in terms of *radiance,* that is, the power per unit of solid angle (steradian) per unit area (W/sr-cm^2). As an example, if the radiance is specified to be 60 W/sr-cm^2, the emitting surface area is 75 μm in diameter, and the fiber core diameter is 50 μm, the total power emitted into a solid angle of 2π steradians is

$$(60)(2\pi)(\pi)(25 \times 10^{-4})^2 = 7.4 \quad mW$$

If a Lambertian source is used with a half-angle of 14°, the power coupled into the fiber is

$$P_c = 7.4[1 - (\cos 14°)^2] = 0.43 \text{ mW}$$

The NA loss is

$$NA \text{ loss} = 10 \log \frac{0.43}{7.4} = -12.3 \quad dB$$

The source/fiber compatibility is important. Glass fibers have two areas of minimal attenuation: 800 to 850 nm and 1050 nm (see Figure 10-7). Most available LEDs have peak wavelengths of 800 to 850 nm and power outputs on the order of 1 mW. Some sources are available at 1050 nm, but the low power (0.1 to 0.2 mW) restricts their use. Solid-state injection laser diodes (ILDs) are very well suited for digital fiber optic systems. The beam is more directional than that of the LEDs, and typical NA losses are on the order of

6 dB. The power output is on the order of 5 to 10 mW. Some disadvantages of ILDs are temperature dependence and nonlinearity of the power output with respect to the drive.

10.5.4 Receivers

The most commonly used receivers are photodiodes, either PIN or avalanche types. These diodes convert the light to an electrical signal which is then applied to a preamplifier. A PIN photodiode consists of a large intrinsic region sand-wiched between p and n doped semiconducting regions. Photons absorbed in this region create electron–hole pairs which are then separated by an electric field, thus generating an electric current. The efficiency of the conversion process is specified by the photodiode's *quantum efficiency* η. This is a measure of the average number of electrons released by each incident photon.

The *responsivity* of the diode is a performance parameter which is more meaningful. It is related to quantum efficiency by

$$r = \frac{\eta\lambda}{1.24} \text{ amperes/watt} \qquad (10.16)$$

The signal current i_s generated when P watts of optical power is incident on the photodiode is given by

$$i_s = rP \qquad (10.17)$$

For typical PIN diodes, the peak responsivity is less than 1 ampere/watt.

The *avalanche photodiode* (APD) is used in applications requiring greater sensitivity. The APD uses rather high voltages (300 V) to generate a strong electric field to create the avalanche effect. Both the APDs and PINs must be reverse-biased in the receiver circuit.

In the receiver circuit, the signal current must contend with the noise current. The dominant noise component in PIN photodiodes is caused by fluc-tuations in the *dark current,* the current that flows through the diode-biasing circuit when no light is incident on the photodiode. Dark-current shot noise power varies linearly with temperature. As a general rule, it doubles for every 10°C increase in the operating temperature.

The *noise equivalent power* (NEP) is also a figure of merit that is related to noise performance. It is usually expressed in units of $W/\sqrt{Hz}$. When the NEP is multiplied by the square root of the detector noise bandwidth B, an absolute unit of power called the *minimum detectable signal* (MDS) is obtained. The MDS defines the optical power incident on the photodiode that is required to generate a photodiode current equal to the total photodiode noise current. This, then, is equivalent to a 0 dB SNR out of the photodiode.

Circuit performance of the receiver is also affected by the response time. The response time in the photodiode is limited by the RC time constant as-sociated with the diode's series resistance plus the load resistance into which it operates, and by the junction capacitance of the diode. The 10 to 90% rise time of PIN diodes is typically a few nanoseconds. The rise times of APDs are somewhat faster.

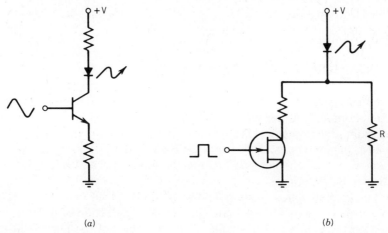

(a)   (b)

Figure 10-11   Transmitting circuits. (*a*) Transistor circuit used for analog intensity modulation. (*b*) FET circuit used for pulsed modulation.

10.5.5 Receive-and-Transmit Circuits

The basic drive circuits for light sources are straightforward (see Figure 10-11). In Figure 10-11*a*, the control current is applied to the base of the transistor. Maximum forward currents are specified for the LED by the manufacturer. Typical values are a few hundred milliamperes. Resistor R in the FET circuit is used to adjust the biasing current to just below the lasing threshold of the ILD. This action increases switching speed since it is not necessary to begin with zero current.

Two common receiver circuits are shown in Figure 10-12. The transimpedance amplifier in Figure 10-12*a* is designed for current sources such as PIN or APDs. The voltage gain circuit in Figure 10-12*b* uses an op-amp to amplify the voltage developed across R_L. A typical value for R_L is on the order of 5 megohms. This large value reduces thermal-noise currents from the resis-

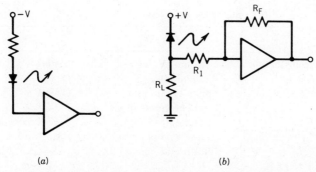

(a)   (b)

Figure 10-12   Basic receiver circuits. (*a*) Transimpedance amplifier. (*b*) Voltage amplifier.

tor, thus lowering the NEP. However, this in turn increases the RC time constant and thus reduces system response time.

10.6 FIBER OPTICS MEASUREMENTS

Fiber optics system measurements are in a constant state of improvement. Two of the most important measurements are the system bandwidth and an optical time domain reflectometer (OTDR) measurement. We previously mentioned a scheme for bandwidth measurement (see Figure 10-4). The reflectometer measurement is a key technique for testing the quality of an optical system and also for troubleshooting.

Typical fiber-optical system field tests are itemized in Table 10-1. These tests are usually run before and after installation. Selection of the test method depends on the equipment available. The continuity of the fiber itself can be checked with a light source as simple as a flashlight. The attenuation is low enough that several kilometers can be checked this way. Splice and connector attenuation requires the use of an optical time domain reflectometer.

10.6.1 Time Domain Reflectometer Measurements

Fiber optic cable testing is greatly facilitated with the use of an optical time domain reflectometer. The model TD 9920 by Laser Precision Corporation is an excellent example of these instruments (see Figure 10-13). The ''real time'' mode allows the user to spot faults instantaneously and to see immediately the effects of cable/connector manipulation. This feature is also of great value in checking fiber alignment prior to splicing.

Dual microprocessor-controlled cursors measure decibel loss between any two points and provide absolute and relative distance measurements between any two points. Other applications include cable length measurements and bending loss. A typical strip chart recording using this instrument is shown in Figure 10-14. This instrument is also capable of performing signal averaging to minimize backscattering effects.

Table 10-1 **Fiber optic cable field tests**

	Test	Method
Prior to installation	Continuity	Visual
	Attenuation	OTDR
Splicing	Splice attenuation	OTDR
	Connector/end splice attenuation	1. OTDR 2. Insertion loss
After installation	End-to-end attenuation	Insertion loss
	End-to-end bandwidth	1. Swept frequency 2. Pulsed

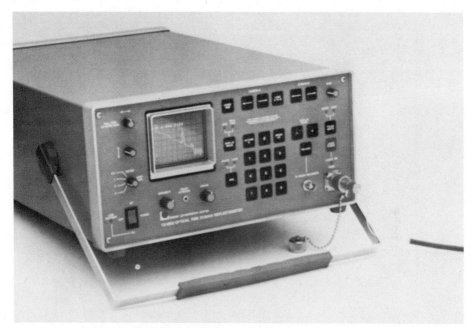

Figure 10-13   Model TD-9920 time domain reflectometer.

10.6.2 Troubleshooting and Maintenance

The field test results (Table 10-1) provide the data for cable plant maintenance. These tests ensure that in tolerance transmission specifications (attenuation and bandwidth) are met. Generally speaking, the fiber optic transmission system consists of the following: (a) electro–optical (E/O) interface, (b) the fiber cable system, and (c) the optical–electrical (O/E) interface. The E/O and O/E interfaces and their common equipment are often referred to as the optical line-terminating equipment (OLTE). The fiber cable system consists of one or more fiber cables joined with splices and terminated at the ends with optical fiber connectors. Often, the OLTE and the fiber cable system are connected by a fiber distribution frame (FDR), which is an optical patch panel.

The minimum optical measurements for the E/O and the O/E interfaces are the output power (power coupled into the fiber) and the minimum received input power. The only equipment required for these measurements is an optical power meter and a variable optical attenuator. Power meters are designed to make the following measurements:

- Fiber cable transmission
- Connector or splice loss
- Absolute fiber output intensity
- Efficiency of receivers
- Absolute intensity of sources

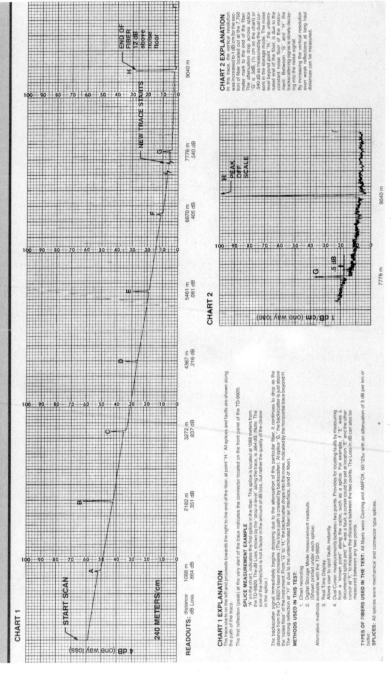

Figure 10-14 TD-9920 time domain reflectometry test results.

Suggested troubleshooting tests are shown in Table 10-2 for a fiber optic communications system transmitting data.

10.7 FIBER OPTICS TEST STANDARDS

As of this writing, a number of optical waveguide test procedures and techniques have been approved, or are in review, for the Electronic Industries Association (EIA). The EIA is the focal point for optical fiber standards in the United States. This work is being performed by the EIA Committee for Fiber Optic Standardization and is designated 10-6. In the following, we discuss some of the EIA test procedures that have been or will be distributed under the designation of a Fiber Optic Test Procedure (FOTP).

With the exception of the *Glossary of Terms,* the standards are published as *Recommended Standards (RS)* in RS 455. The *Glossary of Terms* is published as RS 440. The EIA has adopted the *National Bureau of Standards Handbook 140 (Optical Waveguide Communications Glossary)* as the basic glossary of terms and has published that document as RS 440. New terms are added or deleted on a continuing basis as conditions dictate.

Hundreds of fiber test procedures are currently in various stages of development. The most important of these are procedures for measuring attenuation, bandwidth, refractive index profile, and cone diameter. For example, FOTP 46 describes a procedure for measuring the spectral attenuation of long (greater than 1 km) graded index multimode fibers. This procedure is intended to yield results that are meaningful in predicting the performance of long-haul telecommunication links. These links frequently have several fibers in tandem. The fundamental problem is to define a launch condition that replicates the model power distribution that prevails at the end of a relatively long piece of fiber. High-order, high-loss modes must be eliminated at launch to remove effects that will yield erroneous values of attenuation. Restricted and carefully controlled launch conditions are defined in FOTP 50.

FOTP 50 defines two allowable launch conditions. The first prescribes an

Table 10-2   **Troubleshooting chart**

Problem	Test
No output	Check O/E interface
	Measure received power intensity
	Check E/O interface
	Check cable continuity
	Measure cable attenuation
High BER	All of the above
	Measure system bandwidth
Lossy cable	Perform OTDR measurement to isolate fault-damaged cable, connector, splice, etc.

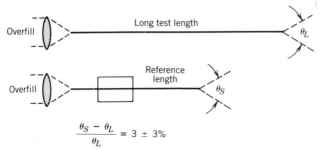

Figure 10-15   Mode filter requirements.

overfilled launch condition with subsequent filtering to eliminate higher-order modes (Figure 10-15). In Figure 10-15, overfill means that the launch spot diameter exceeds the fiber core diameter and also that the launch core angle exceeds the fiber's numerical aperture. Mode filtering is evaluated based on the far-field radiation pattern that it produces. The specification is as follows: A reference length of fiber (1 to 2 meters) is overfilled and subjected to filtering, and the far-field radiation pattern is measured. The angle θ_s at the 5% intensity points must be $-3 \pm 3\%$ (0 to -6%) of the 5% intensity angle measured on the long test fiber without the filter. If this criterion is met, the filter is considered to be adequate.

The second launch condition eliminates the need for filtering. The launch spot diameter and the launch angle are $70 \pm 5\%$ of the fiber cone diameter and the fiber numerical aperture, respectively. This technique is termed the *beam optics launch*. Both techniques yield a measured value of attenuation that scales linearly with distance. The two launch methods do not produce identical modal power distributions but do produce results that are comparable. A popular filter that is used is shown in Figure 10-16.

10.7.1 Bandwidth

Multimode fibers exhibit both differential modal attenuation and differential modal delay (DMD). The EIA allows either a frequency domain or a time domain technique in measuring fiber bandwidth. The frequency domain technique is described in FOTP 30, and the time domain technique is defined in FOTP 51. Both techniques call for controlled launch conditions, as described in FOTP 54. This ensures that the light launched will be independent of the source spatial distribution. Basically, the requirement is that the fiber be over-

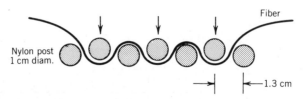

Figure 10-16   Serpentine bend mode filter.

filled and subjected to a mode scrambler. The spot size that results must be uniform to within 30%.

The -3 dB points are considered to represent the bandwidth for either technique. If the -3-dB frequency is not unique, the bandwidth is considered to be defined by the frequency where the magnitude of the frequency response has decreased by 3 dB.

The mechanics for measuring bandwidths are analogous to those for attenuation. The fiber response is compared to that of a short reference length of the same fiber. This serves to calibrate the system. The ratio of the magnitudes of the two frequency response functions determines the magnitude of the transfer function. The -3-dB optical bandwidth is the lowest frequency for which the transfer function magnitude is half the value of the transfer function at the low-frequency reference.

In the interpretation of bandwidth measurements, it is necessary to establish a relationship between RMS pulse duration and optical bandwidth. The theoretical relationship is as follows (FDHM = full-duration half-maximum):

$$K_1 \quad \text{(MHz-nsec)} = 3 \text{ dB optical bandwidth} \tag{10.18}$$
$$\times \text{ RMS pulse duration}$$
$$K_2 \quad \text{(MHz-nsec)} = 3 \text{ dB optical bandwidth} \tag{10.19}$$
$$\times \text{ FDHM pulse duration}$$

For a Gaussian pulse, $K_1 = 187$ MHz-nsec and $K_2 = 440$ MHz-nsec. Experimentally, K_1 has been found to range from 169 to 174 MHz-nsec. The values for K_1 and K_2 depend on the fiber, the cabling, and the amount of mode mixing.

10.7.2 Refractive Index Profile and Core Diameter

The FOTPs that refer to the refractive index profile and core diameter are related primarily to the manufacture and utility of connectors and splices. An accurate determination of the refractive index profile and the core diameter is necessary. The core diameter is usually defined in terms of the refractive index profile as shown in Figure 10-17. The value of n_3, on which the core diameter is based, is defined as

$$n_3 = n_2 + k(n_1 - n_2) \tag{10.20}$$

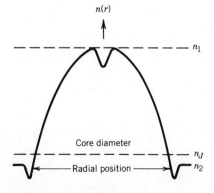

Figure 10-17 A possible refractive-index profile showing how core diameter is defined. Note that absolute values need not be known.

The quantity n_1 is the maximum index and n_2 is the index of the homogeneous cladding. EIA working groups have suggested a value of 0.025 for k. This value for k was arrived at by considering three different candidate methods for determining core size. The FOTPs relating to these measurements are FOTP 29, 43, 44, and 58.

As of this writing, an addendum to RS 455 has been released as RS 455-5. This addendum elevates roughly 24 fiber optic test procedures already published by EIA to the level of standards. Eighty other test procedures are being developed for future addenda to RS 455. Fifteen of these are nearing final approval. Copies of RS 455-5 and other addenda to RS 455 are available from:

Electronic Industries Association
Standards Sales Department
2001 Eye St. N.W.
Washington, DC 20006

10.8 REFERENCES

Buckler, M. J., Measurement of Bandwidth Versus Impulse Response Width In Multimode Fibers. In *NBS Publ. 641, Symposium On Optical Fiber Measurements*, 1982, pp. 33–35.

Flatau, C., OTDR Puts Fibers To The Test. *Photonics Spectra*, March 19, 47–50.

Gallawa, R. L., and Franzen, D. L., Progress in Fiber Standards. *Photonics Spectra*, April 1983, 62–68.

Kaiser, P., Young, W. C., and Curtis, L., Optical Connector Measurement Aspects, Including Single Mode Connectors. In *NBS Publ. 641, Symposium on Optical Fiber Measurements*, 1982, pp. 123–126.

Kummer, R. B., Judy, D. F., and Charin, A. H., Field and Laboratory Transmission and OTDR Splice Loss Measurements of Multimode Optical Fibers. In *NBS Publ. 641, Symposium on Optical Fiber Measurements*, 1982, pp. 109–117.

Matsumoto, T., Bandwidth Estimation of Multispliced GRIN Fibers. *Applied Optics*, **21**, June 1, 1982.

Metcalf, B., and Kleckamp, C., *Designers Guide to Fiber Optics*. Cahners Publishing, 1978.

Reitz, P. R., Predicting Fiber Performance. *Fiber Optics*, February 1983, 35–37.

Szentesi, O. I., Field Measurements of Fiber Optic Cable Systems. In *NBS Publ. 641, Symposium on Optical Fiber Measurements*, 1982, pp. 37–42.

Versluis, J. W., and DeWert, H. P., Prototype System for Automated Measurements of Transmission Properties of Graded Index Fibers. In *NBS Publ. 641, Symposium on Optical Fiber Measurements*, 1982, pp. 47–50.

PART THREE

DIGITAL SYSTEM TROUBLESHOOTING

CHAPTER 11

Digital Communications

Dean Lance Smith
Engineering Consultant Houston, Texas

This chapter starts with a review of digital communications terminology and techniques. It is followed by a more detailed discussion of common data communications channels and their properties. The operations of typical FSK and PSK modems are explained. The common data communications interconnect standards are listed in tabular form. Special data communications test tools are discussed along with component isolation strategies. The chapter concludes with a checklist of common intermittent error problems and solutions.

11.1 GENERAL INFORMATION

Digital or *data communications* involves the transmission of discrete signals. *Discrete signals* can have two or more predefined levels or states. Digital communications can be contrasted to analog communications. In analog communications, continuous signals are transmitted. These continuous signals can take a wide range of values.

If only two discrete states are used, the signal is a *binary signal*. If *m* states are used, the signal is an *m-nary signal; m* is usually an even number. Quaternary (four-level) and octonary (eight-level) signals are encountered in some sophisticated communications systems. However, binary signals are the most common digital signals encountered.

One word of caution. A signal that enters and leaves a digital communications channel in binary form may not maintain that form throughout the entire channel. In some systems, the binary signal may be converted to an *m*-nary signal for ease of transmission and then reconverted to binary at the receiver.

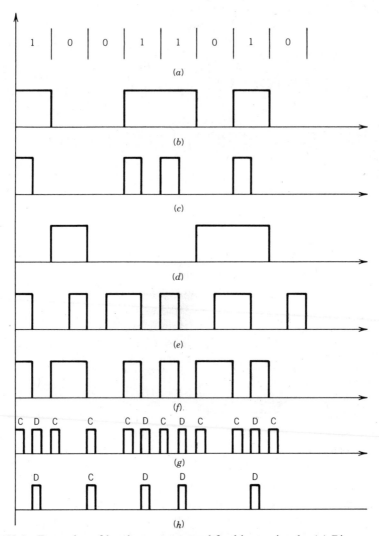

Figure 11-1 Examples of levels or states used for binary signals. (*a*) Binary values.
(*b*) NRZ (non-return to zero) waveform. (High during mark, low during space.) (*c*)
RZ (return to zero) waveform. (Part of a pulse for mark, low for space.) (*d*) NRZI
(Non-return to zero invertible) waveform. (Transition for space, no transition for a
mark.) (*e*) Manchester waveform. (Positive transition for a space, negative transition
for a mark.) (*f*) Biphase mark or Manchester II waveform. (Each bit begins with a
transition. Mark has an additional transition. Space has no additional transition.) (*g*)
Single-density, or FM, waveform. (Clock pulse and data pulse for a mark, clock
pulse and no data pulse for a space.) (*h*) Double-density, or MFM, waveform. (Data
pulse for each mark, clock pulse for each space unless the previous cell was a mark.)
(*i*) FSK (frequency shift keying) waveform. (*j*) PSK (phase shift keying) waveform.

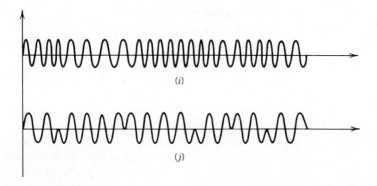

(i)

(j)

11.1.1 Levels, or States

The levels, or states, permitted in a digital communications channel will vary from one system to another or even from one part of a channel to another. The simplest states are voltage or current levels. These are usually used in binary channels. Figure 11-1 illustrates several level formats used.

All the formats shown in Figure 11-1 are *positive logic* formats. *Positive logic* means that a positive voltage or current corresponds to a *mark* (logic 1 or true). No current or voltage or a negative current or voltage corresponds to a *space* (logic 0 or false).

Inverted or *negative* logic is just the reverse of positive logic. No current or voltage or a negative current or voltage corresponds to a mark or logic 1. Positive voltage or current corresponds to a space or logic 0.

Ground (or no signal) need not be one of the states. Some systems, for example, use a positive voltage for one state and a negative voltage for the other.

Signal levels are not the only states used in digital communications systems. Frequencies can be used. *Frequency shift keying* (FSK) uses two or more frequencies to represent different states. Phase shift keying (PSK) uses different phases of the same frequency to represent different states.

Figure 11-1 illustrates FSK and PSK for a binary system. In some sophisticated systems, combinations of amplitude and PSK or amplitude and FSK are used to define the states. *Vestigial sideband amplitude modulation* (VSAM) is also used. VSAM is AM (*amplitude modulation*) with most of one sideband and the carrier filtered off. It requires less bandwidth than an AM signal. The presence of the carrier makes it easier to detect than SSBAM (*single-sideband AM*).

FSK and PSK generation and detection are discussed in more detail below.

11.1.2 Codes

A *bit* is the smallest unit of digital information. It represents a logic 1 or a logic 0 (mark or space). The bit is used as the unit of information in binary and most *m*-nary systems.

Bits by themselves are meaningless unless a code is used to interpret the bit stream. Probably the most common information code is ASCII (American Standard Code for Information Interchange). ASCII is an ANSI (American National Standards Institute) standard. It is also a subset of CCITT V.3. CCITT is an abbreviation for the French equivalent of the International Telephone and Telegraph Communications Conference, an international standards organization. ASCII is also a subset of the proposed NAPLPS (North American Presentation Level Protocol Syntax), a text and graphics code.

ASCII is a seven-level code. However, it is usually sent in binary, as shown in Figure 11-2. The least significant bit is sent first. The most significant bit is sent last. ASCII is sent in parallel on some systems. (The IEEE-488 bus is a common example.)

The start-and-stop bits shown in Figure 11-2 are optional. They are required if the data are sent asynchronously. Characters sent *asynchronously* have an indeterminate delay between characters. Stop bits fill the void between characters. There must be at least one stop bit. Some systems require at least one and one-half or even two. In asynchronous transmission, the system remains in the mark (stop bit) state until the next character is sent.

Start and stop bits are not used if the data are sent synchronously. Characters sent *synchronously* have no delay between characters. The SYN character is used to fill between characters in synchronous systems if fill is necessary.

An optional *parity* bit can be transmitted between the MSB (most significant bit) of the ASCII character and the stop bit. The parity bit is used for error detection. It is an example of an *error detection code;* the parity of a character is the number of ones or marks in a character. Two parity codes are used. *Odd parity* means that the parity bit is set (i.e., a logic 1 or mark is transmitted) when necessary to ensure the parity of the character is odd. *Even*

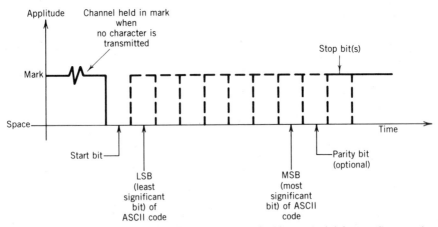

Figure 11-2   Transmission of an ASCII character in binary serial form. (Start and stop bits required only for asynchronous transmission.)

parity means the parity bit is set to ensure the parity of the character is even. Other error detection or correction codes are discussed in the section on noise.

Codes can also be used for data encryption or secrecy.

11.1.3 Channel Capacity

The *data rate* is the rate at which data are transmitted. The data rate is usually measured in *bits per second*.

Channel capacity is measured in *baud*. Baud is the highest rate at which data can be transmitted. It is always greater than or equal to the data rate. For example, data may be transmitted asynchronously in ASCII at 300 baud. However, the data rate is probably less than 300 bits per second because of the variable length of the stop bits.

11.1.4 Equipment Terms

Several special terms are used in data communications. A *data terminal* is a device for sending or receiving a message. If it can only receive a message, it is a RO (receive only) device. Printers are examples of RO devices. Most data terminals have a keyboard (or KB) and are capable of transmitting and receiving. VDTs (video display terminals) and CRTs (cathode ray terminals) are examples.

An interface that converts level codes to FSK or PSK codes, or vice versa, is called a *data set,* or *modem*. Modem is a contraction for modulator–demodulator. Some data sets can be used only to initiate cells. These data sets are called originate-only modems.

Other data sets can be used only to receive calls. These are called answer-only modems. Some data sets have both originate and answer capability.

The terms originate and answer should not be confused with the terms transmit and receive. The former terms apply to calls. The latter terms apply to data.

11.1.5 Transmission Modes

Three modes are used for data transmission. *Full duplex,* or FDX, means that both ends of the channel can transmit and receive simultaneously. *Half duplex,* or HDX, means that only one end of the channel may transmit at a time. However, both ends of the channel may transmit and receive. *Simplex* means that only one end of the channel may transmit and the other receive.

11.1.6 Noise

Noise is any unwanted signal present at the receiver. Noise is introduced by all communications channels. The success of any communications system rests on whether the *signal-to-noise ratio* at the receiver is high enough for the signal to be recovered with little or no loss of information.

An advantage of digital over analog communications is that the signal can be recovered without loss, provided the signal-to-noise ratio is high enough. This concept is illustrated in Figure 11-3 for binary-level signals.

Most digital communications channels are designed to take advantage of this property of digital communications. If the signal-to-noise ratio is too low, error detection and even correction schemes are used.

Echo back is a technique used on FDX channels. The transmitter compares the echo signal to the original and retransmits if an error is detected.

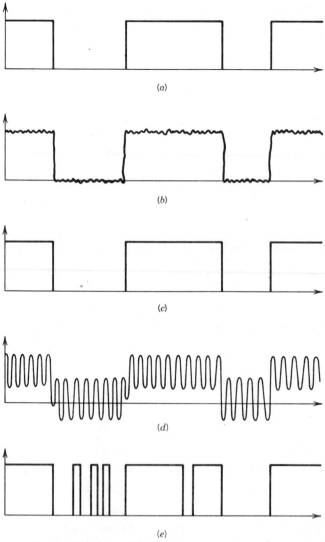

Figure 11-3 An illustration that a digital signal can be received without loss of information if the signal/noise ratio is high enough. (*a*) Transmitted signal. (*b*) Received signal that can be restored without loss of information. (*c*) Restored signal. (*d*) Received signal that cannot be restored without loss of information. (*e*) Restored signal with errors.

Parity bit error detection as used with ASCII codes was mentioned earlier. Strings of ASCII codes, bytes of data, or long blocks of bits can be error-checked using *arithmetic, longitudinal parity,* or *cyclic redundancy* check codes. These codes are sent at the end of the block or string of data. The receiver calculates the check code, compares it to the code transmitted with the message, and requests a retransmission if an error is detected. There are numerous *protocols* or rules for determining the format and length of the blocks.

11.2 DATA COMMUNICATIONS CHANNELS

11.2.1 Wire

The simplest data communications channel is two pieces of wire. Two pieces of wire running parallel to each other with or without spacers is sometimes called *twin lead*. Sometimes, the second, or return, wire is common to several different channels.

If two parallel wires are twisted together, the resulting channel is called a *twisted pair*. Twisting two wires together reduces the noise introduced by *crosstalk*. Crosstalk is an unwanted signal from an adjacent channel introduced by magnetic or electric coupling. A twisted pair also reduces the *hum* introduced by power systems and *hash* from relay banks.

In more severe noise environments, a *coaxial cable,* or a *shielded twin lead,* will be used. Shielded cable is even more immune than twisted pair to stray coupling such as crosstalk, hum, and hash.

Open cable, such as twin lead and twisted pair, can be balanced to reduce stray coupling even further. In an *unbalanced* circuit, one conductor is assumed to be at ground potential. *Ground* is the common reference point in the circuit to which all voltage or potential measurements are referenced.

In a *balanced* circuit, neither conductor is treated as ground. Signals are measured as the difference between the voltages on the two wires. Since most stray coupling is an unbalanced signal, a balanced circuit greatly reduces coupling noise.

Coaxial cable is rarely balanced, because it is essentially self-shielding. Shielded twin lead is always balanced. Unshielded twin lead and twisted pair are used both balanced and unbalanced.

All data communications channels have wire as part of their transmission path, usually in the connections to the transmitter or receiver. Short runs in low-noise environments are usually unbalanced, or *single-ended* twin lead. Twisted pair and/or balanced circuits will be used in noisier environments. Shielded cable will be used in the most severe noise environments.

However, suppression of coupled noise is not the only criterion for selecting cable. Cost is another. Twin lead is the cheapest cable to use. Twisted pair is almost as inexpensive as twin lead. Using balanced transmitters and receivers usually costs more than using unbalanced transmitters and receivers. Shielded cable, especially shielded twin lead, is usually the most expensive channel to construct and maintain.

Another factor important in selecting wire data channels is dispersion.

Dispersion causes pulses to smear together on a communications channel. Dispersion distorts the waveform so that it is difficult to distinguish between allowed states.

Two factors contribute to dispersion. One is unequal attenuation or signal loss at all frequencies. This is called *amplitude distortion*. The other factor is unequal phase shift as a function of frequency, or *phase distortion*.

Open wire line is rarely used to transmit pulses more than a few hundred meters or yards. The longer the line, the lower the channel capacity of the line. Increasing the diameter of the wire helps reduce loss. Increasing the spacing between the wires helps reduce attenuation at lower frequencies, but increases attenuation at higher frequencies.

Short-haul modems can be used to increase transmission distance over open wire line. But modems rarely will increase the distance–channel capacity product by more than a factor of 50.

Higher capacity over longer distances can be achieved with shielded cable. However, the high capacitance of the cable precludes rapid transmission of pulses over long distances. Modems are usually used on cable to achieve fairly long distance transmission.

11.2.2 Carrier Current

Carrier current is a technique for sending information over a power system. The data are sent on a carrier above the power line frequency. Special carrier current modems are required.

The carrier current technique is also used on phone lines. Special modems are used to carry data above the speech frequencies. This technique is sometimes called *speech plus*.

Carrier current works well for short-haul communications, particularly in an office or factory. The technique requires no special wiring. Terminals can readily be moved from one location to another, provided both terminals share the same power distribution transformer or telephone switch.

Power distribution transformers adsorb most of the carrier energy. Telephone switches or exchanges will frequently do the same. However, fused capacitors can be used to bypass power transformers and feed carrier current energy from one distribution circuit to another. A telephone switch can be bypassed with additional modems.

11.2.3 Radio Link

Most radio links used in data communications are dedicated point-to-point microwave links. Some newer systems use Earth satellites as repeaters.

11.2.4 Direct Distance Dial Network

The direct distance dial (DDD) network is a very popular data communications channel. It connects to virtually all major business establishments in the world. It is relatively easy to have a new service drop installed. For short messages in particular, it offers a very cost-effective digital communications channel.

The DDD does have a very significant disadvantage. It was designed to transmit voice, not data. It will rarely pass a signal lower than 300 Hz or higher than 2700 Hz. Voice signals outside the 300- to 2700-Hz passband of the DDD carry no significant information.

The human ear is insensitive to phase distortion. Thus, phase distortion was ignored when the DDD was designed. While the worst frequency response of the DDD can be anticipated, no specifications exist for phase distortion. The phase distortion will vary significantly from one connection to another. The phase and amplitude distortion of the DDD limits the type of digital signals that can be used. It also limits the capacity of the channel to a relatively low rate.

Pulse signals cannot be used because dc is not transmitted by the DDD. FSK and PSK are usually used for transmitting data on the DDD. VSAM is sometimes used.

Table 11-1 lists the properties of several modems commonly used on the DDD. AT&T (Bell Telephone) developed the original modems for the DDD. The specifications they used for these modems have become de facto standards. Independent vendors usually refer to the Bell-type number in their sales literature.

Short-haul modems are available from non-Bell suppliers that operate at higher baud with the same modulation type and frequencies. These higher-baud

Table 11-1   **Properties of common industry standard modem series that transmit digital signals over voice-grade lines**[a]

Bell series	Capacity (baud)	Channel mode	Frequencies (Hz)	Modulation	Comments
103	300	FDX/HDX	1270/2225 (Mark) 1070/2025 (Space)	FSK	DDD
201	2400	FDX/HDX	1800	4DPSK	Conditioned lines, fixed data rate
	75		390 (Mark) 450 (Space)	FSK	Secondary channel
	150		370 (Mark) 470 (Space)	FSK	Secondary channel
202	1200	FDX/HDX	1200 (Mark)	FSK	DDD
	1800		2200 (Space)		Conditioned lines
	5		387	AM	Secondary channel
208	4800	FDX/HDX	1800	8DPSK	Fixed data rate
212	1200	FDX	2400 (Answer) 1200 (Originate)	4DPSK or QAM	DDD only

[a]Capacity shown is maximum for series. QAM is four-level PSK. DPSK is differential PSK (number gives number of levels). Some higher-speed FDX modems require two lines.

Source: Racal-Vadic Specification Sheets.

modems will usually work within an exchange and sometimes across a city, but rarely between cities. They are also used for long runs on wire channels.

Several independent carriers are also operating DDD networks in competition with the phone company. Some of these are long distance only. Others are based on the old telegraph networks. The Bell-type modems will usually work on these independent networks, but the error rate is sometimes higher.

11.2.5 Dedicated Lines

Dedicated lines are available from both the phone company and several independent suppliers. While these lines do not have the switching flexibility of the DDD network, they do offer improved line quality. Some services are merely voice-grade lines with improved amplitude and/or frequency response. Others are wide-band audio lines. Some services even offer true digital communications channels. In all cases, higher capacity is guaranteed.

11.2.6 Local Networks

A *local network* is a wire network connecting several transmitters and receivers in an office or factory. A local network is usually restricted to a length of 1000 m (about 1000 yards) or less.

As this is written, the IEEE is attempting to formulate a standard for local area networks. However, two standards appear to be dominant in at least part of the market. One is the IEEE-488 standard. The other is the Ethernet, a proprietary standard of Xerox, that is supported by Intel and Digital Equipment Corporation.

The IEEE-488 standard was originally intended to be a standard for interconnection of computer-controlled instrumentation. Some of its features are listed in Table 11-2.

The Ethernet was intended for interconnection of office equipment. Some of its features are also listed in Table 11-2.

11.2.7 Fiber Optics

Thin plastic or glass cables are also used for transmitting data. Pulse waveforms are usually used. Semiconductor *LASERs* are used to transmit the data. Photo detector diodes are used for receivers.

Fiber optic technology has superior electrical-noise immunity. Fiber optic cable can handle high data transmission rates.

11.3 FREQUENCY SHIFT KEYING (FSK)

Several techniques are used to generate and detect FSK for modems. Figure 11-4 shows a block diagram of one technique. Mosts newer modems use essentially the FSK generation technique shown in Figure 11-4. A crystal-controlled oscillator ensures more frequency stability (hence higher performance) than any other oscillator circuit. Digital circuitry (even VLSI circuitry in some

Table 11-2   **Features of the IEEE-488 and Ethernet Standard**

Characteristics	IEEE-488	Ethernet
Capacity (baud)	4.0 M with three-state or 2.0 M with open collector	10 M
Cable type	Wire	50-Ohm coaxial
Cable length	Lesser of 20 meters or 2 meters/device	1500 meters/segment 1000 meters point-to-point link
Driver logic family	Shottky TTL	10,000 series ECL
Connector	24-Pin Series 57	Type N
Number of devices on bus	15	1024
Transmission mode	Asynchronous byte serial	Synchronous serial by frame
Priority	Controller	Collision detection and avoidance

Sources: The Ethernet, A Local Area Network, Data Link Layer and Physical Layer Specification, Version 1.0. Intel Corporation, Santa Clara, CA, Sept. 30, 1980; *Tutorial Description of the Hewlett-Packard Interface Bus.* Hewlett-Packard Corp., Palo Alto, CA, 1980.

of the latest modems) is used because of its low cost, stability, and low maintenance. The oscillator frequency is selected to be a multiple of both transmit frequencies. The divider network ensures that both transmit frequencies are synchronous. This improves modem performance.

The data to the modem drives the select lead of the digital data selector (or multiplexer) circuit, determining which frequency will be transmitted. The transmit filter reduces the harmonics to the bandwidth of the channel. Most filters are active filters. An integrated-circuit operational amplifier (IC op amp) with a resistor–capacitor feedback network acts as a bandpass filter.

DDD modems have fairly complex filters that convert the square wave input to almost a pure sine wave. Noncommon carrier modems sometimes have no filter or a simple low-pass filter.

The diplexer connects either the transmit or receive section to the line for a HDX channel. The diplexer blocks the transmit signal from the receive section for a FDX modem. Most diplexers are made from IC op amps. Carrier current modem diplexers also block the power line or voice frequencies from entering the transmitter and receiver. The receive filter improves the signal-to-noise ratio. It reduces the bandwidth of the signal fed to the detector to the minimum necessary. It is usually an active filter. Sometimes, it is combined with the diplexer.

The detector circuitry still varies from one manufacturer to another. One popular technique is to use active filters tuned to each frequency. The filter with the greatest output determines the logic level of the modem output. The comparator compares the filter outputs and sets the modem output logic level.

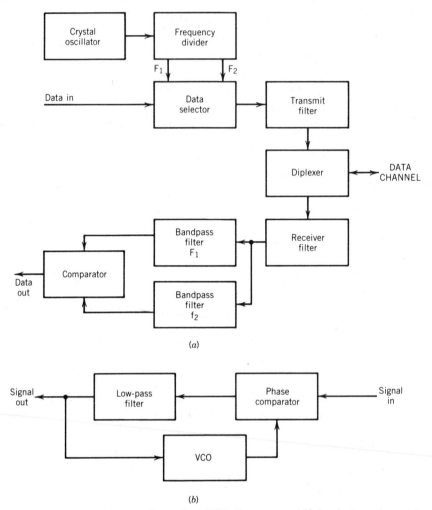

Figure 11-4   Block diagram of a typical FSK (frequency shift keying) modem. (*a*)
Block diagram of modem. (*b*) Block diagram of PLL (phase lock loop).

Some modems replace the filters with phase lock loop (PLL) circuits.
Each PLL is tuned to one of the modem frequencies. In essence, each PLL
acts as an active filter.

Other modem receivers replace the two filters with a single PLL. The
PLL is tuned to approximately halfway between the two modem frequencies.
In essence, the PLL acts as a frequency-to-voltage converter. The output of
the PLL drives a comparator that determines the modem output logic level.

11.4 PHASE SHIFT KEYING (PSK)

The block diagram of a typical PSK modem is shown in Figure 11-5. Note that
the PSK and FSK modems are very similar. There are some significant dif-

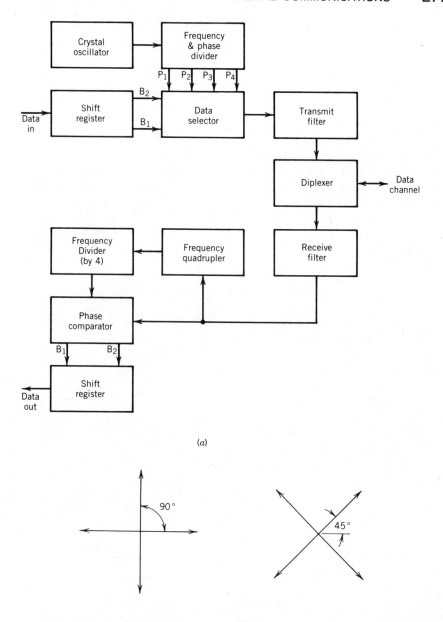

Figure 11-5 Block diagram of a typical PSK (phase shift keying) modem. (*a*) Block diagram of modem. (*b*) Phasor diagrams of PSK signal.

ferences between the FSK and PSK modems shown in Figures 11-4 and 11-5, respectively. First, the PSK modem transmits quaternary signals. Figure 11-5*b* shows a phasor diagram that illustrates the four states. A phasor diagram can be thought of as arrows (or vectors) drawn from the origin, or center, of the graph. The length of the vector is proportional to the root mean square

(RMS) or effective value of a sinusoidal waveform. The angle the vector makes with the abscissa, or horizontal axis, is equal to the phase angle of the sinusoidal wave.

Since the modem transmits quaternary signals, a shift register acts as a two-bit serial to the parallel converter. Two bits of data are shifted into the shift register. The shift register contents are then latched by a two-bit buffer that holds them for the data selector while two more bits are clocked into the shift register. A two-bit parallel to the serial converter on the output of the detector reverses the process in the receiver.

Note that the data selector, or multiplexer, has two select inputs. The divider network has four outputs. All four outputs are at the same frequency but differ in phase by a quarter-period.

A variety of PSK detection techniques are used. The frequency quadrupler and divider circuits are used to regenerate the carrier signal for modems that need it. (The output of the quadrupler circuit is in phase with the carrier,

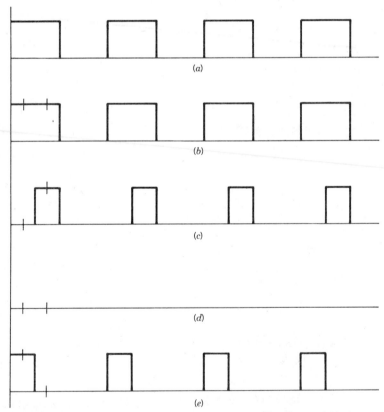

Figure 11-6   Phase detector waveforms. (*a*) Carrier. (*b*) Carrier and 0° phase signal. (*c*) Carrier and −90° phase signal. (*d*) Carrier and 180° phase signal. (*e*) Carrier and 90° phase signal.

regardless of the modulation level. The same is true for the output of a frequency doubler if two-level PSK is used.)

The phases of the carrier and input signal are then compared to determine the data output. A multiplier (or some other nonlinear) circuit and an integrator (or low-pass filter) are usually used as the phase detector.

Figure 11-6 shows typical demodulation waveforms that would result in a digital detector. The multiplier would be an AND gate. Sampling the product waveforms twice per cycle as shown in Figure 11-6 would produce a digital code corresponding to the phase. In essence, the sampling would be a digital integration.

Table 11-3 **RS-232C Connector pinout and functions**

Pin	Symbol RS-232C	Symbol CCITT	Logic	Function
1	AA	101		Protective ground
2	BA	103		Transmit data to DCE (data communication equipment)
3	BB	104		Receive data from DCE
4	CA	105	RTS	Request to send to DCE
5	CB	106	CTS	Clear to send from DCE
6	CC	107	DSR	Data set ready from DCE
7	AB	102	SG	Signal ground (common return)
8	CF	109	LSD	Receive line signal detect sometimes labeled DCD (data carrier detect) from DCE
9				Reserved for data set testing
10				Reserved for data set testing
11				Not assigned
12	SCF	122		Secondary channel receive line signal detector from DCE
13	SCB	121		Secondary channel clear to send from DCE
14	SBA	118		Secondary channel transmit data to DCE
15	DB	114		Transmitter signal element timing from DCE
16	SBB	119		Secondary channel receive data from DCE
17	DD	115		Receiver signal element timing from DCE
18				Not assigned
19	SCA	120		Secondary channel request to send to DCE
20	CD	108.2	DTR	Data terminal ready to DCE
21	CG	110		Signal quality detector from DCE
22	CE	125	RI	Ring indicator from DCE
23	CH/CI	111/112		Data signal rate selector to/from DCE
24	DA	113		Transmit signal element timing to DCE
25				Not assigned

Source: EIA Standard RS-232C, Interface Between Data Terminal Equipment and Data Communication Equipment Employing Serial Binary Data Interchange. Electronic Industries Association, Engineering Department, 2001 Eye Street N.W., Washington, DC 20006, August, 1969, 202/457-4900.

11.5 INTERCONNECT STANDARDS

11.5.1 RS-232C

This standard was originally developed for connecting data terminals to data sets. It is now used for a wide variety of digital communication interconnects.

The RS-232C standard is an EIA and ANSI standard. A similar standard in Europe is called CCITT V.24. A 25-pin subminiature D connector is specified for the connector. Table 11-3 shows the role assigned to the various pins. Table 11-4 lists some other properties of the standard.

Note that a mark is a negative voltage for an RS-232C interface. A space is a positive voltage. However, the control signals are inverted logic. Therefore, true control signals have a positive voltage, false control signals have a negative voltage. The voltage polarities cause much confusion.

A timing diagram in Figure 11-7 shows the relationship of the control signals for both FDX and HDX. A word of caution. Many manufacturers violate the standard. A very common problem is that the receive thresholds are higher than 3 V or lower than −3 V. Another common violation is inverting (using conventional instead of inverted logic) for some or all of the control signals. Some manufacturers tie protective and signal ground together or fail to use a protective ground.

11.5.2 RS-449, RS-422A, and RS-423A

The RS-449 standard is an EIA standard for mechanical connectors only. The electrical specifications for the connectors are given by either the RS-422A or RS-423A EIA standards.

Table 11-4 **Summary of RS-232C features**

All signals unbalanced
Transmit mark (logic 1) from −5 to −15 V
Transmit space (logic 0) from +5 to +15 V
Receive mark from −3 to −25 V
Receive space from +3 to +25 V
Format of data unspecified (usually sent asynchronously in ASCII)
Data rates up to 19,200 baud permitted; common rates include 75, 110, 150, 300, 600, 1200, 1800, 2400, 4800, 9600 and 19,200 baud
Cable length 50 ft maximum, longer lengths permitted
Connector 25-pin subminiature D (MIL-C-24308)
Source resistance greater than 3000 Ohms
Load impedance 3000 to 7000 Ohms resistance, less than 2500 pF capacitance

Source: EIA Standard RS-232C, Interface Between Data Terminal Equipment and Data Communication Equipment Employing Serial Binary Data Interchange. Electronic Industries Association, Engineering Department, 2001 Eye Street N.W., Washington, DC 20006, August 1969, 202/457-4900.

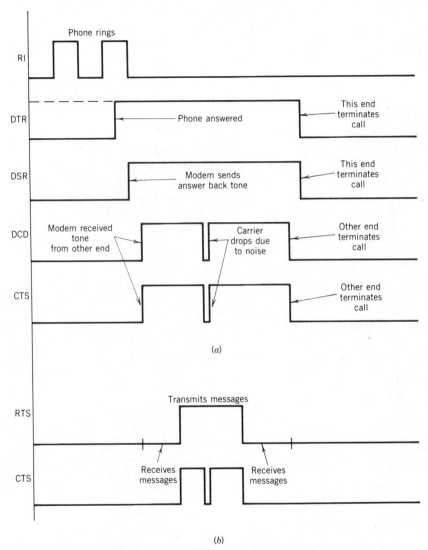

Figure 11-7 Timing diagram of the RS-232C interface control signals. (*a*) FDX channel. (*b*) HDX channel (effect on RTS and CTS only).

Two connectors are specified in the standard. One is a 36-pin subminiature D connector. The other is a nine-pin subminiature D connector. Table 11-5 shows the functions assigned to the pins of each connector. The auxiliary (nine-pin) connector is rarely needed in practice.

The RS-422A standard covers transmission of balanced signals over a longer distance than the RS-232C standard. The RS-423A standard covers the transmission of an unbalanced signal. Tables 11-6 and 11-7 show the properties of the standards.

Table 11-5 **RS-449 Connectors pinout and functions**[a]

Pin	Symbol	Interchange points	Function
37-Position Connector Shield			
1			
2	SI	A–A′	Signaling rate indicator from data communication equipment (DCE)
3			Spare
4	SD	A–A′	Send data to DCE
5	ST	A–A′	Send timing from DCE
6	RD	A–A′	Receive data from DCE
7	RS	A–A′	Request to send to DCE
8	RT	A–A′	Receive timing from DCE
9	CS	A–A′	Clear to send from DCE
10	LL	A–A′	Local loopback to DCE
11	DM	A–A′	Data mode from DCE
12	TR	A–A′	Terminal ready to DCE
13	RR	A–A′	Receiver ready from DCE
14	RL	A–A′	Remote loopback to DCE
15	IC	A–A′	Incoming call from DCE
16	SF/SR	A–A′	Select frequency/signaling rate selector to DCE
17	TT	A–A′	Terminal timing to DCE
18	TM	A–A′	Test mode from DCE
19	SG	C–C′	Signal ground
20	RC	C–B′	Receive common from DCE
21			Spare
22	SD	B/C–B′	Send data to DCE
23	ST	B/C–B′	Send timing from DCE
24	RD	B/C–B′	Receive data from DCE
25	RS	B/C–B′	Request to send to DCE
26	RT	B/C–B′	Receive timing from DCE
27	CS	B/C–B′	Clear to send from DCE
28	IS	A–A′	Terminal in service to DCE
29	DM	B/C–B′	Data mode from DCE
30	TR	B/C–B′	Terminal ready to DCE
31	RR	B/C–B′	Receiver ready from DCE
32	SS	A–A′	Select standby to DCE
33	SQ	A–A′	Signal quality from DCE
34	NS	A–A′	New signal to DCE
35	TT	B/C–B′	Terminal timing to DCE
36	SB	A–A′	Standby indicator from DCE
37	SC	C–B′	Send common to DCE
Nine-Position Connector Shield			
1			
2	SRR	A–A′	Secondary receiver ready from DCE
3	SSD	A–A′	Secondary send data to DCE
4	SRD	A–A′	Secondary receive data from DCE
5	SG	C–C′	Signal ground

Table 11-5   (Continued)

Pin	Symbol	Interchange points	Function
6	RC	C–B'	Receive common from DCE
7	SRS	A–A'	Secondary request to send to DCE
8	SCS	A–A'	Secondary clear to send from DCE
9	SC	C–B'	Send common to DCE

[a]A, Source primary output; A', differential-receiver primary input; B, balanced-source reference output; B', differential-receiver reference input; C, source common; C', receiver common.

Source: EIA Standard RS-449, General Purpose 37-Position and 9-Position Interface for Data Terminal Equipment and Data Circuit-Terminating Equipment Employing Serial Binary Data Interchange. Electronic Industries Association, Engineering Department, 2001 Eye Street N.W., Washington, DC 20006, November 1977, 202/457-4900.

11.5.3 20-mA Current Loop

The 20-mA current loop standard is a de facto standard that is based on an interconnect first used by Teletype Corporation on their mechanical teleprinters. Table 11-8 lists some properties of the standard. Note that the logic levels are based on current, not voltage. This resulted from the electrical properties of the relays used to detect the signal.

The emphasis on current is an advantage in noisy environments or where long cable runs are required and low data rates can be tolerated. Induced noise

Table 11-6   Summary of RS-422A features

Transmitter and receiver balanced
Transmit mark (logic 1) from -2 to -6 V, lead A negative with respect to lead B
Transmit space (logic 0) from $+2$ to $+6$ V, lead A positive with respect to lead B
Source resistance 100 Ohms or less
Short-circuit current less than 150 mA
Multiple loads permitted, up to 10
Termination resistance permitted
Total load (including cable) greater than 90 Ohms with termination resistance, 400 Ohms without termination resistance
Data rates up to 10 Mbaud
Cable lengths up to 1.2 km with proper wave shaping, lengths less than 10 m require no wave shaping
Receivers must accept voltages from -7 to $+7$ V
Test load 100 Ohms balanced (50 Ohms each side to ground)

Source: EIA Standard RS-422A, Electrical Characteristics of Balanced Voltage Digital Interface Circuits. Electronic Industries Association, Engineering Department, 2001 Eye Street N.W., Washington, DC 20006, December 1978, 202/457-4900.

Table 11-7   **Summary of RS-423A features**

Transmitter unbalanced, receiver balanced
Transmit mark (logic 1) from -4 to -6 V
Transmit space (logic 0) from $+4$ to $+6$ V
Source resistance 50 Ohms or less
Short-circuit current 150 mA maximum
Multiple loads permitted, up to 10
Data rates up to 100 kbaud
Cable lengths up to 1.2 km with proper wave shaping, under 10 m require no wave
  shaping
Receiver must accept -7 to $+7$ V
Nominal test load 450 Ohms

Source: EIA Standard RS-423A, Electrical Characteristics of Unbalanced Voltage Digital Interface Circuits. Electronic Industries Association, Engineering Department, 2001 Eye Street N.W., Washington, DC 20006, December 1978, 202/457-4900.

currents are usually considerably less than 20 mA. Most losses in cable are series losses. Series loss will affect received voltage levels but not received current levels.

No connector is specified for the standard. Occasionally, a manufacturer will use the primary or auxiliary channels of an RS-232C connector. This is a violation of that standard.

11.5.4 Phone

There are several FCC-approved connectors for directly connecting certified equipment (such as direct connect modems) to the phone system. These connectors are specified in Part 68 of the Federal Communications Commission (FCC) Rules and Regulations.

The most common connector has six pins. This connector is used with single- and two-line handsets. A different USOC (Universal Service Order Code) is used for each application. The meaning and use of each pin is shown in Table 11-9. Note that all wiring should conform to the NEC (National Electrical Code) Section 800.

11.6 SPECIAL TOOLS

Standard test equipment, such as an oscilloscope or electronic multimeter, can be used to troubleshoot digital communications systems. However, there are

Table 11-8   **Summary of 20-mA current loop features**

Transmit signal levels 20 mA (mark), 0 mA (space)
Receive signal levels 15 to 20 mA (mark), 0 to 3 mA (space)
Load assumed to be approximately 300 Ohms, neither lead grounded
No connector, cable length, handshake signals, data rate, or data code specified

Table 11-9 **Summary of six-position phone plug pinout and function**

Pin	Symbols	Function
1		Reserved for phone company use, not always present
2	A	For phones with hold, shorted to A1 for off-hook
	T2	Tip for second line for two-line nonkey phones
3	R	Ring
	R1	Ring for first line for two-line nonkey phones
4	T	Tip
	T1	Tip for first line for two-line nonkey phones
5	A1	For phones with hold, shorted to A for off-hook
	R2	Ring for second line for two-line nonkey phones
6		Reserved for phone company use, not always present

Source: FCC Rules and Regulations, Part 68, *Code of Federal Regulations,* Vol. 47. Superintendent of Documents, U.S. Government Printing Office, Washington, DC 20402, October 1, 1982.

several pieces of special test equipment that simplify and speed up trouble-shooting. Some of this equipment is simple enough to be easily assembled from widely available components. Most or all of this equipment is available from a number of suppliers.

11.6.1 Loopback Connector

A loopback, or feedback, connector can easily be fabricated by connecting the data-out lead to the data-in lead on an interface connector such as an RS-232C connector. The loopback technique is described below.

Some of the control signals may need to be tied together on some loopback connectors. For example, data terminal ready may need to be connected to data set ready on some RS-232C interfaces. Request to send may need to be tied to clear to send on other RS-232C interfaces.

11.6.2 Null Modem

A null modem is an RS-232C cable that has the data-out and data-in leads interchanged on one end. It permits, for example, two data terminals with RS-232C interfaces to be connected together for testing.

The most common null modem has plugs on both ends. Occasionally, null modems with jacks on both ends or a plug on one end and a jack on the other find use.

11.6.3 Breakout Box

A breakout box is connected between a cable and an interface. It exposes the data and control leads of the interface and permits test equipment probes to be conveniently connected to a data channel.

Some breakout boxes have a set of pins and plugs that permit cross connections to be made across the breakout box. Others have an internal power supply (usually a battery) that permits control leads to be connected to one logic level or the other.

11.6.4 Data Communications Testers

A number of data communications testers are available from several sources. Some models will sense pulse waveforms and display the result in ASCII or some of the less common data communications codes. The logic level inputs can be RS-232C, TTL, or 20-mA current loop. Special data testers are available for the IEEE-488 interface.

Other data communications testers will analyze the spectrum of FSK or PSK signals. These testers can be connected to a modem or a data channel. The results are displayed in a number of different formats ranging from indicator lights or LEDs to cathode ray tube displays similar to the phasor diagram shown in Figure 11-5b.

Data protocol testers are also available. These are especially useful for analyzing certain synchronous data communications systems.

11.7 COMPONENT ISOLATION STRATEGIES

Figure 11-8 shows a block diagram of a digital communications channel in its most general form. One or both of the terminals shown in the figure can be a computer. In short-haul channels, the modems will probably be absent, and the two terminals will be connected together by wire. The item identified as the communications channel in Figure 11-8 could be any of the channels discussed previously. Some newer systems have the modem combined with the data terminal.

The main strategy is to isolate the defect to a module, then get the module repaired. One technique of implementing this strategy is to trace a signal from its source to the defect. Breakout boxes can be used on connectors to expose the signal leads.

This technique does require special servicing tools, such as an oscilloscope and, in some cases, data channel analyzers. It is not always easy to identify a good or bad signal with an oscilloscope. This is particularly true with

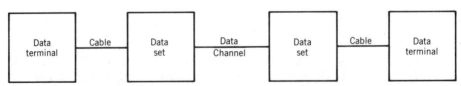

Figure 11-8   Block diagram of a general-purpose digital communications channel.

most PSK and some FSK signals. Nor is it always easy to use correctly some of the data system analyzers.

11.7.1 Loopback

A technique that requires simpler special equipment and frequently is quicker to use is called loopback. A test signal is injected at one end of the system. The signal is wired or looped back to the input so that the channel can be tested in both directions. The loopback point is moved down the system from one end to the other until the defect is isolated.

An advantage of loopback is that a terminal display is used as an indicator. The technique works best on FDX systems. However, it can be applied to many parts of most HDX systems.

Many digital communications system components have a built-in loopback test switch. For example, a terminal that has a FDX/HDX switch can be checked by placing the switch in the HDX position. A character typed onto the keyboard will be shown on the unit's display. This tests all of the terminal, except for the connector and possibly the drive circuitry for the connector.

The remainder of the terminal can be checked by plugging a loopback connector into the terminal. The FDX/HDX switch should be placed in the FDX position. If it is not, each character typed on the keyboard will be displayed twice. One display will be from the internal loopback. The other display will be from the loopback connector. A loopback connector can be used on terminals that don't have a FDX/HDX switch.

Many modems have loopback switches. Where the loopback is made varies from one manufacturer's model to another. It is best to consult the manufacturer's literature for more details. Some modems permit loopback at the connector. The data terminal driving the modem is looped back at the connector. The terminal on the other end of the channel is looped back through the modem receiver and transmit circuitry. This latter type of loopback is available only in FDX modems.

Some modems will also permit loopback at the diplexer. This permits the transmit and receive circuitry to be tested from the modem end of the channel.

Some modems even have built-in test pattern generators. This makes troubleshooting much more convenient.

11.7.2 Substitution of Components

Standard interfaces and the modularity of both data sets and terminals make module swapping easy on most digital communications channels. The only drawback may be the availability of a spare module.

A known good module, such as a terminal or connecting cable, is substituted for a suspect module. If the system still doesn't work properly, then the original module is replaced and another module is swapped.

An assumption in module swapping sometimes causes problems. The connectors or interconnection cables are assumed to be good. These items should also be checked.

11.8 INTERMITTENT ERRORS

Isolating and correcting intermittent failures in a digital communications system is probably the most difficult task there is. The component isolation strategies outlined above are probably the best systematic approach to isolating intermittent problems. Several common sources of intermittent problems are outlined below.

11.8.1 Loose Connections

All connectors should be checked to make sure they are snug. The solder or crimp joints between cables and connectors should also be checked. Joints between cables and connectors tend to fail from repeated flexing. Wires break at the joints. Solder connections sometimes fracture. A partially insulating oxide layer sometimes forms between wires and crimp connectors, particularly in humid and polluted environments.

11.8.2 Vibration

The acoustic-coupled modems commonly used over the DDD network are a common source of noise. The microphones built into these data sets are sensitive to loud sounds, such as those produced by printers, loud conversation, music, or falling objects. The microphones are also sensitive to vibration.

Mounting the modem on a foam pad will reduce interference from vibrations that cannot be eliminated. Seating the telephone hand set more securely on the modem will eliminate many sound interference problems. Surrounding the modem with foam padding will reduce many interfering sounds that cannot be eliminated.

11.9 DATA CHANNEL ERROR SOURCES

11.9.1 Common Carriers

Common carriers include the DDD and leased or dedicated lines. The only solution possible is to complain to the common carrier. If the common carrier cannot, or will not, correct the problem, then complain to the agency regulating or overseeing the common carrier.

Several points should be considered before filing a complaint:

1. Many common carriers use radio links. There is little or nothing (at least short of rebuilding most channels) that can be done about error bursts caused by adverse weather (see section on radio links below).

2. The DDD is not designed to handle data, nor do most tariffs require it to do so. Don't expect any action on your complaint unless conversation traffic is degraded also.

3. The more documentation you can generate supporting a complaint, the better. The common carrier may not be impressed with your documentation. Their regulatory body probably will.

Note the time, location, and any other pertinent information in a signed, dated log. Follow up all verbal complaints in writing. Keep copies of all correspondence about a problem. Keep reports on how it adversely affects your business, including the money lost.

Most troubleshooters find documentation tedious. However, it impresses bureaucrats and (if it gets that far) judges and juries.

11.9.2 Radio Links

Most radio links are very susceptible to error caused by heavy rain. There is usually no cure for this problem (short of increasing the power of the transmitter). The best solution is to wait for the rain to cease.

High humidity or fog will cause false propagation error bursts on many radio links. The solution, again, is to wait for the weather to improve.

11.9.3 Wire Channels

Check for breaks or nicks in the wire. Look for loose connections in splices. Look for water or heavy moisture, particularly near breaks in the sheathing or conduit.

11.10 SUGGESTED READING

Barlin, D., Overcoming the Confusion of Connecting to Bell. *Data Communications,* December 1981, 137–138, 140, 142, 144.

Bell System Technical Reference Catalog. AT&T, New Brunswick, NJ, June 1979.

Data Communications Buyer's Guide. McGraw-Hill, New York, 1982. (Published annually as a special issue of *Data Communications.*)

Data Communications Testing. Hewlett-Packard, Delcon Division, Mountain View, CA, 1980.

Data Products Technical Manual. Western Electric Company, Indianapolis, IN (Select Code 500-926).

DCS-2A Multiplexed Data Carrier System, General Description. Teletone Corporation, 10801 120th Avenue N.E., Kirkland, WA, 1982.

Digital Data Transmission with the HP Fiber Optic System. Hewlett-Packard, Palo Alto, CA, Application Note 1000, 1978.

DeLaune, J. M., *Low-Speed Modem System Design Using the MC6860.* Motorola, Phoenix, AZ. Application Note AN-747.

Doll, D. R., *Data Communications.* Wiley-Interscience, New York, 1978.

Gregg, W. D., *Analog and Digital Communication.* Wiley, New York, 1977.

Lindsey, W. C., and Simon, M. K., *Telecommunication Systems Engineering.* Prentice-Hall, Englewood Cliffs, NJ, 1973.

Martin, J., *Telecommunications and the Computer,* 2nd ed. Prentice-Hall, Englewood Cliffs, NJ, 1976.

McNamara, J. E., *Technical Aspects of Data Communication.* Digital Equipment Corporation, Maynard, MA, 1977.

Nash, G., *Low-Speed Modem Fundamentals.* Motorola, Phoenix, AZ. Application Note AN-731.

National Electrical Code, Article 800—Communications Circuits. National Fire Protection Association, Boston, MA, 1981.

Techo, R., *Data Communications, An Introduction to Concepts and Design.* Plenum, New York, 1980.

Telecommunications—An Introduction to the Network. Western Electric Company, Indianapolis, IN, 1982 (Select Code 500-933).

The Ethernet, a Local Area Network, Data Link Layer and Physical Specifications, Version 1.0. Intel Corp., Santa Clara, CA, 1980.

TMS99532 Single-Chip Bell 103 Compatible MODEM. Texas Instruments, Houston, TX. Application Note MP057, October 1982.

Tutorial Description of the Hewlett-Packard Interface Bus. Hewlett-Packard, Palo Alto, CA, 1980.

Westman, H. P., ed., *Reference Data for Radio Engineers.* Howard W. Sams, Indianapolis, IN, 1972.

CHAPTER 12

Robot Equipment

Robert B. Thorne
Texscan Corporation

12.1 INTRODUCTION

This chapter is about the electronics and peripherals that constitute the robot's nervous system, senses, and muscles, and the way that they all communicate with each other. The Heath Company HERO 1 robot will be used frequently for examples, since it was designed as a teaching tool. You do not need to be familiar with it; the information here is self-explanatory.

We begin with a section on the robot's basic electronic structure. This section will provide the framework for subsequent sections on basic diagnostics procedures, communications (I/O), and operators (things that perform operations as directed by the microprocessor).

12.2 BASIC ELECTRONIC STRUCTURE

Figure 12-1 is one representation of the Heath Company HERO 1 robot. This figure shows how the different parts of the robot are connected electrically to each other; it shows how you might isolate problems in different areas by observing the type of malfunction and tracing signals from the microprocessor to the targeted circuit.

Microprocessor. All instructions come from the microprocessor portion of the robot. The microprocessor section decides what it wants its operators to do, based on its program and what its sensors tell it. The instructions must be sent to the proper *address* (just like an address in memory) to reach the selected device. The micro sends two pieces of information—the address on

Note: Reading Appendix B, Robotic Introduction, may be useful for readers unfamiliar with robots.

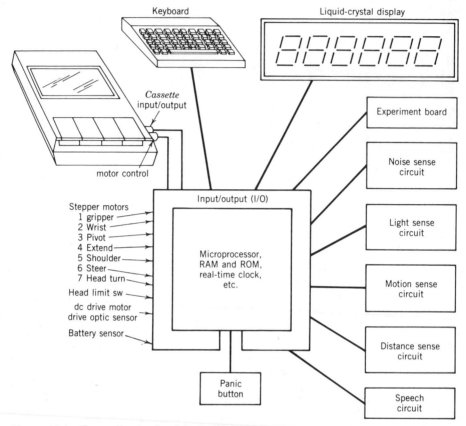

Figure 12-1 Generalized robot block diagram.

the address bus and the data on the data bus. In order to read data coming from a sensor, the microprocessor sends the address indicating the location of the data and then reads returning data from the data bus. There is more explanation below and in Section 12.4.

Address decoder. The address decoder receives the information, in binary written on the address bus, about what I/O device (address) the micro wishes to communicate with. Then, at the selected instant, the decoder sends an "enable" signal to the selected I/O device.

Input/output (I/O). All I/O chips are connected to the micro's data bus and receive the data. But all except the "enabled" device will act like open circuits. If an output (from the micro to an operator) device has been enabled, it will pass the signal from the data bus along to an operator. If an input (from an operator to the micro) device has been enabled, the device will pass the signal from the operator onto the data bus for the microprocessor to read.

Motive operators. Motive operators is used here to signify those devices that supply the robot's motion. The motive operator begins the operation when it receives the instruction from the micro. It is possible that the device driver circuit may make certain "decisions" based upon the input and perform some control actions.

Reactive operators. Reactive operators is used here to signify those operators that react to the robot's movement and to changes in its environment. Most reactive operators are sensors and feed their information back to the micro through the I/O. HERO 1 has sensors for sound, motion, distance to an obstruction, noise, and light.

Feedback. The final steps in the robot's action cycle are the transfer (feedback) of the sensor's information to the I/O, where it waits for the micro to sample it and make further decisions on what to do next.

Troubleshooting. Based on the above information, the basic approach to troubleshooting becomes apparent. You want to know: (1) if there is an output from the micro to the right address on the I/O, (2) if the I/O sends the data on to the operator, (3) if the operator receives it, (4) if the sensors provide information for the micro, (5) if the information gets to the I/O, and (6) if the I/O provides the information correctly to the microprocessor.

12.3 PRELIMINARY DIAGNOSTICS

Before you get bogged down in hardware and software, it is always good to review what is happening, what is not happening, and what the cause could be. This is *diagnostics* and could save you considerable unnecessary work by narrowing the area you have to examine. Begin diagnostics by separating the problem into one of four types: Type 1—the robot fails to move at all. Type 2—one or more particular parts of the robot do not move at all. Type 3—part(s) of the robot operate, but not correctly. Type 4—everything seems to work fine until the whole robot suddenly stops ("locks up").

Type 1. If all of the robot's motive devices fail to respond, you may logically assume that they either cannot move or have not been told to do so. If all movers fail to move, it is likely that they are not receiving power, something that you can normally discover easily and correct. The other possibility, not being told to work, is less likely but also possible. First, the microprocessor and its associate circuitry could have failed, or the program could have been damaged. Consequently, no worthwhile information would be reaching the proper addresses in the I/O. If this is true, you must refer to Chapter 13. Second, if the entire I/O circuit is isolated, it may not be operating correctly. Check for proper input from the microprocessor, proper supply voltage (clean, without noise), and proper grounding.

Type 2. When a single portion of the robot (or just a few portions) fail to operate, assume that it's an isolated problem. If two or more portions fail simultaneously, look first at anything they have in common: power source, wires (or pneumatic/hydraulic lines) that run in the same space, and ultimately the I/O. For single failures, check that particular operator for power, grounding, and mechanical problems. Check also for the signal from the I/O; check that portion of the I/O.

Type 3. If an operator moves when instructed to do so but fails to perform as it should, it may have either of two problems. If the operator starts more or less as it should but fails to stop, often the movement sensor's report is not being recognized by the microprocessor. This could be caused by the sensor, the continuity back to the I/O, the I/O, or even the microprocessor.

If an operator moves randomly, the problem is different. Expect it to be getting incorrect instructions. The I/O could be sending it too many signals (some being meant for other operators), the microprocessor and its support circuitry could be sending incorrect addresses to the I/O, or power and grounding problems could exist. One more likely problem is noise. Either external or internal noise can cause the low-level logic circuitry to generate erroneous data bits, changing instructions in mid-action. Look for proper filtering of the power supply, grounding throughout (especially any coaxial that could be acting as an antenna), and any other way that spurious signals could be getting into the robot's circuitry.

Two other items should be mentioned. If the robot moves erratically as soon as it is turned on, look for flip-flops in the area of the I/O that are in the wrong state at power-up. Some flip-flops have resistors or capacitors to the power supply that are to ensure that the flip-flop is initially in the right state. An open in that circuit can change the condition of the flip-flop and cause power to be sent directly to a mover. Similarly, other flip-flops are designed to initialize correctly when right power is applied. If an expensive IC is replaced with a less expensive equivalent, the new chip might not have the ability to start in the right initial state. Second, if supply power to a microprocessor falls below the required input, some microprocessors tend to operate erratically, sending out erroneous commands. Check the voltage at the microprocessor.

Type 4. Sometimes, a robot will freeze in a position (lock up) after starting normally. The problem may be in the microprocessor or the feedback signal associated with the movement. Often, commands in a robot's program are similar to HERO 1's "move, wait" commands. The command tells the microprocessor to run some device and to perform no other action until the first is complete. If the first action is not completed (or the microprocessor doesn't think it is completed), then the microprocessor will not send the next command.

With these basic considerations in mind, you can now select from the following sections the areas of information you need for any given robot service problem.

12.4 COMMUNICATIONS

If a robot microprocessor is to be useful, it must be able to receive information from a wide number of sensors and send instructions to many operators. As Figure 12-1 shows, the HERO 1 microprocessor receives inputs from its keyboard, motion sensor, sound sensor, light sensor, tape program, programming pendent (control handle), speech option, experimental circuit board, and drive wheel sensor. The microprocessor sends instructions back to all these sensors (to turn them on or off) and to the keyboard's display, the speech option, and eight motors.

Managing all this information at one time could be a tremendous problem were it not for two things. First is the microprocessor's speed, which allows it to multiplex (switch between the many tasks) very rapidly and so appear to be doing all of them simultaneously. Second, the I/O circuits (under instructions of the micro) switch the input and output data through a single data bus as shown in Figure 12-2. This simplifies the physical problems of connecting to all the operators and makes the microprocessor control more practical.

12.4.1 Output

The ICs between the microprocessor circuitry and the operators are the I/O circuitry. When the microprocessor wishes to send instructions (data), it puts the binary data on its eight data lines (the data bus) and signals the address decoder (on the address bus) where the data should go. The address decoder then sends an enable signal to the selected I/O device.

When an output chip is enabled from the address–microprocessor circuitry, it takes the digital data from the data bus lines and locks it on its output. When it is disabled, the output circuit acts like a high resistance to ground, not generating either high or low states.

12.4.2 Input

When an input circuit receives an "enable," it takes the digital highs and lows from the operator and locks them into its output to the data bus. After the micro reads the data, the I/O circuit is disabled and changes state to the high-resistance state, almost like an open circuit, so that no information passes.

12.4.3 General Review

The input and output devices present a high impedance to ground when they are not selected; this allows the rest of the circuitry to operate without being loaded down (on either the input or output side) by the chips. Note that the timing between these operations is critical. The data must be accepted and passed onto the data bus before the microprocessor begins to read it, and the data must be removed from that device before the next read or write operation. With microprocessor clocks running at rates of 1 MHz and higher, it is possible for stray capacitance to throw off the timing between the enable pulse and the arrival of output data. Check the requirements of the chips in the I/O against the actual timing at the chip. Note that the input and output ports are addressed just as the memory is, and are written into and read from in the same manner.

To check operation at the I/O, first check to see that the enable line at the appropriate I/O chip toggles when the microprocessor is writing to, or reading from, that chip. An output chip should show a high impedance to ground when not selected, and should provide an output voltage that is determined by

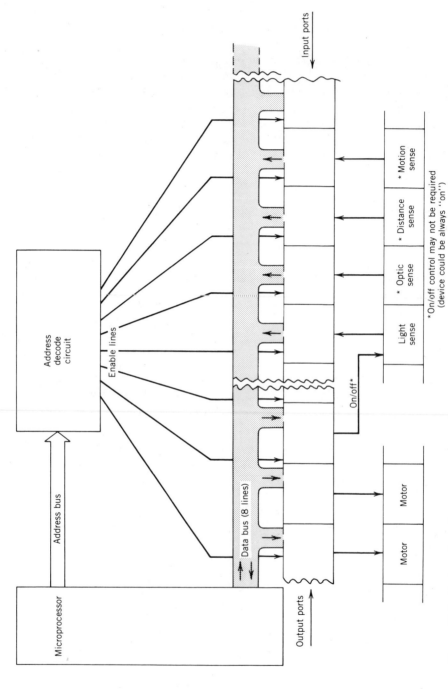

Figure 12-2 Robots input/output block diagram.

the corresponding input line when selected. Some chips invert their output; others do not.

If the signal is getting to the correct I/O chip and is being properly outputed on the other side of the chip, then the problem must be farther along the lines, or at the operator.

12.5 MOTIVE OPERATORS

In this chapter, an operator is considered to be anything that is instructed by the microprocessor; and *motive operators* are operators that cause motion such as electric motors, solenoids, and pneumatic or hydraulic actuators. *Reactive operators* such as sensors or other special devices (such as HERO 1's speech synthesis) are discussed in Section 12.6. The instruction to an operator might be a simple direction to turn on a motor or sensor. Often, the instruction is more complex. For instance, the drive motor for HERO 1's front wheel needs to be told whether to run forward or backward and which of three speeds to use. This section of the chapter is about the different devices that the robot's microprocessor may be directing and how they work.

12.5.1 Motive Operators With Sensing

Usually, the microprocessor needs information about the status of the robot, and about its environment, to know what additional instructions to issue. Typical information for HERO 1's microprocessor includes how far it has rolled, if there is an obstruction in its path, and how bright the light is. The microprocessor gets this information from the sensors. So, if a motor or other operator acts erratically, the most likely place to start troubleshooting is the sensor (or other feedback) loop that affects that device. (Sometimes, you need to check the program to see which sensors affect a particular motive operator; it isn't always the most obvious sensor.)

12.5.2 Motive Operators Without Sensing

There may be some devices that do not have any type of feedback to the microprocessor. The HERO 1 makes extensive use of stepper motors. The direction and speed of these motors is determined by the pattern of pulses (data) that the microprocessor sends to them. If the microprocessor sends 100 sets of pulses, it must assume that the motor has moved 100 "steps." A problem occurs if some force stops the motor from positioning the robot's arm correctly. Then, the microprocessor would continually assume that the arm was somewhere that it was not.

12.5.3 Zeroing

The solution for the last problem mentioned above is to *zero* the position of each stepper-driven robot appendage (arm, leg, or other). This should be done whenever the robot is turned on and whenever movement seems inaccurate.

Zeroing means driving the motor in one direction until a limit switch tells the microprocessor that the appendage has arrived at that end of its travel and the microprocessor knows for certain where it is. A second method of zeroing is possible without the end point limit switches if it will not damage the robot. With the second method, the motor is run for such a long time in one direction that the microprocessor assumes that the appendage *must* have reached a mechanical stop at that end of travel (just as it assumed that it knew where the appendage was before). This method can only be used with low-power robots that allow for its use in their programs. Any time you do not find a movement feedback sensor, such as with stepper motors, it is likely that this type of motor may need to be zeroed for its most accurate operation.

Motor positioning without sensors is relatively inaccurate and should not be used for exacting tasks or under loads that are too great. If you encounter a job where accuracy is a problem and this type of positioning is used, either the requirements are too great for the system or the motors have more load than they can position accurately.

The following sections deal with many of the typical types of movers found in a robot.

12.5.4 Electrical Motive Operators

A fairly easy check to begin with is the motive operator itself. If the drive voltage reaches the motor and matches the requirements shown on its nameplate but the motor does not respond, then probably the robot circuitry is good and the motor has an open circuit. If the voltage at the motor is low and the current is high, expect that either the motor cannot move (is blocked physically), or there is a short in the motor.

One other possibility is in the power switching. A fundamental property of electrical motive operators is that most applications need a fair amount of energy. In virtually any operation, the energy required is more than the microprocessor and its I/O device can supply. This means a secondary power circuit is required for switching the currents and voltages that the operator needs. Locate this circuit and verify that the microprocessor's signals are reaching it and that the SCRs, triacs, or other switching devices are operating correctly. Check also for sufficient drive to allow proper switching in solid-state devices.

12.5.5 Pneumatic and Hydraulic Movers

For the purposes of troubleshooting, we begin by considering these two systems together. Figure 12-3 shows, in generalized terms, how either a pneumatic or a hydraulic device provides the motive force needed to extend the robot's arm or close its gripper. The motor provides the medium (air, hydraulic oil, or other suitable substance) under pressure to push on the piston. The total force at the piston is the pressure, in pounds per square inch, times the number of square inches on the piston face ($F = psi \times area$). Figure 12-3 shows 1 psi pushing on a 15-in.^{2} piston to produce 15 lb of force.

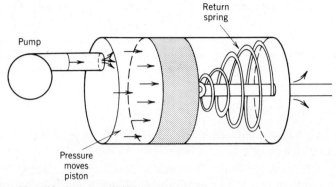

Figure 12-3   Simplified hydraulic/pneumatic actuator.

To accomplish anything, you must be able to control the force, so a valve is required as shown in Figure 12-3. A piston that only moves in one direction won't be much good for a robot arm, so some return force is needed. This could be supplied by a spring or by moving the pressure medium by a valve so that it presses on the other side of the piston. Finally, you must vent (remove) the pressure from the other side of the piston, or it will impede movement. With these basics in mind, let us examine the pneumatic and hydraulic systems.

Hydraulic Systems   The medium in a hydraulic system, Figure 12-4, is an incompressible liquid. This provides the feature of constant pressure for the piston without bouncing or oscillating. The advantages are: very exact positioning (if the sensing and control equipment is good) and the ability to handle

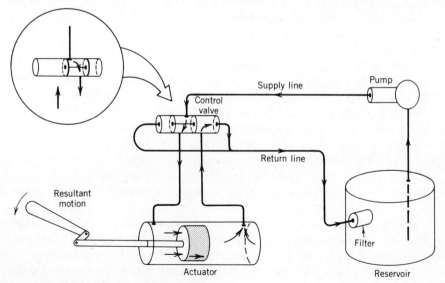

Figure 12-4   Simplified hydraulic system.

large loads without "sagging" (because the liquid is incompressible and doesn't give).

Most hydraulic systems provide the medium to either side of the piston, as required to move it, usually through a single four-way control valve. Hydraulic systems reuse the medium, so there must be lines that return the medium to its reservoir.

When troubleshooting such a system, look for the following:

1. THE ELECTRICAL–HYDRAULIC INTERFACE. Do the electrical control voltages reach the valves and do they appear to operate correctly?

2. THE MEDIUM. A hydraulic system is a closed system. Since the medium is reused, it may accumulate foreign particles which could abrade valves and other parts and degrade performance. Note also that air may get into a hydraulic system. This can cause a loss of motive power and a springy, soft operation, since air is compressible. This, in turn, can result in less accurate positioning than is normally obtained. Air bubbles may also be identified by the noise they cause moving through an otherwise quiet system. If there is air, bleed the system to remove it.

3. WATER IN THE SYSTEM. This problem will not normally make itself known until later, when it damages system components (which were designed for operation in hydraulic oil, not water).

4. THE SUPPLY SYSTEM. The reservoir and pump must supply all the oil needed at the different actuators (especially when more than one part of the system is moving simultaneously), and the pump must operate properly both electrically and mechanically.

5. LEAKS AT ACTUATORS. This may identify where an overpressure occurred (perhaps due to a problem in a valve). Leaks may also mean that air or moisture has entered the system.

Pneumatic Systems The main difference between a hydraulic system and a pneumatic system is the medium. Air is compressible and expendable. Since it is compressible, an actuator will be somewhat less exact in its positioning, even with a good sensing and control system. (One way pneumatic systems overcome this is by working at high air pressures, so that other resistive forces are small in comparison.) Since air is expendable, no return lines or reservoirs are needed and a buildup of foreign materials is less likely. However, it is very important to keep the source air clean and dry; when compressed, the air can lose its moisture into the system, causing damage.

As in the case of hydraulic systems, check the valves for proper electrical characteristics and mechanical operation.

Pneumatic systems normally have pressure regulators between the compressor and the actuators, because compressors tend to work at fairly high pressures and actuator systems work at comparatively lower pressures. Use pressure readings to verify that the regulators are maintaining the prescribed pressures.

12.6 REACTIVE OPERATORS

A robot with a level of sophistication has the ability to react to itself and to its environment.

12.6.1 Sensors

Reacting to itself means that a robot monitors its own motions, the results of its motive devices, its battery voltage levels, internal temperatures, whether some operation (such as synthesized speech) is done, or any variable that might be important to a specific application. You will find most sensors attached to anything that moves (remember the exception of stepper motors, which the microprocessor runs on faith).

Limit switches are the simplest devices used to monitor movement. They indicate when a device has reached the end of a movement. They are useful to prevent a robot from damaging itself by overextending a movement. Limit switches are normally open/closed switches; troubleshoot them with an ohmmeter.

An extension of the limit switch is a series of switches, or a single switch that operates once during each rotation of the drive motor, to provide position information or count rotations, respectively.

Another sensor is a variable resistor. Figure 12-5 illustrates its use as a position sensor for a rotary joint. If a linear, one-turn, 10-k Ω potentiometer were attached to the neck of HERO 1, it would have 0 Ω resistance at one end of the head turn and 10,000 Ω at the other end. Each degree the head turned

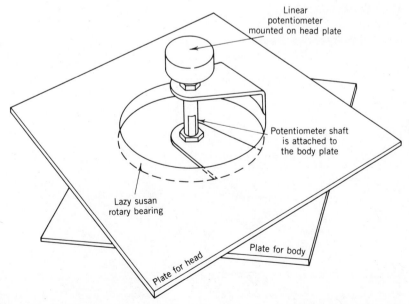

Figure 12-5   Application of a one-turn potentiometer to indicate rotary movement.

would cause about $10,000/360 = 28\ \Omega$ of change. This sensor is an analog device. If the sensor checks out alright, look for problems in the analog-to-digital converter that is necessary to interface with the digital electronics of the microprocessor.

An optical device is a more complex sensor but provides the digital output (with pulse shaping from a Schmidt trigger) that the microprocessor needs. HERO 1's front (drive) wheel uses this type of optical sensor. Figure 12-6 shows a disk mounted on the wheel and the optical sensor mounted above the axle so that it turns left and right with the wheel. The sensor uses its own source of infrared light that reflects off the light and dark parts of the disk to produce high and low resistance as the wheel rolls.

Two precautions are usually taken to avoid having other light cause false readings from the photodetector in the device. The first is to place the optical device as close to the disk as possible and to shroud it so that other light is excluded. The second is to select a wavelength of light for the source and sensor that will not be commonly encountered. After checking the isolation from extraneous light, also check the circuitry that is there to shape the waveforms produced by the sensor.

As usual, begin troubleshooting by the easiest method. Check to see if the photoresistor changes resistance as the wheel is driven (make sure the sensor's light source is turned on). If the photoresistor is not responding, or is responding weakly, the light source may be bad; check for power to the light, then replace. If the light source is good, perhaps the distance to the disk is wrong. These devices are designed to focus and operate at specific distances from the disk (often 0.1 or 0.2 in.). The manual provided with the robot should specify the proper distance. Of course, a dirty disk or dirt on the light source could be the problem.

12.6.2 Other Reactive Devices

There are sensors that are concerned with the robot's external environment. HERO 1's external senses include ultrasonic motion detection and distance

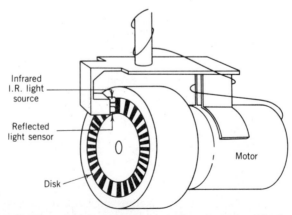

Figure 12-6 Application of optical sensing to indicate rotation of wheel.

measurement to an object. There are also detectors for the level of noise and light. Here are things to consider in these type of devices.

For ultrasonic distance-measuring devices, locate the transmitter and receiver. You can usually tell if the transmitter is operating by listening very closely for a sizzling sound. Transmitter and receiver are usually supplied as matched pairs, because they must operate at the same frequency (around 35,000 to 50,000 Hz). Changing only one of the two devices probably would cause the system to fail to operate unless it is tuned when it is installed.

Distance-measuring devices operate by sending out a burst of ultrasonic waves, then counting until the echo of the burst reaches the receiver. The count between the transmitted and received burst is proportional to the distance. Normally, the receiver must be off during the transmission or it will react to leakage from the transmitter. Some noises can fool an ultrasonic receiver: jingling keys, breaking glass, or the whine of a compressor on a refrigerator. Acoustically absorbent material, or very smooth material that bounces the sound off in another direction, may mean that the receiver won't receive an echo from the transmitted burst. Test this with a board set up to bounce the sound squarely back to the receiver. Check that the circuit's counter is operating correctly to measure the time between transmission and echo reception.

Ultrasonic movement detection operates with continuous transmission or a transmitted series of pulses. The phase angle at the receiver is compared to that at the transmitter. If the relative phase difference changes, something has altered the reflective characteristics of the room.

Light and noise detectors are devices that accept an input via a photocell or microphone, respectively. Measure the strength of that input. HERO 1 breaks that input voltage down into a seven-bit digital representation and can be programmed by the microprocessor to trigger at any relative level from 1 to 256. Verify that the sensor is responding to the input, then check the comparator and the A/D converter. *Note:* either type of device can be made to respond to only one small band of frequencies in its spectrum; remember this when testing the device.

Finally, there is a group of sophisticated special-purpose devices such as TV cameras that recognize shapes and voice synthesis circuits that provide a robot with speech. A detailed discussion of these could fill (and has filled) a book. Still, the initial steps of troubleshooting are the same for sophisticated systems as they are for simpler systems. Be certain that they have power. Check to see that they receive their digital instructions from the microprocessor. Check to see if they are receiving and reacting to the external inputs they are designed for. Finally, verify that their data are received by the microprocessor.

CHAPTER   13

Troubleshooting a Robotic Control System

John Dynes
Educational Media Designer
Heathkit/Zenith Educational Systems

13.1 INTRODUCTION

Robotic control systems are rather unique in that they can contain almost every type of circuitry. Most of the circuitry can be classified into either digital or analog categories, but there are many instances where it is a combination of both.

There are three techniques that apply very well to troubleshooting the control systems of robots. Each of these is discussed separately as the type of control systems on which they are used is explained. The material will be written around detailed functional block diagrams. This allows you to use the techniques on both digital and analog circuits.

Although the discussions mention several areas where a fault may occur, the major thrust of this chapter is to explain the techniques of troubleshooting the circuitry. A major part of the discussion will be based on the electronics of the HERO 1 arm and SONAR system and how they interface with the robot's central processing unit (CPU). Figure 13-1 shows a photograph of the Heath HERO 1 robot. Descriptive names have been assigned to all functional blocks and signals. This was done to readily identify the item in question.

This chapter has several very specific objectives designed to make your troubleshooting task easier and quicker. When you have finished studying this chapter, you will be able to define functional blocks and identify them on a schematic diagram. You will be able to use the *Divide and Conquer, Breaking the Loop,* and *Dividing the Block Diagram* methods of troubleshooting. You

Figure 13-1   HERO 1 Robot. (Reprinted with the permission of Heath Company.)

will understand how the microprocessor of the HERO 1 robot controls its arm and SONAR system. And, most importantly, you will understand the importance of having a good working knowledge of the system you are troubleshooting.

13.2 IDENTIFYING FAILURES

The first step in troubleshooting is to recognize a failure when it occurs. It is important to know the operational characteristics of the system to be able to identify the nature of the failure.

The more you know about a system, the easier it will be to identify and locate its various functional blocks and to recognize when their inputs and outputs are not correct. A thorough understanding of the system and its components is crucial to good troubleshooting.

The best sources for obtaining the necessary detailed information on a system are the operation instructions, the theory of operation, the circuit descriptions, and the schematic diagrams. You can become familiar enough with a system to start troubleshooting it after a short study of these items. You will continue to refer to them as your troubleshooting proceeds.

13.2.1 The Program

A robot is nothing more than a computer with several very specialized peripherals. A computer, of course, needs a program to run. If the programmer does not carefully consider the future uses of the control program, it may not contain all the necessary capabilities to direct the robot to complete a task properly under changing conditions. If the programmer does include features intended for future use but the program is not thoroughly tested, there may be mistakes which show up only when these functions are exercised.

Errors in the control program of a working robotic system are not very common. More often, an error will be found in an application program written by the user.

13.2.2 The Environment

The environment can play a large role in the operating characteristics of any equipment, particularly robots. Also, you must be aware that not all adverse conditions are readily apparent to human senses. For example, changes in temperature or humidity can change the values of components. Electromagnetic interference can induce erroneous signals in wires and circuit boards if shielding is inadequate or not properly in place. Ultrasonic noise from sources such as motors or air ducts can confuse SONAR systems.

Environmental conditions can vary greatly from place to place. For example, a robot may malfunction in a seemingly benign environment and yet, when it is taken to a repair or testing facility, it may function perfectly. If it is practical, you should attempt to troubleshoot the system under conditions similar to those in which it failed.

13.2.3 The Machine

When robots are subjected to extreme environmental conditions, the failure of mechanical parts is accelerated. Many robots are equipped to detect some of these failures and report them or take actions to prevent further damage. Where they are not, a mechanical failure can, on occasion, be difficult to detect. Again, there is a need to thoroughly understand the operation and even the construction of the robot.

Most failures in robots, however, are electrical in nature. Parts exposed to the environment, such as switches and wires, and very fragile parts that are subjected to stress or vibration, such as sensors or transducers, are most susceptible to failure.

13.3 LINEAR CONTROL PATHS AND SIMPLE CONTROL LOOPS

To understand the following discussion, you must be familiar with the HERO 1 arm and SONAR system and their supporting circuitry. The best description for our purpose is a functional block diagram. However, first, it is necessary to show how the circuitry is divided into these blocks.

13.3.1 Identifying Functional Blocks

To understand a circuit's function and simplify the troubleshooting process, it is usually necessary to group circuits logically into functional blocks. These blocks are made up of circuits or a group of circuits that perform a specific function.

Figure 13-2 shows the *SONAR receive* circuit board schematic diagram. On it, the receiver's amplifier functional block is outlined. This circuit was chosen to show that a group of components with several signal paths can be considered one functional block. The selection of the boundaries of a functional block is somewhat arbitrary. The remaining circuitry on the board could just as easily have been included and the block renamed *receiver,* but this defeats the purpose of separating the circuitry by making the function of the block unnecessarily complex.

The *amplifier* block clearly has an input and an output. The signal enters through a coaxial cable at point A and exits at TP-1. Inside the block itself, there is a two-stage amplifier. Each stage has a negative feedback loop and frequency compensation circuitry. While this information is important if you have to troubleshoot this amplifier, it is simply unnecessary information if you don't. Grouping these amplifiers as a single functional block eliminates the need of considering most of this circuitry if a problem is elsewhere.

Another thing that can determine the boundaries of a functional block is the way the schematic is drawn. Since the signals between circuit boards are usually easy to access and are generally signals of interest, it makes sense to use these attributes in defining the contents of a block.

The input and output signals of functional blocks may be analog or digital in form. Each has a specific purpose, and it is important to know what form they take, what they do, and how to measure them.

In Figure 13-3, for example, the digital pulses fed to the *SONAR transmit transducer* are converted into analog sound waves. These waves travel to an object and return as an echo to the *receive transducer*. This converts them into analog electrical signals. These signals are amplified and finally converted to a digital pulse by the *hit detector* (discussed later). In studying the figure, you note that the same signal, the transmit pulse, can take many forms. It is important to know these forms at the output of the individual functional blocks.

The input or output of a simple functional block may be made up of more than one electrical signal as long as they are closely related. Figure 13-4 shows

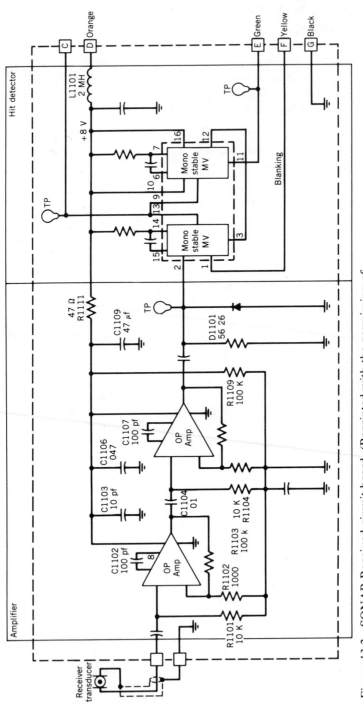

Figure 13-2 SONAR Received circuit board. (Reprinted with the permission of Heath Company.)

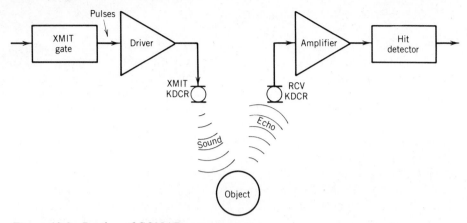

Figure 13-3   Portion of SONAR system.

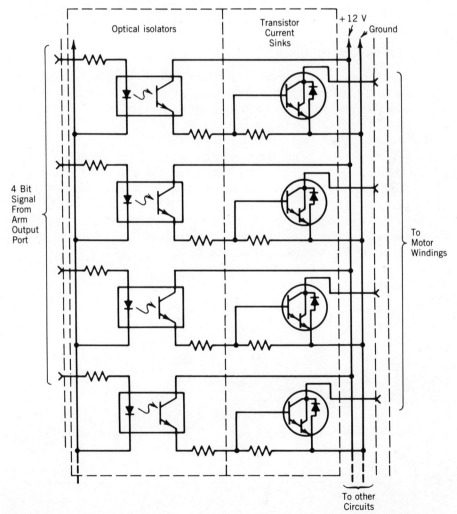

Figure 13-4   Portion of arm drive circuit board. (Reprinted with permission of Heath Company.)

a portion of the arm schematic. As you can see, a four-bit pattern determines which of the motor windings will be energized. Inasmuch as these bits are related, they may be considered part of the same signal. The signals of other circuitry on the board are excluded because they are used to operate other motors.

Every functional block has at least one input. For example, the free-running *reference oscillator* shown in Figure 13-5 appears as a functional block with no obvious input, but it is controlled by switching its power supply on and off. In the case of the HERO 1 *SONAR system*, one of the bits of the power port enables the power supply for the SONAR transmit and receive circuit boards. When the power is on, the system is free running. The microprocessor controls the SONAR system power through this digital signal.

Every functional block also has at least one output. As you can see in the SONAR system, an output is produced by the *transmit transducer*: sound. In the arm, the motors produce the outputs: motion.

13.3.2 Linear Control Paths

A linear control path is a series of functional blocks each of which has only one input and one output. They are connected in series so that the input of a block is determined solely by the output of the previous block.

The HERO 1 arm uses a linear path to control its motors. The control path for the gripper motor is shown in Figure 13-6. The control signal path starts with the system timer. When power is applied to the robot, the *real time clock* is initialized to generate a reference frequency of 1024 Hz. This signal is fed to the interrupt circuitry. This generates the clock $\overline{\text{IRQ}}$ (interrupt request) signal and sends it to the microprocessor, which then writes a four-bit pattern to the arm output port. A set of optical isolators transmits the signal to the transistor current sinks, which draw current through the motor windings. This produces motion in the motor output shaft.

13.3.3 Divide and Conquer

Perhaps the most often used systematic technique for troubleshooting is the *divide and conquer* method. It is used on linear control paths like the HERO

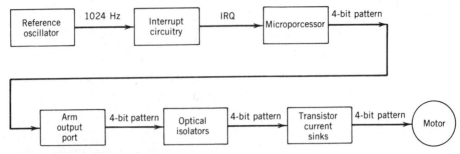

Figure 13-5   Gripper block diagram.

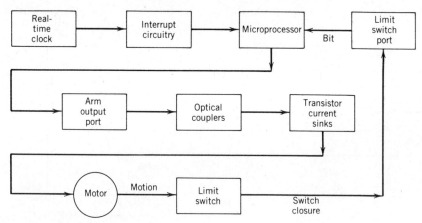

Figure 13-6   Shoulder block diagram.

1 arm. The principle is simple. First, locate one block on the diagram where the output signal is known to be correct and another where the signal is known to be incorrect. Observe the output of a block about halfway between these blocks. If the output of this block is correct, the fault must be between this and the last block. If the output is not correct, then the fault must be between this and the first block.

The process is repeated until the fault has been isolated to the malfunctioning block or component. Each time it is applied, the number of blocks left to examine is reduced by half.

EXAMPLE 13.1   Assume that power is applied to the HERO 1 arm, that the gripper is open, and that a valid command has been issued to make it close. You find, however, that it does not.

You know that power is applied to the real time clock and that the motor output shaft is not turning. On the diagram in Figure 13-6, the point halfway between the real time clock and the motor is the arm output port. Examine the outputs of the arm port to see if the correct bit pattern is present.

If the bit pattern is correct, then the fault must lie farther down the control path toward the motor. Conversely, if the pattern is not correct, the fault is farther up the path toward the real time clock. In either case, the number of blocks left to investigate is reduced by half. With successive applications of this technique, it is easy to locate the faulty block quickly.

13.3.4 Simple Control Loops

In a control loop, the output of one block is fed back to a previous block, modifying its output. In Figure 13-7, a feedback path has been added to the arm block diagram. It consists of a *limit switch* and an input port. This is a representation of the shoulder motor circuitry.

Sufficient movement by the shoulder motor will drive the arm to its limit of travel. When that limit is reached, the limit switch closes. This produces a

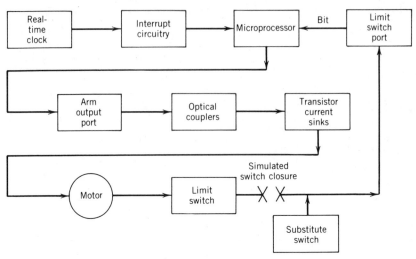

Figure 13-7   Shoulder block diagram with "broken" loop.

digital signal that is fed to the *limit port*. The function of the microprocessor block is changed to include interrogation of the limit port before writing the next bit pattern to the arm output port. If the bit pattern read from the limit port indicates that the limit switch is not closed, the microprocessor will continue; but if the switch is closed, it will not.

13.3.5 Breaking the Loop

To troubleshoot a system with a control loop, it is often necessary to *break the loop*. With this method of troubleshooting, the control loop is interrupted and the signal is simulated. This separates the functional blocks into linear control paths. Once they are separated, the *divide-and-conquer method* can be applied.

EXAMPLE 13.2   To determine whether or not the microprocessor is producing the correct output, it is necessary to control its inputs. This is done by breaking the control loop as shown in Figure 13-7. Here, the output from the limit switch is disconnected from the limit port and another switch is substituted. In this manner, it is easy to simulate the opening and closing of the limit switch and thereby verify the output of the microprocessor.

In this example, the output of the limit switch was chosen because it appears to be the easiest to simulate. Manually moving the motor, or substituting a bit pattern to drive the optical couplers, works just as well but is not as easy to implement.

13.4 COMPLEX CONTROL LOOPS

Until now, the discussion has been limited to only the simplest control systems. The majority of systems however are considerably more complex. Instead of linear paths or single feedback loops, these systems have multiple control loops, parallel paths, modified signals, and various other features that complicate the troubleshooting process. Furthermore, with the more complicated control paths, the previously mentioned troubleshooting methods are no longer directly applicable.

13.4.1 Identifying Complex Functional Blocks

Functional blocks are often much more complicated than those previously mentioned. A good example is the microprocessor block. Its functions in the SONAR system include enabling the power, responding to the SONAR interrupt, reading the range, and acknowledging the fact that it has responded. As shown in Figure 13-8, this block has several inputs and outputs.

Another example is the *range gate,* which must control the flow of pulses from the reference oscillator to the range counter and reset the counter at the appropriate time. The gate opens on the leading edge of the transmit gate signal, closes when it receives the hit signal, and resets the range counter when it receives either the next transmit gate signal or the interrupt service acknowledge.

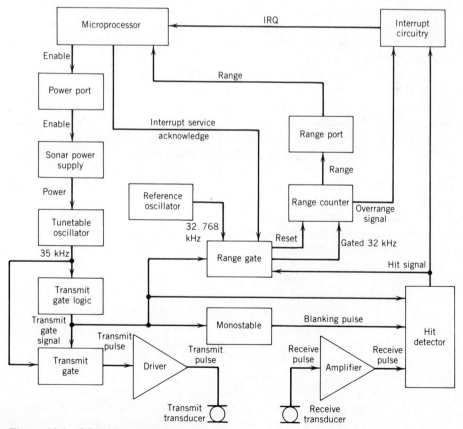

Figure 13-8 SONAR system block diagram.

These blocks are made up of several integrated circuits and discrete components. It is possible however to have a very complex functional block made up of a single IC. For example, the hit detector is a single IC containing two independent monostable multivibrators.

13.4.2 Isolating the Fault to a Specific Area

To locate a fault in a complex system, it is necessary to isolate it to a specific area of the block diagram. This can be accomplished by separating the various portions of the diagram into simpler control paths or loops.

Once the fault has been isolated to one of these paths, it is a simple matter to employ simple deduction, the divide-and-conquer or break-the-loop methods to isolate it further to a functional block and then to the offending component. Demonstration of this technique requires some familiarity with the HERO 1 SONAR system. Refer to Figure 14.7 for the following description. The batteries of HERO 1 are tapped to provide an uninterrupted power source for the real time clock. This clock runs a 32.768-kHz crystal reference oscillator. This very stable signal is used to determine the time between the transmit pulse and its echo.

When instructed by the program, the microprocessor writes a bit pattern to the power port, which enables the SONAR system power. This provides power to an oscillator (as well as other circuits) that runs at approximately 35 kHz. This oscillator is tunable so that its frequency may be matched with the resonant frequency of the transducers.

The transmit gate logic block is essentially a 14-bit divider and an RS-type latch. Internally, these circuits generate two signals. One signal defines the leading edge of the transmit gate signal, and the other defines its trailing edge. The time between these signals is the duration of the transmit gate signal.

The transmit gate signal is very important to the operation of the SONAR system. First, it momentarily opens the transmit gate, allowing several pulses from the tunable oscillator to pass. Also, its leading edge opens the range gate, enables the hit detector in the receiver, and triggers the monostable, which produces a blanking pulse for the hit detector.

The pulses from the transmit gate are amplified and applied to the transmit transducer. The resulting sound, the transmit pulse, travels until it reflects off an object. The echo of this sound is converted to an electrical pulse by the receive transducer and is amplified.

The hit detector is enabled by the transmit gate signal. A blanking pulse is provided by the monostable to inhibit the direct reception of the transmit pulse. The hit detector produces the hit signal only after it is enabled by the transmit gate signal, the blanking pulse has ended, and an echo is received.

As mentioned earlier, the range gate passes pulses from the reference oscillator to the range counter from the leading edge of the transmit gate signal until it receives either the hit signal or the interrupt service acknowledge. It also resets the range counter. The range counter is a 12-bit counter in a single IC. Its most significant bit is used as the overrange signal. The next eight bits are connected to the range port, which transfers the range information to the microprocessor. The remaining three bits are used as a divide-by-eight circuit to reduce the frequency of the reference oscillator pulses to 4096 Hz.

The interrupt circuitry generates the $\overline{\text{IRQ}}$ signal for the microprocessor. A SONAR $\overline{\text{IRQ}}$ is generated when either the hit signal or the overrange signal is received. When the microprocessor receives the $\overline{\text{IRQ}}$ signal, it reads the value at the range port. It then stores this value in a reserved memory location for use by the application program.

13.4.3 Dividing the Block Diagram

Looking at Figure 13-9, the block diagram of the SONAR system, you can see that the functional blocks might be separated into four groups: a transmit group, a receive group, a counting group, and a control group.

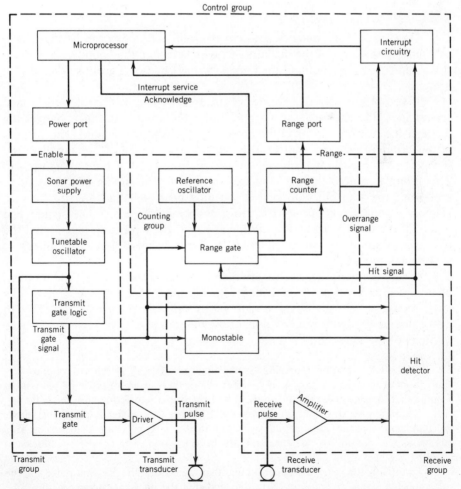

Figure 13-9 SONAR system block diagram "divided" into functional groups.

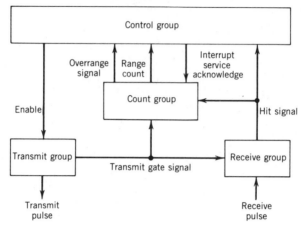

Figure 13-10   SONAR system functional groups.

Dividing a block diagram into larger groups of blocks can be thought of as changing the *level* of the diagram. As with selecting the boundaries of functional blocks, the selection of the boundaries of functional groups is somewhat arbitrary.

If you observe the boundaries of the groups, you will see that they have inputs and outputs similar to those of functional blocks. In this example, the input of the transmit group is the SONAR system power enable. The outputs of this group are the transmit gate signal and transmit pulse. The inputs of the receive group are the transmit gate signal and the echo of the transmit pulse. This group produces the hit signal as an output. The inputs of the counting group are the transmit gate signal, the hit signal, and the interrupt service acknowledge. Its outputs are the range count and the overrange signal. The inputs of the control group are the range count, the overrange signal, and the hit signal. Its outputs are the SONAR system power enable and the interrupt service acknowledge.

When functional blocks are grouped together in this manner, the problem of troubleshooting a complex system is made much simpler. After the blocks have been grouped, they can be redrawn as shown in Figure 13-10. You can now examine the inputs and outputs of the functional groups to eliminate large portions of the diagram at a time.

EXAMPLE 13.3   For this example, assume that the SONAR is reporting impossibly short, nonzero ranges regardless of whether or not there is an object in its path. Also assume that there is no external source of 35-kHz sound interfering with the echo.

Examining the transmit group, you find that the enable signal is present. Next, you determine that the transmit gate signal and the transmit pulse are also correct. This eliminates the possibility that the fault lies in this group. The number of blocks to examine has been reduced by five.

Examining the receive group, you find that the echo of the transmit pulse and the transmit gate signal are present. You also find that the hit signal occurs at the same time as the transmit gate signal. Since the range is counted from the beginning of the transmit gate signal to the beginning of the hit signal, there should be a delay between them proportional to the distance from the object. From this, you can conclude that the fault lies in the receive group. This conclusion reduces the number of blocks to examine to three.

As you can see, with a few tests, almost the entire block diagram is eliminated. Now that the fault is isolated to the receive group, the inputs and outputs of the functional blocks within it can be examined to find the faulty block and the faulty component.

CHAPTER 14

Personal Computers

Dean Lance Smith

Engineering Consultant
Houston, Texas

Any computer costing less than $5000 is considered a *personal computer* by most business analysts. Most personal computers are intended to be used by one user at a time. However, some top-of-the-line models can handle two or more users simultaneously.

The personal computer market is divided into several different segments. Some personal computers are intended for business use while others are intended for recreational or home use. Some personal computers are designed to be portable. Most are not, although the trend is toward more portable personal computers.

Regardless of the type of personal computer, the basic principles of operation are the same. The approaches to servicing are the same. Some of the details, however, may differ.

14.1 THEORY OF OPERATION

14.1.1 Computer System

Figure 14-1 shows a block diagram of a *computer system. Peripherals,* such as disk drives, printers, modems, terminals, and so on, surround the computer. Sometimes, the peripherals are in the same case as the computer.

Some peripherals permit large quantities of data to be temporarily or permanently stored by the computer; disk and tape drives are examples. Other peripherals permit data to be exchanged with the computer. Disk and tape drives can be used to exchange large amounts of data, such as data bases or computer programs. Printers can be used to display printed copies of data. Modems can be used to exchange data over a phone line.

The *console* is used to control the system. A display on the console shows information from the computer. The display is usually similar to a TV picture tube. This type of display is sometimes called a CRT (cathode ray tube), or

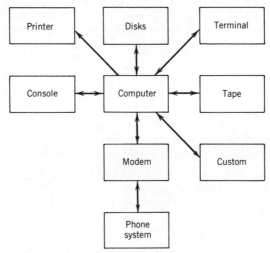

Figure 14-1   Block diagram of a computer system.

VDT (video display terminal). *Hard copy,* or printing displays, are sometimes used.

A keyboard on the console permits the user to type commands and data. Some newer systems use a *mouse,* (a small, rolling, movable data entry device) or a touch-sensitive display in conjunction with a keyboard.

14.1.2 Computer

Figure 14-2 shows a block diagram of the computer. The heart of the computer is the CPU. CPU is an abbreviation for *central processing unit.* The CPU contains the hardware that performs the logical and arithmetic operations built into the computer.

The CPU contains a *clock* to pace the other logic in the computer. The clock is usually crystal-controlled.

If most or all of the CPU is built into one integrated circuit, that circuit is called a *microprocessor.* Most personal computers are based on a micro-

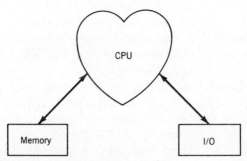

Figure 14-2   Block diagram of a computer.

processor. Most eight-bit machines are based on the 6502, 8080, 8085, 8088, or Z80 microprocessor. Most 16-bit machines are based on the 8086 or 68000 microprocessor. Machines with 32-bit microprocessors are now being introduced.

A *memory* is a device that can store information. Several types of memory are found in personal computers. A *random access memory (RAM)* is a semiconductor read/write memory. RAM is used to store data that may change as the computer processes information. The data can be information that is supplied by the user. The data can also be calculated by the computer.

All RAM is *volatile*. Once the power to the RAM is turned off, the contents of the memory are lost. RAM comes in two forms. *Dynamic RAM,* or *DRAM,* must be refreshed, its internal capacitors recharged, or its contents are lost. *Static RAM* needs no refresh. Once data is written into static RAM, the data remain until the power is turned off or new data is written over the old data.

Refresh occurs whenever a DRAM cell is read. Refresh is usually accomplished by special refresh controls on the chip. These controls permit many cells to be internally read at once. Each bit in a dynamic RAM must be refreshed at least once every 1 or 2 msec. In some systems, the processor is halted every time RAM is refreshed. In other systems, part of each instruction cycle is used for refresh. This is called *cycle stealing*. Usually, only $\frac{1}{128}$ of the RAM is refreshed each refresh cycle. Different parts of the DRAM are refreshed each refresh cycle until the entire RAM is refreshed.

Another type of memory found in all personal computers is *ROM*. ROM stands for read only memory. The user usually cannot write into ROM once it is installed in the computer. ROM comes in several different forms. True ROM, or *mask-programmable* ROM, has its contents permanently determined when the chip is manufactured.

PROM, or *programmable ROM,* can be programmed or written into by the user of the chip. PROM is usually programmed, or burned, by the manufacturer of the computer. Once it is programmed, it usually cannot be reprogrammed.

EPROM is PROM that can be erased and reprogrammed. This type of ROM is erased by exposing a window in the top of the chip to ultraviolet light. The light must be strong enough, and the length of exposure long enough, to erase the chip completely. Lower-intensity ultraviolet light, or shorter exposures, can cause partial erasure of the chip. Partial erasure is called *dropping* (or adding) bits. X-rays can also cause partial erasure.

A disk is also used to store data on some personal computers. A disk is a circular metal or plastic plate that is coated on one or both sides with magnetic material. A *disk drive* is the holder for the disk.

Two types of disks, or *media,* are used on personal computers. A *hard disk drive* uses a rigid disk. It is faster and is used on more expensive business systems. *Floppy,* or *flexible, disk drives* use nonrigid plastic media inside a square plastic or paper holder. The disk and holder are inserted and removed as a unit from the drive. Floppy disks are slower and have less storage capacity than hard disks. They are used on lower-performance and less expensive systems.

Some home computers use magnetic *cassette tape* for *bulk*, or *mass*, *storage*. Tape is slower and less expensive than floppy disks.

14.1.3 Programs

The *program* is the series of binary numbers (ones and zeros) that actually run the machine. This type of program is sometimes referred to as *machine code*, or *object code*, to distinguish it from a more English-like form of a program called the *source program*, or *source code*. The programmer writes the program in source code and a computer program is used to convert it to machine code.

If the source program is written in *mnemonics*, the program is called an *assembly program*. A mnemonic is an alphanumeric symbol for the binary instruction wired into the CPU. A program that converts the mnemonics to object code is called an *assembler*.

If the source program is written in English-like statements, then it is written in a *higher-level language*. A program that converts the higher-level language statements to machine code is called a *translator*, or *compiler*.

There is another type of program that processes higher-level language statements. This type of program is called an *interpreter*. An interpreter does not produce object code. Rather, it uses the higher-level statements to link up parts of the interpreter. Once the linkages are complete, the higher-level program can run on the computer.

A *control program* permits the user to load an application program and its data into the computer. The *application program* is the program used to perform a desired task. Word processors, accounting systems, data base management systems, editors, assemblers, and *BASIC* compilers are examples of application programs.

Two styles of personal computers are available. The all RAM system has mostly RAM for its semiconductor memory. The control program for this system is usually called an *OS*, or *operating system*.

One or two ROMs in an all RAM system contain a program called the *bootstrap*, or *bootstrap loader*. The bootstrap loads the operating system into RAM when power is first applied to the system. Depressing the reset button on the computer will also reload the operating system. Loading the operating system from the bootstrap ROM is called a *cold start*.

The system user initiates a *warm start* by depressing a special key on the console. A warm start is quicker than a cold start because it uses part of the OS in RAM to reload the remainder of the OS.

A *ROM-oriented* personal computer has part of its RAM replaced with ROM. It may use cassette tape or disk for bulk storage.

The control program for a ROM-oriented system is called a *monitor*. Usually, the monitor contains a programming language like BASIC.

14.1.4 Input/Output

The computer communicates with the peripherals in the system through *input ports* and *output ports*. Many personal computers use special-purpose I/O (input/output) chips designed to interface the computer to a specific peripheral.

These special I/O chips are called *programmable peripheral* chips. Common programmable peripheral chips are floppy disk controllers, CRT displays, UARTs (universal asynchronous receiver transmitters), and programmable parallel I/O ports.

14.1.5 Machine Cycles

All machines execute an instruction with a series of machine cycles. All microprocessors have at least two types of machine cycles, a read and a write cycle.

During a *read cycle,* the CPU reads a memory location. The CPU places the address of the location in memory on a series of lines called the *address bus*. This occurs during the first part of the read cycle.

After receiving the address, the memory chip places the data stored at that address on another series of lines called the *data bus*. The CPU latches this information at the end of the cycle.

During a *write cycle,* the CPU writes data into a memory location. The address is placed on the address bus during the first part of cycle. At approximately the same time, the data are placed on the data bus. The RAM latches up the data at the end of the cycle.

One or more *control lines* tell the memory and I/O chips if the machine is doing a read or write cycle. The control lines also tell the peripheral chips when data can be latched on a write cycle. The details of how these control lines work vary greatly from one microprocessor to another.

Some microprocessors have more than two machine cycles. Some machine *architectures,* or designs, have a special read cycle called an *op code fetch* cycle. This cycle is used when the CPU reads the first word of an instruction. (Instructions in most microprocessors can be one, two, three, and even four words long.) The first word is called the *operation code,* or *op code.* A special control line usually indicates when a read cycle is an op code fetch cycle.

Other architectures have a special address space for I/O. This feature is called *separate I/O.* The read and write cycles to the separate I/O space are called *input* and *output cycles,* respectively. One or more control lines permit the peripherals to tell the difference between memory and I/O cycles or between read and input or write and output cycles.

Other special machine cycles are available on some machines. Special control lines or combinations of control lines tell when the CPU is executing one of these special cycles.

14.1.6 Other Integrated Circuits

Some LSI (large-scale integrated) circuits do not have enough power to drive a data or address bus. These chips drive the bus through bus drivers. *Bus drivers* have a high *fan-out,* or drive, capability. *Bus transceivers* are used not only for their high drive capability but also to combine two one-directional buses into a bi-directional bus.

Upper address decoders are chips that decode the upper address lines and drive the *chip select* or *enable* pins of memory and I/O chips. The upper address decoders prevent two or more memory chips from responding to the same address placed on the lower address lines.

The control signals of some microprocessors need to be combined before driving the memory and I/O chips. Simple or *random logic* chips (NAND, NOR, INVERTER, AND, OR, etc.) are usually used for this purpose.

14.2 TYPES OF FAILURE

Most failures in integrated circuits occur because the connections, or *bonds,* between the silicon wafer and the package become damaged. From the outside of the chip, it appears a pin has opened, shorted to ground, shorted to the supply voltage, or shorted to another pin. Some failures occur because a cell in the chip becomes damaged. From outside the package, these failures appear as a loss of part of the chip's function. An example would be when an EPROM drops or adds a bit because a cell has been erased by X-rays.

Hard failures are consistent failures. An example would be a data input or output pin on a chip that shorts to logic 1 or logic 0. Hard failures are relatively easy to diagnose.

Soft failures occur occasionally or in a seemingly random pattern. An example would be a data output pin on a chip that opens. The chip inputs connected to the defective output will float. The logic level taken by these inputs will often be a function of such secondary effects as stray magnetic fields, data patterns on nearby pins, and so on. Soft failures are usually much more difficult to diagnose.

System failures can also occur as a result of errors in the *software,* or program, running on the system. The software error can result from a mistake in writing the program. The error can also result from unintentional alteration of correct code. The alteration can result from a hardware failure or from user error.

One of the most interesting and most frustrating aspects of computer servicing is the interaction of software and hardware when failures occur. Successful troubleshooters need a thorough understanding of both hardware and software.

A *restart failure* has occurred if the machine fails to respond properly when either the power is turned on or the reset button is depressed. Restart failures are usually tougher to diagnose than failures that are noticed after the machine is reset because the symptoms are general rather than specific in nature.

14.3 SYSTEMATIC SERVICING APPROACHES

There are four systematic approaches to diagnosing personal computer failures.

14.3.1 Traditional

The traditional approach to computer servicing uses traditional electronic instruments, such as an EVM (electronic volt meter) and an oscilloscope, and traditional troubleshooting techniques, such as signal injection, to diagnose the problem.

Advantages The advantage of this approach is that no special measuring equipment is required. No advanced preparation is required either. This approach finds its most common application once one of the other approaches has been used to isolate the problem to a small number of parts. It is also useful in diagnosing some types of restart failures.

Disadvantages There are many disadvantages to this approach. First, it is a long, tedious process. It is very difficult to make it work in digital equipment.

Traditional signal-tracing techniques, such as in-to-out and out-to-in tracing, break down when there is feedback. Computers inherently have a lot of feedback, and the software driving the machine may create even more.

Comparison is a much better technique when feedback loops cannot be broken. Good computer-oriented servicing approaches take advantage of comparison.

An EVM is too slow to measure the key properties of most digital signals. Its only use in computer troubleshooting is to measure power supply voltage levels.

An oscilloscope is relatively useless for computer troubleshooting. Many important buses in personal computers are three-state buses. These buses float when they are not being driven. An oscilloscope usually cannot sense the floating state. Indeed, an oscilloscope may even confuse the floating state with logic 1 or logic 0 state.

Another disadvantage of an oscilloscope is that waveshape is not very important in digital work. As long as a waveform does not overshoot or undershoot too far and settles down rapidly enough within the proper logic levels, just about any waveshape is acceptable. It is also difficult and awkward to determine if a logic signal is within proper logic levels when an oscilloscope is used.

14.3.2 Digital Instruments

Instruments designed for computer servicing are usually designed as pass/fail-type instruments. They sense logic levels and timing relationships and give a true/false indication.

A *logic probe* is one such instrument. It is a relatively inexpensive handheld tool. It senses logic levels. Higher-priced models sense floating nodes. It can detect fast pulses or transitions. It can also sense activity and give a relative measure of duty cycle.

A logic probe is a superior instrument to an oscilloscope for troubleshooting using the conventional approach. It also costs much less.

A *logic pulser* is another simple, relatively inexpensive digital trouble-shooting instrument. It can inject a low-duty cycle pulse train into any node it touches. The pulse train will override any properly operating node and drive it momentarily high and then low. It does so without damaging the chips connected to that node. It is useful for injecting a signal.

A *digital current tracer* is another simple, relatively inexpensive digital troubleshooting instrument. It senses relative current flow. It is useful for finding shorted logic inputs on a data or address bus. It is also useful for finding shorted power supply bypass capacitors, because the current through them is high.

Other instruments are available that can measure frequency, period, rise time, fall time, and other properties of a pulse waveform. They do so digitally with a technology that is more accurate and easier to use than an oscilloscope. These instruments usually cost less than an oscilloscope.

14.4 SINGLE STEPPING

A fault is diagnosed using the single-step approach by comparing the computer's machine cycle behavior to what it should be. A data latch or logic analyzer is connected to the address and data buses of the system under test. The instrument is used to follow the system as it executes a program. The program is usually an application program or the control program. However, a special test program can be traced using single stepping.

There are two types of single-step approaches: single instruction and single cycle. In *single instruction,* the data on the buses are latched at the end of each op code fetch cycle. In *single cycle,* the data on the buses are latched at the end of each cycle.

Single instruction is used to locate a problem rapidly. Single cycle is used at the problem to analyze it. Conventional test equipment such as oscilloscopes are often useful if single stepping is available on a computer.

14.4.1 Data Latch

A data latch is a series of D flip-flops, or latches, connected to the data and address buses of the system under test. The outputs of the latches drive a display. The display is usually in binary or hexadecimal. Some control and status lines may also be latched.

A control network built into the data latch determines when data are stable on the address and data buses. The control network latches the information on the buses and then halts the processor. Switches on the front panel permit the user to select run or single-step operation and the type of single-step operation (single cycle or single instruction).

Some older personal computers have data latches built into their front panel. Data latches are commercially available hardware that can plug into the expansion slot of some personal computers. Other commercial models have a clip that will permit connection to the microprocessor chip. General-purpose latches are also available.

14.4.2 Logic Analyzer

A logic analyzer can also be used to gather single-step information. However, a logic analyzer is a more general-purpose piece of equipment. It also permits the information from several machine cycles to be captured at one time and then displayed at the user's convenience.

A logic analyzer is based on an array of memory. The data in the memory are displayed unless data are being captured and stored in the memory.

If the memory contents are displayed as numbers, the analyzer is called a *logic state analyzer*. This type of logic analyzer is more useful for the single-step approach. If the memory contents are displayed as a pseudotime waveform, the analyzer is called a *logic timing analyzer*. This type of logic analyzer is more useful for diagnosing some types of computer development problems. Some logic analyzers can display information in both forms.

The user can place the logic analyzer in the capture mode. In this mode, data on the data leads of the logic analyzer are stored in memory. A new word (based on the number of logic analyzer data leads) of data is stored in memory every time a predetermined transition (positive or negative) is received on the logic analyzer's clock lead.

To use a logic analyzer for single-step, the logic analyzer's data leads are connected to the computer system's address and data buses. Any extra data leads are usually connected to status and control signals of the system under test. The clock lead is usually connected to a control signal that has a transition when the contents of the address and data buses are stable. Most microprocessors have one such control signal that makes a transition during an op code fetch cycle and another that makes such a transition every machine cycle.

A logic analyzer will start capturing data once it recognizes a predetermined trigger word. It can store data up to the trigger word (very useful for diagnosing some types of random errors) or after the trigger word. A predetermined delay can occur after the trigger word before data are captured. The delay can be based on a predetermined number of clock transitions or on a predetermined number of trigger events.

Some logic analyzers have data qualifier leads. These leads permit the trigger word to be expanded (but not the data word). Some logic analyzers have clock qualifier leads. These leads permit more than one signal to be logically combined to give a clock transition.

14.4.3 Diagnosis

If incorrect data are read from RAM but correct data were written into RAM, then either the RAM is defective or there is a defect in the write control circuitry, the read control circuitry, the upper address decoder, or a data bus driver. Additional connections of the logic analyzer or data latch will give more information.

For example, checking the write-enable pin of the RAM during a write cycle, the read-enable pin of the RAM during a read cycle, the chip-select or chip-enable pin of the data bus driver during a read or write cycle will all give more information. A logic probe or even an oscilloscope can be used as an indicating instrument.

If incorrect data are read from a ROM, then either the ROM, its upper address decode circuitry, or its data bus driver are defective. Additional checks should be made on these circuits.

Sometimes, incorrect data will be read from a RAM or ROM because more than one memory chip is responding to the address. This usually indicates a defect in one of the upper address decoders.

If a correct op code and data for an instruction are read by the CPU and it fails to process the instruction correctly, then the microprocessor has failed. Any bus drivers associated with the microprocessor should be checked also.

14.4.4 Advantages

The main advantage of the single-step approach is that little or no advanced preparation is required.

14.4.5 Disadvantages

There are several disadvantages to the single-step approach. One major disadvantage is the amount of knowledge needed to use the technique successfully. The troubleshooter must thoroughly understand the hardware in the system, all the instructions in the microprocessor, and how the machine cycles work for each instruction. The troubleshooter must be able to read a schematic and a machine code listing of the software used for the test.

Another disadvantage of the approach is that the troubleshooter must have a schematic of the system. Some manufacturers consider this information proprietary and do not make it available.

Another disadvantage is that an assembly code listing of the test software must be available. It is convenient to use the bootstrap loader or control program as the test software. Unfortunately, these programs may be written in a higher-level language. An assembly level listing of the program may not be produced by the compiler. Even if the program was written in assembly language, many manufacturers consider the software proprietary and will not make a listing of the program available.

An assembly listing of the machine code can be obtained by disassembling the machine code. (A disassembler program does the opposite of an assembler program. It converts machine code into mnemonics rather than mnemonics into machine code.) Unfortunately, the comments are unavailable after disassembling the object code, and the disassembled listing may not be entirely accurate. A well-commented assembly program is far easier to read than a program without comments.

Another disadvantage of the technique is that many data latches and most logic analyzers are relatively awkward pieces of equipment to use. Newer models are easier to use. But all models, old or new, always seem to have one or two fewer data leads than what are needed. This requires tedious and clever use of most models. Sometimes, the use of a second logic analyzer or another measuring instrument will overcome the finite word size of the machine.

Another disadvantage of the single-step approach is that it takes a long time to make a diagnosis. The process is slow even with a built-in data latch. Connecting a logic analyzer takes a significant amount of time.

14.5 SELF-TEST (DIAGNOSTIC)

The *self-test* approach is sometimes called the *diagnostic program* approach. In this approach, the troubleshooter runs a computer program that tests the hardware of the system and identifies failures.

Self-test programs are built into some personal computers. Partial or complete diagnostic programs are available from either the manufacturer or independent suppliers for many other personal computers.

14.5.1 Tests Run

RAM Test   The test program usually writes and reads one or more patterns in RAM. All bits are usually set and reset. The patterns selected are usually assumed to be a reasonably good test of pattern sensitivity.

Pattern sensitivity occurs when RAM will pass one pattern test but fail to pass another. Pattern sensitivity can occur even if both patterns set and reset all bits.

Some RAM tests will detect *dropped bits*. Dropped bits occur when one or more bits are set (or reset) after a delay of several milliseconds. Dynamic RAM with defective refresh circuitry is especially prone to dropped bits.

ROM Test   A ROM test program will usually be run. The most common ROM test is to compare a *checksum* of the ROM contents to what it should be. A *checksum* is calculated by adding the contents of every memory location in the ROM and ignoring the overflow. For example, a checksum in an eight-bit machine could be calculated by ignoring all overflows beyond eight bits. A checksum is an example of an *arithmetic check code,* or *modulo* arithmetic. A 16-bit checksum is the most common ROM test, where the sum is loaded into two registers.

Interface Tests   Interfaces are usually checked by a diagnostic program. Output interfaces, such as lights, are illuminated. Input interfaces, such as keyboards, are usually tested after an output interface, such as a display, has been tested. Software causes the data provided by the troubleshooter to be displayed on the output interface.

Interfaces with compatible outputs and inputs are frequently tested together using a technique called loopback. For example, an RS-232C input and output interface will be tested together by connecting the data-out and data-in pins together.

CPU Test   The microprocessor chip is rarely tested directly with a self-test program. The program would be too long and complicated. The assumption is made that if the system passes the preceding tests, the CPU is working. This is usually a reasonable assumption.

14.5.2 Advantages

The main advantage of the self-test is that it quickly isolates a defect to an area. A person with little or no computer background often can use the technique.

14.5.3 Disadvantages

The diagnostic program usually must be prepared prior to a system failure. Writing and debugging a self-test program while the system is defective is difficult and often impossible.

While self-tests will rapidly identify a defective area of a system, most cannot identify a defective component in that area. Usually, some other technique must be used to identify one of several chips that could cause the failure.

Self-tests can fail to spot a defect. It is impossible to write a diagnostic program that will test for all possible defects in a system. A good test program will spot hard failures but may not spot soft failures.

Writing a good self-test program requires a thorough knowledge of the system hardware. It also requires experience in programming. Also, the program usually must be written in assembly language. Many modern high-level programming techniques can result in a poor self-test program.

14.6 SIGNATURE ANALYSIS

In signature analysis, a test program is run over and over in a loop. The waveforms, or signatures, at each node or connection in the system are measured. The signatures are compared to signatures measured when the system was working properly. The signatures are usually recorded in tables. Defective signatures are traced back to the failed component.

Two tests are usually run on the system. The first test is called the free-run, or kernel, test. The free-run test is implemented with hardware. The second test is a software test of the rest of the system.

14.6.1 Signature Analyzer

While an oscilloscope can be used to look at the signature, it is usually inconvenient to do so. The waveforms are usually too long. The preferred instrument is a signature analyzer. The signature analyzer is a relatively inexpensive instrument available from many suppliers.

The signature analyzer has three connections that are made to the system under test. These connections can change for each table.

Start-and-stop leads tell the signature analyzer when to start and when to stop measuring a signature. A clock lead tells the signature analyzer when to sample data. All of these leads are edge sensitive. Switches on the signature analyzer determine whether sensitivity is to the rising or falling edge of a signal.

The signature at a node is measured with a logic probe that is connected to the signature analyzer. The logic probe is useful for distinguishing between signatures that are always at ground or logic 1 potential and those that are always sampled at these two logic levels.

Some signature analyzers have an enable lead. This connection is usually level sensitive. The enable lead minimizes the reconnection of leads in some situations. It also permits more flexibility in the software used for the second test.

14.6.2 Tests Run

Free-Run, or Kernel, Test   The data bus is broken at the microprocessor chip. The microprocessor side of the break is hardwired with the NOP (no operation) instruction. (Actually, any instruction that works internal to the CPU and takes one machine cycle can be used.)

Systems with signature analysis designed into them can usually be free-run by throwing some switches on the data bus. Free-run adapter sockets are available for retrofitting signature analysis. Free-running the system converts the microprocessor into a binary counter. The microprocessor counts from address 0 to its highest address over and over again. The address lines act as outputs from the counter and show the circuit.

The free-run test is a good test of the address lines, upper address decoders, and some of the control logic of the system. It is a fairly good test of the CPU. It also checks any ROMs in the system.

At least two tables are generated for the free-run test. The first table is general free-run signatures. An additional table is generated for each ROM in the system. The general free-run table is generated using the most significant address line for the start-and-stop signal. The triggering edge is not critical for most systems as long as it is the same for both the start and stop leads. Different signatures will result from selecting one transition rather than the other.

The signatures for each ROM table are started on the falling edge of the chip select signal for the ROM. The measurement is stopped on the rising edge of the chip select signal. Each ROM table checks the output of the ROM, the data bus, and any data bus drivers between that ROM and the CPU.

Second Test   The second test checks the remainder of the system. Patterns are written and read from RAM. Output ports are stimulated. Input ports are read. Loopback tests are run on compatible output and input ports.

Care must be taken when writing a second test program to ensure that the program behaves the same way regardless of the data read. The program should not decide if the pattern read is correct because this will change the program's behavior for defects in the system. The signature analyzer will spot incorrect data.

The program should be written with the effect of hardware failures on the program taken into consideration. For example, the second test should start immediately on system reset to minimize the effect of restart failures on the test. Subroutine call and return instructions should be avoided. A RAM defect could destroy the subroutine return address stored on the stack.

At least two tables are generated for the second test. One table lists write signatures. (If patterns are not written correctly into RAM and I/O chips, correct patterns will not be read from those chips.) The second table gives the read signatures.

In some processors, such as the 8085 and Z80, the read and write control lines can be used to clock the signatures for the second test. For processors such as the 6502 that have a multiplexed read/write line, the clock signals are a little more difficult to obtain.

A level-sensitive enable signal can be used to sort out the write and read signals. This assumes that the signature analyzer has a level-sensitive enable signal and that the proper level can be selected. Sometimes, combinations of control signals are available in the system that give the equivalent of separate write and read signals. Sometimes, the equivalent of a separate write signal is available, and both read and write signatures can be read for the second table. (This works because the write signatures were checked on the first table.)

The read table will usually have several subtables for the data bus signatures. If two or more chips are driving the data bus at the same time in the read signature table, it is impossible to determine which chip is causing a defective signature.

If signature analysis is designed into the system, switches are usually placed between the upper address decoder and chip select pin of most of the RAMs and peripheral chips. Only one bank of RAM is permitted to drive the data bus for the read table. Additional banks of RAM and peripheral chips are switched in, one at a time, while observing the data bus signatures. Switching in a chip that produces a defective signature isolates the defective chip.

When retrofitting signature analysis, shorting clips can be used to short the chip select inputs to the inactive state. (Most drivers can survive driving a short circuit.) These clips serve the function of the switches when signature analysis is designed in. Removing chips from sockets is another retrofit technique for isolating data bus drivers. Sometimes, multiple second-test programs and tables, one for each data bus driver, are written for retrofits. Isolation is achieved with software rather than hardware.

Obtaining start and stop signals for the second test requires some cleverness. Sometimes, a memory or port address is written into or read only once during the program. The decoder output for this address can be used as the start or stop signal.

Sometimes, an unused memory or port address can be used for the start and stop signal. The data bus should have pull-up resistors so that a known value is read from the location. An unused upper address decoder output or even an unused address line can be connected to the start and stop lead.

14.6.3 Advantages

The main advantage of signature analysis is that a relatively low-skilled person can trace the defect to a node. Usually, only two or three chips are connected to a node. Frequently, enough information is available to uniquely identify the defective chip connected to the node. Additional tools, such as a logic pulser or a digital current tracer, are sometimes useful for identifying the defective chip.

14.6.4 Disadvantages

There are several disadvantages to signature analysis. First, the technique is relatively slow at isolating a defect to an area. Once the area is identified, however, signature analysis is a good technique for isolating the defect to a node and even a chip. Self-tests are often used as a complement to signature analysis to overcome this disadvantage.

Another disadvantage is that all the test programs and signatures must be available before servicing the system. Some free-run test signatures can be predicted by referring to previously published tables. However, these universal signatures are rarely enough to permit isolation of a defect.

A highly skilled person knowledgeable in hardware, computer architecture, and software at the assembly code level is needed to specify the signature analyzer connections and write the test software. Signature analysis test software is often less difficult to write than a self-test program. However, the interactions of software and hardware are more subtle than for the self-test. A signature analysis package can be poor at identifying some restart failures unless properly prepared.

A fairly large amount of time is required to prepare the documentation for signature analysis. Test software must be written and debugged. Signatures for all tests must be measured, documented, checked, and rechecked.

14.7 SIMPLE CHECKS

The following preliminary checks can be performed by just about any computer user. They require no special skills. Many personal computer failures can be traced to the following problems.

Connectors   Make sure the connections on all cables are tight. Check the power cord, cables to the RS-232C connectors, disk drive cables, keyboard cables, and any special interface connections.

Plug and unplug all signal connectors a few times. In humid and polluted environments, dirt or corrosion can accumulate on connector contacts. The dirt and corrosion can act as an insulator on signal connectors. This is particularly true if the equipment has been idle. Plugging and unplugging the connectors often will remove enough dirt and corrosion to permit good contact.

In severe cases, the dirt and corrosion will have to be removed with a cleaning solvent or, as a last resort, an eraser. Inspect and clean the connectors if necessary.

Boards   Make sure that all the boards of a multiboard system fit tightly in their motherboard sockets. This is especially true if the equipment is portable or has been moved. Remove any dirt or corrosion that may have accumulated on the sockets as described above under connectors.

Power   Apply power to the equipment for a few minutes. Feel the case. If the equipment is not warm, check the power plug to make sure it is plugged in. Many personal computers have a power cord that plugs into both a wall outlet and the equipment. Check both plugs.

Turn off the equipment and remove the fuse. The fuse is located near the power cord on most personal computers. The fuse can be removed from most equipment by twisting the holder counterclockwise. Inspect the fuse and replace it if it is defective.

Internal Diagnostics   Some personal computers have a built-in diagnostic or self-test program. Run this program. Correct any defect spotted by the program.

Portable Radio Test   Apply power to the computer and operate a portable radio next to it. The computer should cause interference with the radio, particularly if the radio is operated on the FM band. Reset the computer a few times. The clock in the computer is defective if no interference is detected. The new EMI requirements of the FCC may soon render this test invalid except with older computers.

14.8 SYSTEM FAILS TO RESTART

A personal computer can fail to restart if one of the control or status lines to the CPU has failed. A restart failure can also occur if a RAM, ROM, or CPU failure causes the computer to crash during the restart routine. The restart routine brings the system up in an orderly fashion after reset. The restart routine is usually part of the monitor or bootstrap loader.

A RAM, ROM, or CPU failure can be diagnosed using the single-step approach. A good signature analysis package will also detect the defective chip. The conventional approach can also be used. A logic probe or oscilloscope can be used to check for activity on the address and data buses, the chip select outputs of the upper address decode networks, the dynamic RAM refresh circuitry, the console interface, and so on. However, prior to using one of these approaches, the following checks should be made.

Power Supplies   Check the power supplies. A computer cannot restart unless the proper voltage is supplied to each chip. Use a VOM (volt ohmmeter) or EVM. Check the voltage at several key chips such as the microprocessor, RAM, ROM, and programmable I/O chips.

Check not only the 5 V supply that supplies most of the power to these chips but also any other power supplies in the system. Many chips require -5 V, 12 V, and even -12 V dc supplies. Check these special supplies at several of the key chips they power.

Clock   Check the clock at the microprocessor. The system can't work without the clock to pace it. Check the clock at the memory and programmable peripheral chips in those systems, such as the 6502, that use the clock signal as a control signal.

Use a logic probe or an oscilloscope to check the clock and other control or status lines. An indication of activity on a logic probe is usually sufficient indication that the clock is operating properly. Rarely will a crystal-controlled clock change frequency enough to cause a restart failure.

Reset Line   Push the reset button while looking at the reset input to the microprocessor. Repeat the procedure with programmable peripheral chips that are tied to the reset line. If these chips are not reset properly, the system will not restart.

Some microprocessors require the reset line to be held in the active state several milliseconds, particularly on power up. If the system under test does not power up properly but does restart with the reset button, check the delay circuitry on the reset line.

Ready or Wait Lines   Most microprocessors have a *ready* or *wait* control line. It is used to slow the processor for slow memory and I/O chips. Placing this line in the active state forces the processor into wait states or cycles. The processor will wait permanently if this line is hung in the active state.

Check this line at the microprocessor. Trace a fault back to its source near the memory or the I/O chips.

Halt or Hold Lines   Many microprocessors have a line that is called halt, hold, bus request, and so on. Activating this line causes the processor to complete an instruction and then halt. Usually, the address and data buses are floated when the halt occurs. This line is used by another processor, such as DMA (direct memory access) controller for fast disk operation, or a CRT refresh controller, to gain control of memory.

If this line hangs in the active state, the processor cannot fetch an instruction. Check this line at the microprocessor and trace a hung line back to its source.

Interrupt Lines   All microprocessors have at least one interrupt line. Check the interrupt lines at the processor and make sure they stay in the inactive state during restart. If they don't, trace the problem to its source.

A hung interrupt line can cause the processor to service an interrupt either during or after the restart routine. Some systems will hang up if no peripheral has requested the interrupt.

14.9 SYSTEM RESTARTS

Run any built-up diagnostic or self-test program. This will rapidly isolate the defect to an area.

Run a retrofit self-test program if one is available for your system. The most effective retrofit diagnostic programs are supplied in ROM or EPROM. They can be plugged into the ROM socket that normally contains the restart program. Sometimes, a socket adapter is needed. Less effective self-test programs are supplied on tape or disk. Most of these run off an industry standard operating system.

An *emulator* is a convenient way of running a retrofit self-test program. An *emulator* plugs into the microprocessor socket. The emulator contains a known working CPU. A control program is included in the emulator along with enough RAM to service the control program.

The emulator permits the user to look inside the SUT (system under test). Many emulators have provisions for loading test programs into the SUT. Some emulators have general-purpose diagnostic programs as part of the package.

If time permits and talent is available, write a self-test program or a series of self-test routines. Use the loader features of the control program or an emulator to load and run the diagnostics. An assembly language program can always be assembled by hand if the system can't run its assembler program.

Next, use signature analysis to isolate a defect to a node. If signature analysis does not give enough information to determine which chip connected to the node is defective, use a logic pulser and digital current tracer, if necessary, to find the chip. If the chips are in sockets, replace the suspect chips connected to the node one at a time until the defect is corrected (shotgun technique).

If signature analysis is not built into the system, use a retrofit package if it is available. The most effective retrofit packages have the second test program in a ROM that plugs into the ROM socket containing the restart program.

Some emulators have a signature analyzer built into the package. Signature analysis test programs for major components, such as RAM, may be included with the package. The software can be used to generate signatures for a working system. The software can frequently be of help even if signatures have not been measured on a good system.

If a suspect chip is one that can be tested by the free-run test, then free-run the system. Use the universal free-run signatures given in Tables 14-1 through 14-11 to help trace the defect. These tables are based on values calculated by a computer program called *signature generator*. Most of the entries were verified by measurement. The tables use the "funny" hexadecimal code from Hewlett-Packard (0 to 9, A, C, F, H, P, and U). They are also based on the Hewlett-Packard feedback taps for the signature analyzer.

Table 14-1 Free-run address signatures for an eight-bit, 64K memory space microprocessor (lower 32K first, then upper 32K)

Connections[a]			Check signatures
Start—A15 (falling edge)			Ground—0000
Stop—A15 (falling edge)			V$_{cc}$—0001
Clock—RD* (rising edge)—Z80			
RD* (rising edge)—8085 with demultiplexed address lines			
SYNC (falling edge)—6502			
A0—UUUU	A4—5H21	A8—HC89	A12—HAP7
A1—5555	A5—0AFA	A9—2H70	A13—3C96
A2—CCCC	A6—UPFH	A10—HPP0	A14—3827
A3—7F7F	A7—52F8	A11—1293	A15—755U

[a]An asterisk after a logic symbol means active low.

Source: Smith, D. L., Free-Run Signature Tables. In *Proceedings of the 4th Annual Conference on Computer Developments.* University of Houston at Clear Lake City, Clear Lake City, TX, November 5, 1982.

Table 14-2 Free-run address signatures for an eight-bit, 64K memory space microprocessor (upper 32K first, then lower 32K)

Connections[a]			Check signatures
Start—A15 (rising edge)			Ground—0000
Stop—A15 (rising edge)			V$_{cc}$—0001
Clock—RD* (rising edge)—Z80			
RD (rising edge)—8085 with demultiplexed address lines			
SYNC (falling edge)—6502			
A0—UUUU	A4—5H21	A8—HC89	A12—HAP7
A1—5555	A5—0AFA	A9—2H70	A13—3C96
A2—CCCC	A6—UPFH	A10—HPP0	A14—3827
A3—7F7F	A7—52F8	A11—1293	A15—755P

[a]An asterisk after a logic symbol means active low.

Source: Smith, D. L., Free-Run Signature Tables. In *Proceedings of the 4th Annual Conference on Computer Developments.* University of Houston at Clear Lake City, Clear Lake City, TX, November 5, 1982.

Table 14-3 Free-run address signatures for an eight-bit, 64K memory space microprocessor that pipelines (lower 32K first, then upper 32K)

Connections			Check signatures
Start—A15 (falling edge)			Ground—0000
Stop—A15 (falling edge)			V$_{cc}$—0003
Clock—THETA2 (falling edge)—6502, 6800			
A0—UUUU	A4—1U5P	A8—7791	A12—4FCA
A1—FFFF	A5—0356	A9—6321	A13—4868
A2—8484	A6—U759	A10—37C5	A14—9UP1
A3—P763	A7—6F9A	A11—6U28	A15—0001

Source: Smith, D. L., Free-Run Signature Tables. In *Proceedings of the 4th Annual Conference on Computer Developments.* University of Houston at Clear Lake City, Clear Lake City, TX, November 5, 1982.

Table 14-4 **Free-run address signatures for an eight-bit, 64K memory space microprocessor that pipelines (upper 32K first, then lower 32K)**

Connections			Check signatures
Start—A15 (rising edge)			Ground—0000
Stop—A15 (rising edge)			V_{cc}—0003
Clock—THETA2 (falling edge)—6502, 6800			
A0—UUUU	A4—1U5P	A8—7791	A12—4FCA
A1—FFFF	A5—0356	A9—6321	A13—4868
A2—8484	A6—U759	A10—37C5	A14—9UP1
A3—P763	A7—6F9A	A11—6U28	A15—0002

Source: Smith, D. L., Free-Run Signature Tables. In *Proceedings of the 4th Annual Conference on Computer Developments.* University of Houston at Clear Lake City, Clear Lake City, TX, November 5, 1982.

Free-run adapter sockets are available commercially. One can also be made from two integrated circuit sockets compatible with the microprocessor chip. One socket should have the data bus pins wired with the no-operation instruction. That socket should be plugged into the second socket, making sure the data bus pins from the top socket do not contact the bottom socket.

The microprocessor chip is removed from its socket. (If it is soldered in, unsolder it and replace it with a socket.) The free-run adapter socket is plugged into the empty microprocessor socket. The microprocessor chip is plugged into the free-run adapter socket.

Table 14-5 **ROM test address signatures for an eight-bit, 64K memory space microprocessor (2K ROM)**

Connections[a]			Check signatures[b]
Start—CS* (falling edge)			Ground—0000
Stop—CS* (rising edge)			V_{cc}—7A70
Clock—RD* (rising edge)—Z80			
RD* (rising edge)—8085 with demultiplexed address lines			
SYNC (falling edge)—6502			
A0—H62U	A4—P030	A8—9635	A12—[c]
A1—C21A	A5—4442	A9—1734	A13—[c]
A2—HA07	A6—4U2A	A10—8P54	A14—[c]
A3—H0AA	A7—0772	A11—[c]	A15—[c]

[a]An asterisk after a logic symbol means active low.
[b]B after a signature indicates that there is activity on the node and that the logic probe will be blinking.
[c]0000-B or 7A70-B depending on the position of the ROM in the address space.
Source: Smith, D. L., Free-Run Signature Tables. In *Proceedings of the 4th Annual Conference on Computer Developments.* University of Houston at Clear Lake City, Clear Lake City, TX, November 5, 1982.

Table 14-6 ROM test address signatures for an eight-bit, 64K memory space microprocessor with pipelining (2K ROM)

Connections[a]				Check signatures[b]
Start—CS* (falling edge)				Ground—0000
Stop—CS* (rising edge)				V_{cc}—826P
Clock—THETA2 (falling edge)—6502, 6800				
A0—2A1F	A4—3319	A8—19H6	A12—[c]	
A1—A206	A5—7C47	A9—HP66	A13—[c]	
A2—C133	A6—C25F	A10—7A70	A14—[c]	
A3—8P3U	A7—5H21	A11—[c]	A15—[c]	

[a]An asterisk after a logic symbol means active low.
[b]B after a signature indicates that there is activity on the node and that the logic probe will be blinking.
[c]0000-B or 826P-B depending on the ROM's position in the address space.
Source: Smith, D. L., Free-Run Signature Tables. In *Proceedings of the 4th Annual Conference on Computer Developments.* University of Houston at Clear Lake City, Clear Lake City, TX, November 5, 1982.

If signature analysis is not available, then run the diagnostic routine that spotted the defect in your system in a loop. Use a logic probe or oscilloscope to trace the signal to the defect.

If none of these techniques can be used, then do single stepping on proven software. If a logic analyzer, data latch, or an assembly listing of the software is not available, then use the conventional approach.

Table 14-7 ROM test address signatures for an eight-bit, 64K memory space microprocessor (4K ROM)

Connections[a]				Check signatures[b]
Start—CS* (falling edge)				Ground—0000
Stop—CS* (rising edge)				+5 V—826P
Clock—RD* (rising edge)—Z80				
RD* (rising edge)—8085 with demultiplexed address lines				
SYNC (falling edge)—6502				
A0—7P25	A4—8P3U	A8—5H21	A12—[c]	
A1—2A1F	A5—3319	A9—19H6	A13—[c]	
A2—A206	A6—7C47	A10—HP66	A14—[c]	
A3—C133	A7—C25F	A11—7A70	A15—[c]	

[a]An asterisk after a logic symbol means active low.
[b]B after a signature indicates that there is activity on the node and that the logic probe will be blinking.
[c]0000-B or 826P-B depending on the ROM's position in the address space.
Source: Smith, D. L., *Testing and Servicing Microprocessors.* Dean Lance Smith, P.E., Houston, TX, 1983.

Table 14-8 **ROM test address signatures for an eight-bit, 64K memory space microprocessor with pipelining (4K ROM)**

Connections[a]				Check signatures[b]
Start—CS* (falling edge)				Ground—0000
Stop—CS* (rising edge)				+5 V—P254
Clock—THETA2 (falling edge)—6502, 6800				
A0—FA11	A4—46HC	A8—1U5P		A12—[c]
A1—3HUA	A5—65CA	A9—AAHU		A13—[c]
A2—12U0	A6—8AUC	A10—U665		A14—[c]
A3—C75A	A7—9241	A11—826P		A15—[c]

[a]An asterisk after a logic symbol means active low.
[b]B after a signature indicates that there is activity on the node and that the logic probe will be blinking.
[c]0000-B or P254-B depending on the ROM's position in the address space.
Source: Smith, D. L., *Testing and Servicing Microprocessors.* Dean Lance Smith, P.E., Houston, TX, 1983.

Table 14-9 **Short table of V$_{cc}$ (+5 V supply) signatures**

Sample	Signature	Sample	Signature	Sample	Signature
1	0001	14	3U9F	2K	7A70
2	0003	15	7U39	4K	826P
3	0007	16	UP73	8K	P254
4	000U	17	UFP6	16K	1180
5	001U	18	U9FF	32K	755U
6	003U	19	U399	64K	0001
7	007U	20	P733	128K	0003
8	00UP	32	3951	256K	000U
9	01UF	64	A70F	512K	00UP
10	03U9	128	6PCP	1M	UP73
11	07U3	256	CC34	2M	3951
12	0UP7	512	4596	4M	A70F
13	1UFP	1K	8P54	8M	6PCP
				16M	CC34

Source: Smith, D. L., Free-Run Signature Tables. *Proceedings of the 4th Annual Conference on Computer Developments.* University of Houston at Clear Lake City, Clear Lake City, TX, November 5, 1982.

Table 14-10 **Free-run decoder output signatures for a 3 to 8 decoder driven by an eight-bit, 64K memory space microprocessor (most significant decoder input connected to A15, outputs active low, start and stop off the falling edge of A15)**

DO0—4P0A	DO2—PC01	DO4—6H49	DO6—U3H5
DO1—12U3	DO3—F2A6	DO5—0996	DO7—P255

Source: Smith, D. L., Free-Run Signature Tables. In *Proceedings of the 4th Annual Conference on Computer Developments.* University of Houston at Clear Lake City, Clear Lake City, TX, November 5, 1982.

Table 14-11 **Free-run decoder output signatures for a 3 to 8 decoder driven by an eight-bit, 64K memory space microprocessor that pipelines (most significant decoder input connected to A15, outputs active low, start and stop off the falling edge of A15)**

DO0—C9U1	DO2—F9CF	DO4—5FUA	DO6—64HF
DO1—534H	DO3—2302	DO5—29A4	DO7—1183

Source: Smith, D. L., Free-Run Signature Tables. In *Proceedings of the 4th Annual Conference on Computer Developments*. University of Houston at Clear Lake City, Clear Lake City, TX, November 5, 1982.

14.10 POWER SYSTEM PROBLEMS

There are two types of power line problems that can cause havoc with some personal computers. One type is power outages. The other type is imperfections in the power system voltage.

14.10.1 Outages

Power outages can cause two problems with personal computers. First, obviously the computer cannot be used during the outage. Even worse, data can be lost when the outage occurs.

Since RAM is volatile, its contents are lost when power is lost. Any data not saved on tape or disk will be lost when the power fails.

Data can also be lost if the failure occurs when disk or tape is being accessed. Power failures during bulk memory writes will almost always scramble part of the data on the media. The disk directory for most personal computers resides on disk. Scrambling the directory is especially troublesome. Power failures during disk or tape reads can sometimes destroy data on the media.

Some personal computers are designed to sense power outages and shut the system down in an orderly fashion. These systems usually sense a drop in the power line voltage before the outputs of the logic supplies decay significantly. Power fail routines save the state of the processor in RAM. The RAM is powered by a rechargeable battery. A charger keeps the battery charged. Thus, the RAM contents are not lost during the power failure. When the power system returns to normal, the processor restarts and checks the battery-powered RAM to see if the state of the processor should be restored.

Most personal computers do not have power fail circuitry. If uninterrupted operation is essential, such a system can be run off storage batteries. An inverter is usually needed to drive a CRT or VDT display and the disk drive motors. A battery charger can keep the batteries charged whenever the power system is functioning.

Sometimes, the battery can be connected into the existing power supply. Frequently, the battery and supply voltage are incompatible, and a completely new power is needed.

Commercial noninterruptible power supplies are also available. These have an internal storage battery that is charged whenever the power system is functioning. If the power system fails, an inverter supplies power to the computer from the battery. Some commercial noninterruptible power supplies supply power through the inverter at all times. The power system only charges the battery.

Small diesel- and gasoline-powered generators are also available. These suppply power at standard power line voltages and frequency.

The least expensive way to avoid lost data is to use a backup. Data entered into RAM should frequently be copied onto disk or tape. This will minimize lost RAM data.

Disk or tape data losses can be minimized by frequently backing up data on another disk or tape. For example, on systems with three disk drives, the operating system file copy program can be used periodically to transfer data from a second drive to a third drive containing the backup disk.

A special backup disk can be made for one and two drive systems. The backup disk is like any other copy of the operating system, except that most of the system utilities are left out. Only the basic operating system and the file copy program are left on the disk. The backup disk is periodically inserted into the drive normally used for the operating system. Data on the second disk drive are copied onto the backup disk.

In single-drive systems, the data are first copied to RAM. Then, the backup disk is inserted and the data are copied to the backup disk.

14.10.2 Imperfections

Imperfections in the power system voltage can include overvoltage, undervoltage, spikes (over- or undervoltages lasting a fraction of a cycle), surges (over- or undervoltages lasting several cycles), and transients (sudden changes in voltage levels that decay to a steady value, sometimes in an oscillatory manner).

As a rule, personal computer power supplies are designed to accept these power system imperfections with two possible exceptions. Most systems are not designed to handle chronic or sustained over- or undervoltages outside the standard limits (110 to 125 V for residential or commercial office service in most of North America). If voltages outside these limits cause problems, the issue should be pursued with the local power company. The power company is violating accepted standards and is obligated, in most locations, to correct the problem.

The other exception is nearby lightning strikes. Power systems are designed to safely clear lightning strikes to distribution lines. This frequently results in momentary or prolonged power outages during an electrical storm. However, most residences and small commercial buildings are not protected from an electrical strike to the building or its power drop.

Sensitivity to normal power line variations usually indicates a failure in the power supply. That failure should be corrected.

In those rare cases where a power supply has been improperly designed or abnormal power line variations cannot be corrected, commercially available power line conditioners are available from a number of sources. Some of these will just filter out noise. Better units will limit spikes. The best models will also regulate the power system voltage to the computer.

14.11 SUGGESTED READING

A Designer's Guide to Signature Analysis. Hewlett-Packard, Palo Alto, CA. Application Note 222, 1977. (An excellent discussion of signature analysis. References to other publications.)

A Guide to Testing Microprocessor Based Systems and Boards. Millenium Systems, Cupertino, CA. (A good overall description of emulation combined with self-test and signature analysis.)

A Manager's Guide to Signature Analysis. Hewlett-Packard, Palo Alto, CA. Application Note 222-3, 1980.

Application Articles on Signature Analysis. Hewlett-Packard, Palo Alto, CA. Application Note 222-2, 1979. (An excellent collection of previously published articles on signature analysis. References to other sources.)

Carr, J. J., *Digital Electronic Troubleshooting.* Tab Books, Blue Ridge Summit, PA, 1981.

Coffron, J., *Getting Started in Digital Troubleshooting.* Reston Publishing, Reston, VA, 1979.

Computer, 13 (No. 3) March 1980. (Special issue on fault-tolerant testing. Several survey articles. Excellent bibliographies.)

Hardos, B., *Diagnostic Programming for Microprocessor-Based Systems.* Millenium Systems Inc., 19020 Pruneridge Ave., Cupertino, CA. Application Note 2.

Hardos, B., *Programming with the μSA MicroSystem Analyzer.* Millenium Systems Inc., 19020 Pruneridge Ave., Cupertino, CA. Application Note 1.

Implementing Signature Analysis for Production Testing with the HP 3060A Board Test System. Hewlett-Packard, Palo Alto, CA. Application Note 222-1.

Kneen, J., *Logic Analyzers for Microprocessors.* Hayden, Rochelle Park, NJ, 1980.

Lenk, J. D., *Handbook of Practical Microcomputer Troubleshooting.* Reston Publishing, Reston, VA, 1979.

Lesea, A., and Zaks, R., *Microprocessor Interfacing Techniques.* Sybex, Berkeley, CA, 1978, Chapter 8.

MCS6500 Microcomputer Family Hardware Manual. MOS Technology, Norristown, PA, August, 1975, p. 124. (Description of data latch for 6502 microprocessor. Note erratum for this section.)

New Techniques of Digital Troubleshooting. Hewlett-Packard, Palo Alto, CA. Application Note 163-2.

Operating and Service Manual, Model 1602A Logic State Analyzer. Hewlett-Packard, Colorado Springs, CO, 1978.

Slater, M., and Bronson, B., *Practical Microprocessors, Hardware, Software and Troubleshooting.* Hewlett-Packard, Palo Alto, CA, 1979.

Smith, D. L., Free-Run Signature Tables. *Proc. 4th Annual Conference on Computer Developments,* University of Houston at Clear Lake City, Clear Lake City, TX, November 5, 1982.

Smith, D. L., Limits on Exhaustive Tests of Medical Microprocessors. In *Proc. Frontiers of Engineering in Health Care,* Houston, TX, September 20, 1981.

Smith, D. L., Servicing A Small Computer That Fails To Restart. In *Proc. JAICCC (Joint Applications in Instrumentation, Control and Computing Conference),* Clear Lake City, TX, March 13, 1980.

Smith, D. L., *Signature Generator.* Dean Lance Smith, P.E., Houston, TX, 1982.

Smith, D. L., Techniques for Inserting Test Programs into Single-Board Micro-Computer Systems. In *Proc. Electronics Test and Measurement Conference,* Chicago, IL, October 5–8, 1981.

Smith, D. L., *Testing and Servicing Microprocessors*. Dean Lance Smith, P.E., Houston, TX, 1983.

Standard Patterns for Testing Memory. Micro Control Company, 7956 Main Street N.E., Minneapolis, MN. Application Note 00041.

SYM-1 Self Diagnostic Test Program. Synertek Corporation, P.O. Box 552, Sunnyvale, CA 95052, 1979. Publication SSC PUB MAN-A-260036-A EPS-1.

CHAPTER 15

Peripherals

Dean Lance Smith
Engineering Consultant
Houston, Texas

This chapter reviews the theory and operation of most common system peripherals. It discusses maintenance of each peripheral, when appropriate, and testing and servicing techniques.

15.1 SWITCH AND RELAY CONTACTS

Figure 15-1 shows several techniques for interfacing switch and relay contacts to a computer. The interface performs two functions. First, it converts "dry" contacts to the logic voltage levels required. Second, the interface debounces the contacts.

All switch contacts bounce when they are set to a new rest position. This results in continuity being broken several times after initial contact. Up to four bounces are not uncommon. The period of each bounce is on the order of 1 msec. Some contacts, particularly worn or dirty contacts, will make and break contact several times while leaving a rest position.

Most computer systems operate much faster than the bounce period of switch contacts. Without debouncing, the computer will interpret each bounce as a separate contact throw.

The interface in Figure 15-1a is debounced with software. The computer program checking the contact's position delays a millisecond or two after noting a transition. It then reads the contact position again and accepts the change only if the two readings agree. Otherwise, it will ignore the transition. Some computer programs will make three readings in a row before accepting a change in contact position.

Figure 15-2 shows timing diagrams for the inputs and outputs of the circuits in Figure 15-1. The two NAND gates in Figure 15-1b form an R–S flip-flop. The circuit in Figure 15-1c uses a one-shot (or monostable) and an OR (actually a NAND gate) gate to debounce the contacts. Both of these circuits are available as integrated circuits.

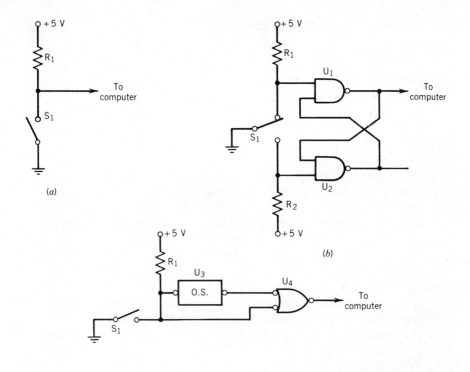

Figure 15-1 Typical switch and relay contact debounce circuits. (*a*) Software debounce circuit. (*b*) R–S debounce circuit. (*c*) One-shot (monostable) debounce circuit.

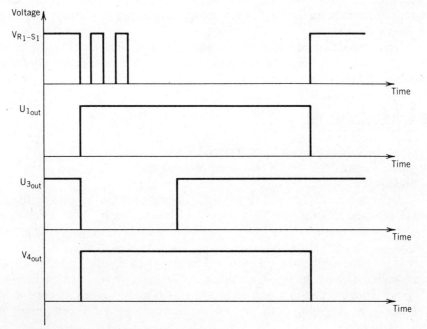

Figure 15-2 Debounce circuit timing diagrams.

15.1.1 Troubleshooting

The majority of contact interface failures are due to damaged or fouled contacts. Many switch or relay contacts are sealed. Replacement is the only fix.

Accessible contacts can be cleaned with a brush and common organic solvents such as acetone or alcohol. Avoid using cotton swabs. Cotton lint may foul the contacts. Commercial aerosol contact cleaners are available; these are usually fluorocarbon compounds.

Some contacts are replaceable. Broken or stretched contact springs can be replaced or adjusted on some switches and relays. Special contact and spring adjustment tools are available to aid maintenance.

Occasionally, a resistor or IC will fail in the interface. A logic probe is the preferred tool for checking the logic levels. An oscilloscope can also be used. An oscilloscope is the best tool for spotting debounce failures. A VOM (volt–ohmmeter) or an EVM (electronic voltmeter) can also be used for checking static logic levels.

MOS (metal oxide semiconductor) technology, especially CMOS (complementary MOS) chips, are susceptible to damage from static electricity discharges. Proper system maintenance will reduce static discharge failures.

15.1.2 Maintenance

Preventive maintenance consists of periodically cleaning loose dirt and lint from the area of the contacts. A vacuum cleaner and soft brush are excellent tools for removing dust and lint.

Banning smoke, food, and drinks in the area of the equipment will prolong contact life. Tobacco smoke tends to foul contacts. Spilled beverages cause shorts. Sugared beverages will foul contacts when they dry. Many beverages are mildly corrosive.

A grounded floor pad near the system will prolong interface life in locations subject to static electricity. Humidifying the air in the area of the equipment will also eliminate static electricity. Optimum relative humidity is 50 to 70%.

15.2 KEYBOARDS

Small keyboards are sometimes interfaced to a computer as a series of individual switches. Figure 15-3 shows two other techniques for interfacing a keyboard to a computer.

An encoder can be used to output a binary code corresponding to each switch selected. Hardware can be used to debounce each switch prior to encoding. Some encoders have debounce circuitry in their inputs. Software can also be used to debounce the outputs.

Most large keyboards are scanned as shown in Figure 15-3b. Each row line is pulled high by the pull-up resistor. A low pulse drives each column, one at a time. The row tied to a depressed switch will go low when its column line is driven low. Knowing both the column and row uniquely determines the depressed key.

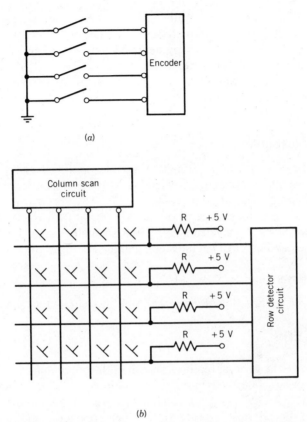

Figure 15-3   Typical keyboard interface circuits. (a) Encoder interface. (b) Scanned interface.

The columns are usually scanned between 1000 and 10,000 times per second. The scanning, debounce, and contact identification functions can be microprocessor-controlled. SSI (small-scale integrated) and MSI (medium-scale integrated) components can also be used to perform the control, scan, and debounce functions.

LSI (large-scale integrated) circuit chips are available that perform the scan, debounce, and control functions in one chip. Most LSI chips have an external resistor–capacitor network that controls the scan frequency. The key closure is usually available at the output of the chip as a binary code. Some LSI chips have an internal ROM (read only memory). The ROM permits the output code to be any binary code, such as ASCII.

All keyboard interfaces must perform one other function: rollover control. Rollover occurs when two or more keyboard switches are depressed at the same time.

Some encoder interfaces have priority encoding. The code corresponding to the highest or lowest input is generated by the interface.

Other encoder interfaces use N key rollover or N key lockout to handle two or more key depressions. With N key lockout, the first key recognized locks the keyboard encoder until all keys are released.

With N key rollover, other key depressions are recognized once the first key is recognized. However, a key is not recognized again until it is released. N key rollover usually requires diodes in series with each switch. Without the diodes, undepressed keys may be recognized because of the current paths through the depressed keys.

15.2.1 Troubleshooting

Troubleshooting keyboards is similar to troubleshooting switches. Keyboard switches are usually not sealed. They can be disassembled and cleaned. Broken contacts and springs can often be replaced.

A logic probe is an excellent tool for looking at the rows and columns of a scanned keyboard. The electronic circuitry is usually working correctly if the probe indicates activity. An oscilloscope can also be used to observe activity.

15.2.2 Maintenance

Maintenance of keyboards is identical to that of switch and relay contacts.

15.3 PRINTERS

All printers have the general features shown in the block diagram in Figure 15-4. Several standard interfaces are used on printers. The most popular serial interface is the RS-232C. This interface is discussed in more detail in Chapter 11. The RS-449 and 20-mA current loop interfaces are sometimes used for serial interfaces. These interfaces are also discussed in Chapter 11.

The most popular parallel interface is the Centronics interface. This is a de facto standard first used by the Centronics Company on its printers. Table 15-1 lists some properties of this interface. Table 15-2 shows the function of each of the 36 pins in the connector used for the interface. Figure 15-5 shows a timing diagram of the Centronics interface handshake signals.

The IEEE-488 interface is also used for parallel data. This interface is discussed in Chapter 11.

The interface circuitry in Figure 15-4 converts the data to the logic levels used inside the printer. Most modern printers use TTL (transistor–transistor logic) levels. Data are converted from serial format to parallel format in most serial interface circuits.

The control circuitry coordinates the movement of data into the printer and to the printing head and carriage. Most printers use a handshake protocol to permit data to be transferred to the printer without loss.

Most printers have some RAM (random access memory) or read/write memory to act as a buffer. The buffer serves several purposes. It permits data to be transferred to the printer at rates greater than the character print rate for brief periods. Some printing heads require that a line or two of data be stored prior to printing.

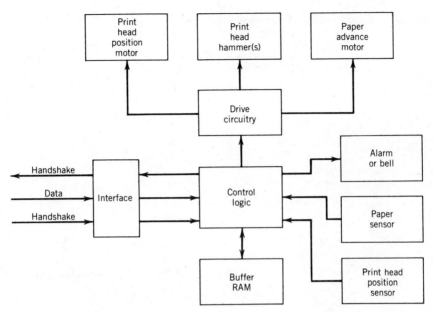

Figure 15-4 Block diagram of a printer.

The drive circuitry converts the TTL logic levels to the voltage and current necessary to drive the paper advance motor, print head position motor, and print head hammer(s). Some paper advance motors are stepping motors. A logic pulse causes a stepping motor to rotate in discrete steps. Other paper advance motors are continuous motors. The amount the motor rotates is determined by the length of the pulse applied to it.

The print hammer drive circuitry will also convert the data code to the print hammer code, if necessary. Most motors and print hammers operate from 12- or 24-V dc supplies.

Most printers have an audio alarm. For historical reasons, the alarm is

Table 15-1 **Summary of Centronics interface properties**[a]

Connector: 36-pin connector (Amphenol No. 57-40360)
Voltage levels: TTL (SN7400 series)
Logic levels: 2.4 to 5.0 V (logic 1)
          0 to 0.4 V (logic 0)
Current levels: source 0.320 mA at 2.4 V
           sink 14 mA at 0.4 V
Line termination: 1000 Ohms to +5 V (DATA 1–DATA 8)
           470 Ohms to +5 V (DATA STROBE* and INPUT PRIME*)
Maximum distance: 10 feet

[a]Asterisk means active low. Adapted from *Interface Specifications*. Centronics Data Computer Corporation, Hudson, NH 03501, 1977.

Table 15-2 **Centronics interface pinout and function**[a]

Pin	Signal name	Function
1	DATA STROBE*	1.0-msec pulse (min.), clocks data
2	DATA 1	Data, least significant bit
3	DATA 2	
4	DATA 3	
5	DATA 4	
6	DATA 5	
7	DATA 6	
8	DATA 7	
9	DATA 8	Data, most significant bit
10	ACKNLG*	Indicates data received by printer
11	BUSY	Printer cannot accept data
12	PE	Printer out of paper
13	SLCT	Printer selected
14	OV	Signal ground (sometimes called SS)
15	OSCXT	A 100-KHz or 200-KHz signal
16	OV	Signal ground
17	Chassis Gnd	Frame ground
18	+5 V	+5 V power bus
19	DATA STROBE*	Return pin 1
20	DATA 1	Return pin 2
21	DATA 2	Return pin 3
22	DATA 3	Return pin 4
23	DATA 4	Return pin 5
24	DATA 5	Return pin 6
25	DATA 6	Return pin 7
26	DATA 7	Return pin 8
27	DATA 8	Return pin 9
28	ACKNLG*	Return pin 10
29	BUSY	Return pin 11
30	INPUT PRIME*	Return pin 31
31	INPUT PRIME*	Clears print buffer, resets printer
32	FAULT*	Problem with printer (no paper, etc.)
33	FAULT*	Return pin 32
34	Line count pulse	One side of line count switch
35	Line count pulse	Other side of line count switch
36		Not used

[a]Asterisk means active low; return signals are part of a twisted pair. Adapted from *Interface Specifications*. Centronics Data Computer Corporation, Hudson, NH 03501, 1977.

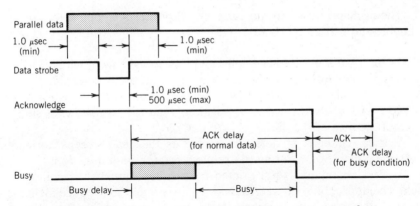

Parallel data

1.0 μsec (min)

1.0 μsec (min)

Data strobe

1.0 μsec (min)
500 μsec (max)

Acknowledge

ACK delay
(for normal data)

ACK

ACK delay
(for busy condition)

Busy

Busy delay

Busy

Figure 15-5   Timing diagram for the Centronics interface. (From *Interface Specifications*. Copyright 1977 Centronics Data Computer Corporation, Hudson, NH. Reprinted by permission.)

called a bell. The bell can be triggered by a data character. The alarm can also be triggered by problems detected by the printer control circuitry.

One problem that will trigger the alarm on most printers is running out of paper. The paper passes through a sensor before passing over the print head. The sensor is usually a switch. The paper closes the switch on some models. On other models, the paper keeps the switch open.

Other errors that may trigger the alarm include running out of ribbon or toner, having the print buffer overflow, receiving a nonprintable control character that is not recognized by the printer, and power failure. Not all these conditions will trigger the alarm on all printer models. These conditions will also inhibit the interface handshake signal on some printers.

Many different types of print heads are used. The features of the more common heads are given below.

15.3.1 Types

Daisy Wheel   Daisy wheel printers produce high quality printing at speeds up to 50 characters per second. The printing is produced by a hammer striking the backside of a piece of type. The hammer is actuated by a solenoid.

The type presses against the paper through an inked ribbon when struck by the hammer. Carbon ribbons are also used. The process is similar to a typewriter except that the type is arranged on a wheel like petals on a daisy.

The daisy wheel rotates continuously in some models. The printer control circuitry tracks the rotation and forces the hammer to strike when the correct character is in the vertical position.

In other models, the daisy wheel does not rotate continuously. Rather, the daisy wheel rotates in discrete amounts. It rotates only as far as necessary to bring the proper character into the vertical position. The wheel then stops until the next character needs to be typed.

The principles used in the daisy wheel printer are similar to those used in thimble-and-ball-type printers.

Dot Matrix   A dot matrix head consists of a vertical row of wires. Each wire strikes the paper through a ribbon when activated by its solenoid. The wires pass through a die when striking the paper. The die is positioned just above the paper and holds the wires in position. Figure 15-6a shows a simplified sketch of the face of the die.

Vertical columns of dots are printed as the head sweeps horizontally across the paper. Some dot matrix printers print in both directions.

Letters and symbols are printed by treating a number of columns as a block. The appropriate dots in each block are printed to produce a character. Figure 15-6b shows how the character A can be created in a 9 × 7 block of dots.

The outer rows and columns in each block are usually not printed. This improves separation of characters. Common block formats are nine by seven (as shown in Figure 15-6b), 9 × 7, and 7 × 5. Other block formats are used to give different types of fonts.

Dot matrix heads can also be used to print graphs or pictures. Each dot

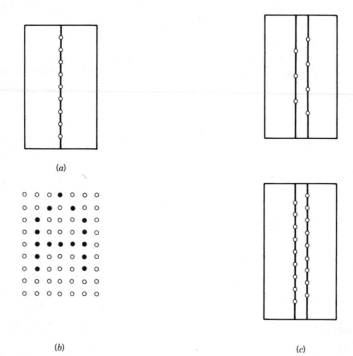

(a)

(b)   (c)

Figure 15-6   Dot matrix printer features. (*a*) Dot matrix print head (simplified front view). (*b*) Character A in 9 × 7 format. (*c*) Letter quality print heads (simplified front view).

can be treated as a pixel, or picture element. Character blocks can also be treated as pixels. The result is a picture or graph that looks something like a picture printed in a newspaper.

Higher-quality characters that closely resemble daisy wheel printing can be produced by advancing the carriage half a pixel and retyping the character. Some dot matrix print heads improve character quality by having an additional set of staggered print wires as shown in Figure 15-6c. The additional set of wires fills in the vertical gaps.

Many dot matrix printers have the capability of mixing several different types of fonts and graphics. The higher-quality print fonts reduce printing speed.

Chain A chain printer has type attached to a chain. The chain forms a loop. The loop runs horizontally across the paper.

A hammer strikes the type and presses it against the paper through a ribbon. High-speed chain printers have a hammer for each character position. Lower-speed printers have one hammer that moves across the page and strikes the chain as the proper character goes by.

Chain printers can operate at up to several thousand characters per second. The darkness of the type tends to be uneven, particularly if the hammers and chain are worn or improperly adjusted. These are seldom used for personal computers.

Thermal Thermal printers require a special heat-sensitive paper. Warm type contacts the paper and produces the character. Some print mechanisms work on the chain principle. Others work on the dot matrix principle.

The result is moderate- to fairly high-quality printing with moderate contrast. The background tends to darken as the paper ages, particularly if stored in a warm environment.

Ink Jet A stream of ink is painted on the paper corresponding to the material to be printed. The stream is electrically charged so that the stream can be deflected electrostatically. The jet is mechanically moved across the page while the jet is electrostatically deflected to produce the characters.

The result is a high-speed printer that can produce fairly good-quality printing. Some models can intermix type and graphics. Type fonts can be changed while printing.

Electrostatic Electrostatic printers work on the same principles as a copying machine. A charged particle beam paints charge on the paper corresponding to the material to be printed. The charge produces static electricity. Toner is applied to the paper. The static electricity causes the toner to adhere to the paper where the charge has been painted. Baking the paper permanently bonds the toner to the paper.

The result is high-quality, high-speed printing. Type and graphics can be intermixed. Type fonts can be changed while printing.

15.3.2 Troubleshooting

Several tests can be performed to isolate problems with a printer:

1. Many newer microprocessor-controlled printers have built-in self-test capability. The test is usually a good functional test of the entire system, except for the interface. Most tests repeatedly type all printable characters in a known sequence. This test can be used to determine if the overall system is working properly. It can also be used to spot broken or misaligned type, ribbon, and paper feed problems.

2. Some computers have a printer test built into them. Other computers can easily be programmed to test a printer. A popular printer test is to type the phrase, "The quick brown fox jumped over the lazy dog." This phrase uses almost all of the characters in the English alphabet. A better test is to follow the phrase with the numbers 0 through 9 and all the other printable characters on the printer. Either of these tests can be used to observe the overall behavior of the printer and the behavior of the major mechanical parts.

3. A printer with a Centronics interface can be tested by plugging a special test plug into the interface. Table 15-3 shows how the test plug should be wired. Connecting the handshake lines together as shown in the table permits the printer to handshake with itself. Wiring the data connections into an alternate 10101010 or 01010101 pattern is a good test of the data circuitry. However, any pattern can be used. This test can be used to check the overall operation of the printer. It can also be used with a logic probe or oscilloscope to trace signals through the electrical parts of the printer. A signature analyzer can also be used with this test, provided the signature tables have been generated beforehand.

   A word of caution. The motor and driver voltages will damage most logic probes and signature analyzers. These instruments are usually designed to be used only on TTL voltage levels.

4. Printer test boxes are available from several printer manufacturers and independent suppliers. These plug into the printer interface and drive it with a test message.

Table 15-3   **Centronics interface test connections**[a]

Pattern	ASCII character	Connections
11111111	Delete	None
10101010	*	8 – 26, 6 – 24, 4 – 22, 2 – 20
01010101	U	9 – 27, 7 – 25, 5 – 23, 3 – 21
00000000	Null	9 – 27, 8 – 26, 7 – 25, 6 – 24, 5 – 23, 4 – 22, 3 – 21, 2 – 20

[a]Pins 1 and 10 (DATA STROBE* and ACKNLG*) should be connected for all test connectors. The automatic line feed option of the printer (if it has one) should be turned on.

5. A terminal can be connected to the printer, provided the printer and terminal have compatible interfaces. The terminal keyboard can be used to type test messages on the printer.

15.3.3 Mechanical Problem Checklist

Most problems that occur with a printer are mechanical problems:

1. Check to see that the printer has adequate paper and that the paper isn't jammed. A scrap of paper will sometimes jam in the paper sensor. The jammed paper can cause the machine to indicate it is out of paper when it isn't, or vice versa. A scrap of paper can jam the tractor that feeds paper to the machine. Scrap paper can jam the print bar or print head.

2. Check the ribbon. A jammed ribbon will eventually wear out in one spot. As a result, the machine will produce illegible characters. Eventually, the print head will cut the ribbon by repeatedly striking it at one point. Make sure the ribbon is in all the guides that route it by the printing head. Make sure the ribbon moves freely through the guides. Make sure the ribbon cartridge is securely positioned. Some ribbon cartridges have a knob that permits the ribbon to be advanced manually. Use it to check ribbon alignment. Check that carbon ribbons are being used properly. These ribbons are intended for one pass through the machine. Sometimes, two or three passes can be obtained before the ribbon wears out or breaks. However, print quality is drastically reduced by repeated use of these ribbons.

3. Make sure the cable from the computer to the printer is plugged in securely. Make sure the power cord is securely plugged into the wall socket. Printers vibrate. The vibrations can loosen the power cord and the interface cable.

4. Misaligned printing heads can be aligned for most machines. Check the service manual for the particular machine. Special alignment tools are available for some machines. Check with the machine's manufacturer or an independent electronic tool supplier. Broken or worn daisy wheels, chains, and dot matrix heads are easily replaceable on most units. Check the operator or service manual for details.

15.3.4 Electrical Problem Checklist

Most electrical problems are due to failures in the motors, driver circuits, and the special power supplies for the driver circuits:

1. Use an EVM or VOM to check the power supply voltages.

2. Use an oscilloscope or digital waveform tester to check the drive waveforms at the solenoids and motors. Stimulate the solenoids and motors with one of the tests described earlier. Consult the equipment service manual for specific instructions for replacing defective hammer solenoids and motors.

3. Check the remaining logic with a logic probe or an oscilloscope. Use one of the tests described earlier to stimulate the machine. Refer to the unit's power or service manual for schematics and specific instructions.

15.3.5 Maintenance

Maintenance for most printers consists of periodically cleaning the unit and changing the ribbon. Some units do require lubrication or other special maintenance. Consult the unit's owner or service manual for details.

Cleaning   Cleaning should be done at least once a year. Once a month would not be too often for high-speed units that are used almost continuously. The lint and small pieces of paper that accompany most printer paper will eventually jam a printer.

Cleaning consists of thoroughly vacuuming the paper feed and the inside of the unit. Disconnect the power to the printer before cleaning. A soft brush is useful for loosening dust and lint that a vacuum cleaner alone cannot remove.

Ribbons   Ribbons should be replaced periodically based on use or wear. An ink ribbon that is used infrequently will dry out before all the ink is used. As a result, ink ribbons should be replaced at least every two months. (Saving the ribbon in a sealed plastic bag when not in use will keep it from drying out.) More frequent replacement is needed for continuous-duty printers. Replacing the ribbon once or twice a week is not uncommon for continuous-duty high-speed printers.

15.4 CRT DISPLAYS

Some CRT (cathode ray tube) or VDT (video display terminal) displays use a standard black and white TV picture tube to display information, though many use CRTs with green or yellow phosphor. Both text and graphic material can be displayed. Color TV picture tubes are used for color displays.

Higher-resolution displays use special CRTs. RADAR display tubes are used in some high-performance displays. Low-cost computers come with rf interfaces to use a television set as a display.

All these displays work on the same basic principles. An electron gun in the tube paints the information on the tube's screen. The screen is coated with a phosphor compound. The phosphor compound emits light in proportion to the number of electrons striking it.

The display must be refreshed periodically or it will fade. TV picture tube displays are usually refreshed 25 to 30 times per second. RADAR-type displays usually are coated with a longer-persistence phosphor compound. These displays need not be refreshed as often.

Storage-tube-type displays do not require periodic external refresh. The tube does the refresh automatically. Material can be added to a storage tube display. However, erasing material requires that the entire display be erased and the new display drawn. Storage tube displays are not used with personal computers.

15.4.1 Types of Display

TV-Type Displays TV picture tube displays usually take advantage of the low-cost high-voltage circuitry available for driving these tubes. This circuitry traces from 525 to 625 horizontal lines every frame. It paints 25 to 30 new frames every second.

Interfaces that drive commercial TV sets modify this format slightly. TV set pictures require interlaced scanning. Half the picture is painted on one frame. The other half of the picture is painted on the next frame. Alternate lines are painted on alternate frames.

Text information is painted using character blocks similar to those created by a dot matrix printer. A common character block format uses nine rows and seven columns. Many VDTs will display 24 or 25 rows of characters with 80 characters per row. The bandwidth of the resulting video signal can be over 15 mHz. This, however, is too high for most TV sets.

TV set displays usually use 12 rows of 54 characters or less. Sometimes, fewer rows are used. A 7 × 5 character format is often used. The resulting video signal is modulated on a carrier frequency corresponding to one of the lower VHF television channels and fed directly to the VHF antenna connection of the set.

Figure 15-7 shows a block diagram of a typical TV picture tube video

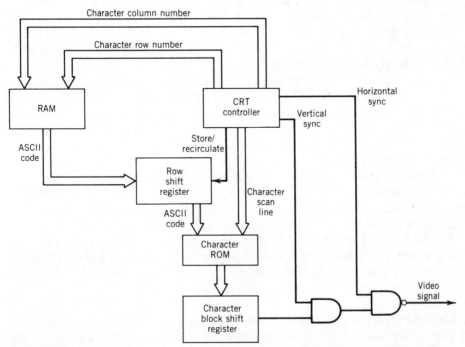

Figure 15-7 Block diagram of CRT (cathode ray tube) display circuit for driving a TV-style CRT.

display. The CRT controller times the display and generates the horizontal and vertical synchronization signals. The characters for each row are stored in the RAM. The characters are usually stored in ASCII code.

At the start of each row, the CRT controller fetches the characters for that row from the RAM and stores them in the row shift register. The row shift register output drives part of the ROM (read only memory) address pins. The remaining ROM address pins are driven by the CRT controller. The CRT controller supplies the character row number to the ROM.

The output from the ROM is the dot pattern for one row of a character. The ROM output is stored in the character shift register. The serial output of the character shift register is the main component of the video signal.

The CRT controller recirculates the contents of the row shift register for each line of the row until all lines of the row have been displayed. Then, the CRT controller fetches a new row from RAM and the cycle repeats for the next character row.

The CRT controller supplies horizontal synchronization pulses at the end of each line. It supplies vertical synchronization pulses at the end of each frame. These are combined with the character shift register output to form the video signal.

If the CRT display is built into the computer system, the character RAM can be part of the system RAM. The CRT controller interrupts the computer every time it needs the characters for a new row. The CRT controller halts the computer's CPU and takes control of the system RAM until it has stored the new row into the 80-character shift register.

Interrupting and halting the computer to share the character RAM reduces the performance of the computer. The computer is halted every time the CRT controller needs to refresh a new row.

The computer performance can be improved by bank sharing. In bank sharing, the CRT controller has its own character RAM. The computer sets a flag and interchanges the character RAM with a bank of its own RAM any time it wishes to modify the display. However, the computer is halted any time it tries to read or write into the character RAM when the CRT controller is refreshing a new row.

Bank switching requires more RAM and more hardware to coordinate the CRT controller and the computer. However, it does improve the computer performance. The computer is halted only when it attempts to modify the character RAM while the CRT controller is refreshing a new row.

If the CRT display stands alone or is part of a terminal, a separate controller stores the incoming data in the character RAM. Data storage is usually halted when the CRT controller needs to refresh a new row.

High-Performance Displays RADAR-type display tubes usually have a phosphor that emits light for a long period of time after an electron strikes it. The techniques used to display information are similar to those for TV picture tube displays. However, the longer light emission of the phosphorus compound permits longer refresh times. This means a finer grid can be used to display information and still keep video signal bandwidth to a reasonable level. The result is a better-quality display.

Some RADAR-type display tubes are what are called storage tubes. The information of the display is stored on the face of the tube rather than in RAM. The display is refreshed inside the tube until a signal is removed from a control grid in the tube. This type of tube can use an extremely fine grid and yet still use relatively low-bandwidth video signals to move the electron beam from one point on the screen to another. Characters are drawn on the screen similar to the way a drafter would draw letters. These displays are used almost exclusively for graphics.

15.4.2 Troubleshooting

Testing CRT displays is similar to testing printers. The object of the test is to check the display's response to all permitted characters.

Most failures in CRT displays are in the high-voltage CRT driver circuits and their power supplies. Troubleshooting these TV picture tube displays is almost identical to troubleshooting similar circuits in a TV set.

An oscilloscope with a high-voltage probe and an EVM with a high-voltage probe are the most useful test instruments. Any of the tests described under printers can be used to generate a test signal.

A word of caution. The high-voltage sections of a CRT display pose a severe shock hazard. Do as much work as possible with the power turned off. If the power must be turned on to make measurements, place one hand behind your back. Use the other to probe the equipment. This will prevent a shock from traveling through the chest (and the heart). Remove all rings, watches, and other metal objects from both hands. Remove all metal chains and pendants. Metal lowers the contact resistance of the skin and can cause an accidental short circuit.

TTL voltage level logic can be probed with a logic probe or an oscilloscope. Take care to avoid contact with the high-voltage section when probing the low-voltage logic circuits.

One other observation. The high-voltage power supplies have little or no filtering of their dc outputs. This means the output voltages of these supplies are very susceptible to line voltage variations. These variations, or transients, will cause the display to shake. They can be caused by nearby heavy equipment turning on and off. The variations can also be caused by lightning. If the shaking of the display is objectionable, then control the source of the transients (if possible) or suppress the transients. The transients can be suppressed by plugging the power cord for the display into a voltage-regulating transformer or line filter.

15.4.3 Maintenance

The only maintenance required of CRT displays is an occasional washing of the screen. Use a soft cloth dampened in a solution of mild soap or detergent and water to remove the dirt. Do so with the power turned off. Avoid spraying water on the unit.

15.5 LED, LC, AND GD DISPLAYS

15.5.1 Description

A LED (light-emitting diode) is a diode that emits almost pure light of one color when excited by electricity. An LC (liquid crystal) device emits no light of its own. Rather, its ability to transmit light changes if electricity is applied to it. LC devices are usually mounted in front of a mirror or light source.

GD (gas discharge) devices emit almost pure light of one color when electrically excited. A GD device is constructed by placing two electrodes into a glass tube containing a special gas under relatively low pressure.

LED, LC, and GD displays come in a variety of forms. LEDs are used as on/off indicators. LEDs, LCs, and GDs are used as 7- and 16-segment displays as shown in Figure 15-8. Seven-segment displays can be used to display the numbers 0 through 9. They can also be used to represent the hexadecimal characters a, b, C, d, E, and F. Some letters, but not all, can be displayed by a seven-segment display. Sixteen-segment displays can be used to represent the numbers 0 to 9, all capital letters of the English alphabet, and some special characters such as +, −, *, and the like.

LEDs are sometimes used in dot matrix displays.

Figure 15-9 shows two typical LED driver circuits for single displays. The resistors limit the current to 10 or 20 mA when the transistor turns on. Integrated circuit drivers are available that require no external components except the LED. Integrated circuits that contain both the LEDs and drivers in one package are also available.

Seven-segment LED displays are available as common anode or common cathode (as shown in Figure 15-9c) integrated circuits. Seven-segment driver circuits are also available as separate integrated circuits or integrated with the displays. The displays also come integrated with latches and counters.

Multiple seven-segment displays are usually scanned to reduce the parts count. Figure 15-9c shows a typical circuit. The same segment of each display

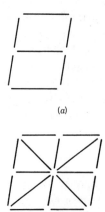

(a)

(b)

Figure 15-8 Seven and 16 segment display patterns.
(a) Seven-segment display. (b) Sixteen-segment display.

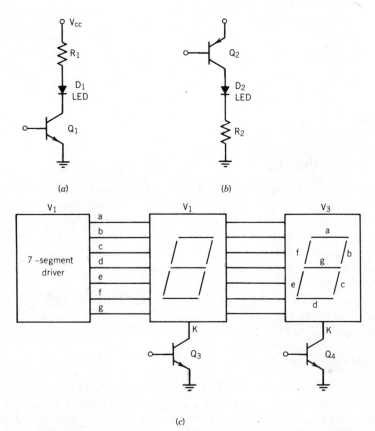

Figure 15-9 LED (light-emitting diode) driver circuits. (*a*) npn Transistor driver cirucit. (*b*) pnp Transistor driver circuit. (*c*) Scanned seven-segment display.

is connected to the same driver. The transistors shown in the figure act as enable drivers. Each display is turned on (with the others off) in succession. Each display is turned on between 100 and 10,000 times per second. A scan rate above 10 to 20 times per second will not be noticed by the human eye. All displays will appear to be illuminated simultaneously.

Some scanned displays are designed so that the average current is under the 10- to 20-mA rating of the LED segments. The peak current however will exceed the average maximum rating of the LEDs. If the scan stops for some reason, the LED segments that happen to be on may burn out.

More conservative designs avoid this problem by restricting the peak current to the average current rating of the LEDs. This does sacrifice brightness for safety if the scan circuitry fails.

Sixteen-segment displays are driven the same way as seven-segment displays. Dot matrix displays usually have the scan and driver circuitry built into the displays. Internal buffers hold the display data once they are written into the chip. Newer 7- and 16-segment LED displays also have built-in scan, driver, and buffer circuitry.

LC and GD driver circuits are similar to those shown for the LED displays. LED displays usually require at least a 5-V source. LC display circuits can operate from supplies as low as 1.4 V. Whereas LEDs are essentially diodes and have low resistance, LCs are fairly high-resistance devices. Their lower power consumption makes them ideal for portable equipment.

GD devices usually require higher-voltage supplies (from 24 to 120 V, depending on the gas and technology used). Some GD devices consume less power than LEDs. GD devices are usually found in more complicated displays.

15.5.2 Troubleshooting

LED, LC, and GD displays are fairly reliable devices. LEDs will burn out occasionally. GDs probably are more prone to burnout than LEDs. Turning all the elements of a display on will show if a display or segment is working.

LC and GD devices are sensitive to cold. Warming them up will usually cure any problems that occur due to cold. The higher-voltage supplies used with GD devices are sometimes unreliable. If none of the display devices works, the power supply is probably the problem.

An oscilloscope can be used to trace the signal to a malfunctioning display while it is being driven. A logic probe can be used on LEDs and LCDs driven by TTL supplies. A VOM or EVM can be used on unscanned displays.

15.5.3 Maintenance

No maintenance is required for LED, LCD, or GD displays except occasionally cleaning the display surface. Wipe the dust from the display with a soft cloth. The cloth can be dampened with a water solution of mild soap or mild detergent. Wipe the solution off with a soft cloth dampened with water.

Avoid cleaning when power is applied to the display. Avoid cleaning while GD displays are warm. A cold cloth may rapidly cool and crack a display. Avoid applying water directly to the displays. The residue may cause a shock hazard.

15.6 TERMINALS

A terminal is essentially a keyboard and display combined in one unit. See the earlier sections on keyboards and displays for more information about theory of operation, servicing, and maintenance.

Chapter 11 discusses the overall troubleshooting of terminals in more detail. The use of the HDX/FDX switch and the loopback connector to test a terminal are discussed in that chapter.

15.7 FLOPPY DISK DRIVES

15.7.1 Operation

A floppy, or flexible, disk is a plastic disk coated with magnetic material. The disk is permanently housed in a square plastic cover. The disk and inside of the cover are permanently lubricated. The disk and housing are inserted into the drive as a unit, and the disk rotates inside the housing.

A spindle clamps the disk through a large center hole when the disk is inserted into the drive. The spindle rotates the disk. A read/write head contacts the disk surface any time the disk drive is selected.

Access to the magnetic surface (on the underside of the disk on a single-sided disk, both sides on a double-sided disk) is obtained through a slot (or slots) in the housing. The disk is flexible and flops or wobbles when it moves in the drive, hence the name. The disk and housing are stored in a protective envelope.

Floppy disks are currently manufactured in two popular sizes, 8 inches and $5\frac{1}{4}$ inches. Table 15-4 lists the physical properties of both the 8- and $5\frac{1}{4}$-inch drives. The properties of $5\frac{1}{4}$-inch drives vary.

Table 15-4 **Properties of floppy disk drives**

	Drive size	
Property[a]	8 inches	$5\frac{1}{4}$ inches
Unformated capacity (kbytes)		
SS SD	409.6	125/250
SS DD	819.2	250/500
DS SD	819.2	250/500
DS DD	1638.4	500/1000
Rotational speed (rpm)	360	300
Inner track recording		
Density (bpi)		
SD	3200	2768/2788/2961
DD	6400	5536/5576/5922
Track density (tpi)	48	48/96
Tracks	77	40/80
Ambient temperature (°F)	40 – 115	40/50 – 115
(°C)	4 – 46	4/10 – 46
Relative humidity (%)	20 – 80	20 – 80
Typical MTBF (hours)	8000	8000
Error rates		
Soft read	1 in 10^9	1 in 10^8 or 10^9
Hard read	1 in 10^{12}	1 in 10^{11} or 10^{12}
Seek	1 in 10^6	1 in 10^6
Media life		
Passes/track	3.5×10^6	3.0×10^6
Insertions	30,000 +	

[a]SD, Single density (FM); DD, double density (MFM); SS, single-sided; DS, double-sided.

Sources: SA 800/801 Diskette Storage Drive Service, Maintenance Manual. Shugart Associates, Sunnyvale, CA, 1977; *SA200 Minifloppy Diskette Storage Drive OEM Manual.* Shugart Associates, Sunnyvale, CA, 1982; *SA410/460 96TPI Single/Double-Sided Minifloppy Diskette Storage Drives OEM Manual.* Shugart Associates, Sunnyvale, CA, 1981.

Table 15-5   **SASI 8-inch floppy disk interface pinout and function**

Pin	Symbol	Function
1		
2		
3		
4		
5		
6		
7		
8		
9		Return pin 10
10	TWO-SIDED*	Two-sided disk in selected drive
11		Return pin 12
12	SIDE SELECT	High for side 0, low for side 1 (signal to two-sided drives)
	DISK CHANGE*	Drive door open (optional on single-sided drives)
13		
14		
15		Return pin 16
16	IN USE*	Turns on activity LED over door (optional)
17		Return pin 18
18	HEAD LOAD*	Loads read/write head on disk (optional)
19		Return pin 20
20	INDEX*	Beginning of track
21		Return pin 22
22	READY*	Drive stable and ready for use
23		Return pin 24
24	SECTOR*	New sector (hard-sector disks only)
25		Return pin 26
26	DS1*	Drive select 1, loads head, turns on LED
27		Return pin 28
28	DS2*	Drive select 2, loads read/write head
29		Return pin 30
30	DS3*	Drive select 3, loads read/write head
31		Return pin 32
32	DS4*	Drive select 4, loads read/write head
33		Return pin 34
34	DIRC	Direction select for read/write head; one is out, zero is in when STEP* active
35		Return pin 36
36	STEP	Rising edge steps read/write head one track
37		Return pin 38
38	WRITE DATA*	Falling edge toggles current in read/write head
39		Return pin 40
40	WRITE GATE*	Permits writing when active, permits reading when inactive
41		Return pin 42

Table 15-5 (*Continued*)

Pin	Symbol	Function
42	TRACK 00*	Selected drive on track 0 (outermost track)
43		Return pin 44
44	WPRT*	Write protected disk in selected drive
45		Return pin 46
46	READ DATA *	Raw (clock and data) read data from selected disk
47		Return pin 48
48	SEP DATA*	Read data separated from clock
49		Return pin 50
50	SEP CLOCK*	Clock separated from raw read data

*a*Asterisk means active low; Return is common or ground. Adapted from *SA 800/801 Diskette Storage Drive Service, Maintenance Manual*. Shugart Associates, Sunnyvale, CA, 1977.

As this chapter is being written, several manufacturers have introduced drives that take disks that are approximately 3 inches in diameter. Most of these drives are incompatible. It is not yet clear which, if any, of these drives will become a de facto standard as was the case with the 8- and 5¼-inch drives. Indeed, an attempt is being made to set a formal industry standard.

The Shugart Associates Standard Interface (SASI) is the de facto interface for both the 8- and the 5¼-inch drives. The SASI interface for the 8-inch drives is given in Table 15-5. The SASI interface for the 5¼-inch drives is shown in Table 15-6. Most manufacturers of the 3-inch drives seem to be using the SASI 5¼-inch interface.

Data on a disk is organized by track and sector. Tracks are concentric rings of data; they are identified by number and are numbered consecutively. The outermost track is track 0.

A sector is a part of a track. Sectors are also identified by number and are normally numbered consecutively. The first sector is sector 1.

Several different techniques are used to identify sectors. The most popular is soft sectoring. A soft-sectored disk has one index hole. The sector following the hole on each track is sector 1. The number of sectors on each track is determined by the recording format used and the length of each format.

Two recording codes are used: FM and MFM. Both are discussed in Chapter 11. FM is usually referred to as SD (single density) in the sales literature. MFM is usually referred to as DD (double density) in the sales literature. Double-density recording permits twice as much data to be stored on a disk as single-density recording.

The protocols used for both single- and double-density recording are shown in Table 15-7. The length of the data stored in each sector is called a record. Several different record lengths are used. Table 15-8 shows the number of sectors per track for each type of record length and the amount of formatted

Table 15-6 **SASI 5¼-inch floppy disk interface pinout and function**

Pin	Symbol	Function
1		Return pin 2
2	DOOR LOCK	Locks door (optional)
3		Return pin 4
4	IN USE	Activates display LED
5		Return pin 6
6	DS4*	Drive select 4
7		Return pin 8
8	INDEX*/SECTOR*	Beginning of track/new sector (hard sectors)
9		Return pin 10
10	DS1*	Drive select 1
11		Return pin 12
12	DS2*	Drive select 2
13		Return pin 14
14	DS3*	Drive select 3
15		Return pin 16
16	MOTOR ON*	Turns drive motor on for selected drive
17		Return pin 18
18	DIRC	Direction select for read/write head; one is out, zero is in when STEP* active
19		Return pin 20
20	STEP	Rising edge steps read/write head one track
21		Return pin 22
22	WRITE DATA*	Falling edge toggles current in read/write head
23		Return pin 24
24	WRITE GATE*	Permits writing when active, permits reading when inactive
25		Return pin 26
26	TRACK 00*	Selected drive has found track 0 on disk
27		Return pin 28
28	WPRT*	Write protected disk in selected drive
29		Return pin 30
30	READ DATA*	Raw (clock and data) read data from selected drive
31		Return pin 32
32	SIDE SELECT	High for side 0, low for side 1 (signal to two-sided drives)
33		Return pin 34
34	DRIVE STATUS*	Undisturbed disk in drive

[a]Asterisk means active low; return is common or ground. Adapted from *SA 410/460 96TPI Single/Double-Sided Minifloppy Diskette Storage Drives OEM Manual*. Shugart Associates, Sunnyvale, CA, 1981.

Table 15-7 **Initial floppy disk formats**[a]

Single density		Double density	
Number of bytes	Hex value of bytes written	Number of bytes	Hex value of bytes written
Beginning of Each Track			
40	FF (or 00)	80	4E
6	00	12	00
		3	F6 (writes C2)
1	FC (index mark)	1	FC (index mark)
26	FF (or 00)	50	4E
Each Sector			
6	00	12	00
		3	F5 (writes A1)
1	FE (ID address mark)	1	FE (ID address mark)
1	(Track number)	1	(Track number)
1	(Side number)	1	(Side number) (00 or 01)
1	(Sector number)	1	(Sector number)
1	(Sector length)[b]	1	(Sector length)[b]
1	F7 (2 CRCs written)	1	F7 (2 CRCs written)
11	FF (or 00)	22	4E
6	00	12	00
		3	F5 (writes A1)
1	FB (address mark)	1	FB (data address mark)
+	Data (E5)	c	Data
1	F7 (2 CRCs written)	1	F7 (2 CRCs written)
27	FF (or 00)	54	4E
End of Each Track			
247*	FF (or 00)	598[d]	4E

[a]Adapted from *FD179X-02 Floppy Disk Formatter/Controller Family, 1983 Components Handbook.* Western Digital Corporation, Irvine, CA, 1983, p. 189.
[b]Sector length codes are: 00 for 128 bytes/sector, 01 for 256 bytes/sector, 03 for 512 bytes/sector, and 04 for 1024 bytes/sector.
[c]Either 128, 256, 512, or 1024.
[d]Approximately.

data that can be stored on both 8- and 5¼-inch disks using different formats and different recording densities.

Single-density recording with 128 bytes/sector is sometimes called the IBM 3740 format. Double-density recording with 256 bytes/sector is sometimes called the IBM System 34 format.

There are hard sector disks that are less commonly used. These disks have several holes equally spaced around the peripheral or center hole of the disk. Each hole defines the start of a new sector.

In addition to single- and double-density recordings, there are also SS

Table 15-8 **Formatted capacity of floppy disks**[a]

Size and density	Tracks	Tracks/inch	Sector	Bytes/sector	Capacity (bytes)	
					SS	DS
8 SD	77	48	128	26	256,256	512,512
			256	15	295,680	591,360
			512	8	315,392	630,784
8 DD	77	48	256	26	512,512	1,025,024
			512	15	591,360	1,182,720
			1024	8	630,784	1,261,568
5 SD	40	48	128	18	92,160	184,320
			256	10	102,400	204,800
			512	5	102,400	204,800
5 SD	80	96	128	18	184,320	368,640
			256	10	204,800	409,600
			512	5	204,800	409,600
5 DD	40	48	256	18	184,320	368,640
			512	10	204,800	409,600
			1024	5	204,800	409,600
5 DD	80	96	256	18	368,640	737,280
			512	10	409,600	819,200
			1024	5	409,800	819,200

[a]SS, Single-sided; DS, double-sided; SD, single density (FM); DD, double density (MFM). Sector count from *Owner's Manual, Model 2422 Multimode Floppy Disk Controller.* California Computer Systems, Sunnyvale, CA, 1980, pp. 3–16.

(single-sided) and DS (double-sided) disks. Single-sided disks can be recorded on only one side. Double-sided disks can be recorded on both sides in special drives that have two read/write heads.

There are also reversible disks. These disks can be recorded on one side of an SD drive, then removed, flipped over, and recorded on the other side.

A double-density-rated disk can be used on a single-density drive, but not vice versa. A single-sided disk can be used on a double-sided drive, but the reverse is usually not true. A double-sided drive usually senses when a single-sided disk is inserted and uses just that side.

The concept of logical sectors also adds problems to diagnosing disk problems. Adjacent physical sectors rarely contain data in sequence. Skew is introduced in recording data to improve the access time to the data. A sector skew of 6 is common but not standard.

For example, a sequence of data may start on sector 1, continue on sector 7, then sector 13, and so on, if the sector skew is 6. Sectors 1, 7, 13, and so on, would be logically adjacent but not physically adjacent. Usually, the skew factor is chosen so that all sectors in a track are included in a sequence before the sequence repeats.

Also, several arrangements are possible of logical sectors on double-sided disks. A common approach is sometimes called the cylinder approach. For speed, all sectors on one side of a disk for a given track are accessed, and then all sectors on the second side of the disk for the same track are accessed before moving the head to the next track.

Another approach for double-sided disks is to move from the outer to the inner track on one side and then from the inner to the outer track on the other side.

Dust, dirt, or lint can easily cause transient errors when writing or reading data with floppy disks. The check codes at the end of each sector are used to spot errors. Most sectors are read at least 10 times before the software controlling the reading assumes that an error is permanent.

Similarly, most sectors are read after being written to ensure that the correct data were written. Most control software will attempt to write the data and read them back at least 10 times before concluding that a permanent failure has occurred.

Figure 15-10 shows a block diagram for a typical floppy disk interface

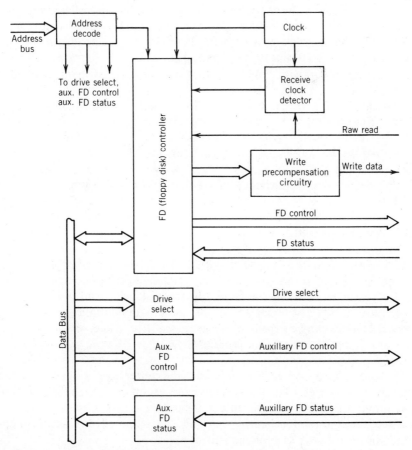

Figure 15-10 Block diagram of a floppy disk controller interface.

circuit. A number of integrated circuit FD (floppy disk) controller chips can be used as part of the interface.

Figure 15-11 shows a block diagram of the disk electronics. The microprocessor software selects which disk drive it wishes to address. Up to four drives are commonly connected to one SASI interface.

If the microprocessor software knows the track location of the read/write head, it tells the controller to increment or decrement the track positioning motor to the desired track. This is usually done by writing a control word into the FDC (floppy disk controller) chip.

If the track location is unknown, the software tells the FDC to move the head to track 0. Then, it tells the FDC to increment the head stepping motor to the correct track. Once the correct track is located, the read/write head is loaded on the disk. The controller is instructed to alert the processor when the index hole goes by. This indicates the location of sector 1.

The processor reads the sector IDs until it finds the correct sector. (For hard sector disks, the number of sector holes is counted.) The processor then reads or writes to the correct sector.

After a sector is read, the processor checks the controller error flag to make sure no error was detected. It then proceeds to read the next logical sector on the same track, if necessary. If an error is detected, the sector is read the next time it spins by the head. The process is repeated at least 10 times before the software concludes that there is a permanent error in the sector.

Writing data to the disk is similar to reading the data. Some disk write software will read the data on the next revolution to ensure it is correct. The write and read sequence will be repeated at least 10 times if an error is discovered before the software gives up and concludes that there is a permanent error.

The head is removed from the disk after the last read or write is performed on a sector.

15.7.2 Troubleshooting

Disk Failures   In spite of the sensitivity of the magnetic recording process to dust and dirt, floppy disk drives are fairly reliable pieces of equipment. Eventually, a sector will become scratched or the disk will bend or tear. On the average, a disk will last about six to nine months in continuous use before a permanent failure can be expected. It is good practice to keep extra (or backup) copies of all important data and programs.

Center Hole Failure   Some disks will fail at the center hole. Rings are available to reinforce the center hole. These seem more useful on $5\frac{1}{4}$-inch than on 8-inch drives. Many users don't bother to use them.

Note that the reinforcing rings are not a recommended repair technique. Rather, they are a way of prolonging disk life.

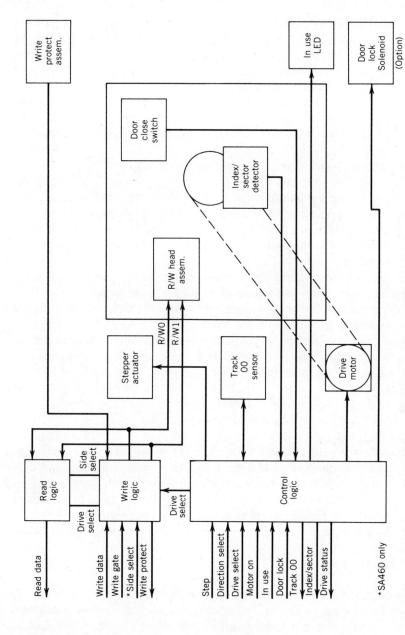

Figure 15-11 Block diagram of a floppy disk's electronics. (From *SA 410/460 96TPI Single/Double-Sided Minifloppy Diskette Storage Drives*, p. v. Copyright 1981 Shugart Associates, Sunnyvale, CA. Reprinted by permission.)

Write Protect Notch   Write protect notches cause many apparent disk failures. An 8-inch disk has a notch on its back edge. If the notch is present, the disk can be read but not written to. A piece of tape placed over the notch will permit the disk to be written to. Many new disks come prenotched. A pad of tape covers is included to permit writing to the disk.

To confuse the issue further, the notch on a $5\frac{1}{4}$-inch disk plays the opposite role. The notch permits the disk to be written to. Covering the notch write protects the disk. Also, many $5\frac{1}{4}$-inch disks containing software come unnotched.

The notch on $5\frac{1}{4}$-inch disks is located on a side edge.

Sector Failures   A sector failure should not be considered permanent until at least 10 attempts have been made to read the sector. Several computer programs are available to recover data from a disk with a failed sector. Some of these programs are included with some systems. These programs are usually very hardware dependent.

Other data recovery programs operate off popular operating systems. They work on a wide variety of systems using the same operating system. They are especially useful on operating systems that store the disk directory on the disk. A failure in the directory of these systems is very trying.

Some of these data recovery programs will also test disks and disk drives. They will also identify defective sectors. In some cases, they will permit defective sectors to be flagged so that the defective sector is ignored and the disk reused.

Software dealers, computer magazines, and software directories are good sources to check for the distributors of these programs.

Disk Initialization   A disk initialization program is supplied with some systems. Programs are sometimes available from commercial suppliers and user groups. Most disk initialization programs are very hardware dependent.

A disk initialization program permits an uninitialized disk to be initialized for a given system. It permits preinitialized disks to be adapted to a system that is out of alignment. It also permits a disk or a sector to be bulk-erased.

15.7.3 Drive Failures

Head Cleaning   Most manufacturers recommend that read/write heads be cleaned only under two conditions: (1) when more than three or four temporary failures occur in one month of continuous use; (2) when a failure will not clear after 10 writes and reads to a known good disk.

Magnetic-recording-head cleaning kits are available in most electronic supply stores. These kits require disassembly of the drive to gain access to the head.

Special floppy-disk cleaning kits are available from several disk manufacturers. These avoid disassembling the drives to clean the heads. All work by inserting one or more special cleaning disks into the drive for a few minutes.

Drive Alignment   Special drive alignment disks are available from most drive manufacturers. Drive alignment is necessary only if a defective head has been replaced or if properly formatted disks from a properly aligned system cannot be read. Alignment might also be considered if the drives have been subjected to severe shock, such as being dropped on the floor from a table.

The alignment instructions supplied by the manufacturer should be followed. Most of the alignment disks permit checking alignment at track 0 and the innermost track. They also permit alignment to be checked at one or more intermediate tracks. An alignment disk sometimes includes tracks recorded with known formats to check the operation of the read circuitry.

Test Programs   Test programs are available from a number of commercial sources. Some manufacturers will supply test programs. Some data recovery programs (described earlier) include drive test programs.

Most drive test programs will force the head to track 0 to do a write and a read, then to the extreme inner track to do a write and read, back to track 0, and then to an intermediate track. The program may also write several different patterns on each track.

Any prerecorded disk can be used to test the read circuitry.

Disk Exercisers   Disk exercisers are available from several manufacturers of drives. These connect directly to the drives through the interface. They permit the drives to be tested without being connected to a computer. Some exercisers supply test signals that can be traced through the electronics of the drives. An oscilloscope or logic probe can be used to trace the signal.

15.7.4 Maintenance

Most drive manufacturers recommend periodic vacuuming of the drives. A soft brush should be used along with the vacuum to remove stubborn dust and lint. Most manufacturers recommend that this be done at least once a year.

The belts that drive the spindle should be inspected and replaced if they are frayed or excessively worn.

Most drive manufacturers do not recommend cleaning the read/write head as part of periodic maintenance. They do recommend inspecting the head. The head should be cleaned only if it has oxide or dirt on it.

15.8 HARD DISK DRIVES

The theory of operation of hard disks is essentially the same as for floppy disks. However, several important details of the technology do differ.

Most hard disk drives use a magnetically coated aluminum disk. The disk is permanently attached to the spindle on most drives. Some disk drives have removable disks. Bernoulli disks use a flexible magnetic-coated medium that is forced against an aluminum disk while both rotate. Most hard disk drives are based on Winchester technology. Hard disks usually rotate at speeds near 1800 or 3600 rpm.

Hard disk read/write heads do not ride on the surface of the recording media. Rather, they float just above the surface on a small cushion of air. The air cushion is generated by the aerodynamics of the rotating disk and the shape of the head. The thickness of the air cushion is usually much smaller than the thickness of a human hair.

Cleanliness is important in hard disk systems. Most hard disk enclosures are pressurized. The pressure prevents dirt, dust, and contaminants from being sucked into the disk enclosure. Special filters are used to clean the air used to pressurize the enclosure. Dirt or dust may cause the head to "crash" into the disk. This frequently results in permanent damage to the disk or the head.

Several standards for hard disk systems exist. Lower-performance systems use the SASI interface described earlier. Several higher-capacity interfaces are also used.

Several different recording formats are used. These are similar to the floppy disk formats mentioned earlier. Cyclic redundancy checks are often used for the check codes. Sector lengths of 128, 256, 512, and 1024 bytes are common.

The recording capacity varies from model to model and from manufacturer to manufacturer. Capacities of over 80 megabytes per disk surface are available. Several manufacturers make double-sided disks. Many manufacturers stack several disk assemblies on one spindle. Some high-performance systems even use a separate head for each track.

Higher-performance disks use DMA (direct memory access) to read or write data. DMA uses a special controller to transfer data between read/write memory and the disk. The main processor initiates the DMA processes by telling the DMA controller the starting address and length of the block of data. The controller either halts the main processor when the disk is ready or uses a separate bank of read/write memory for the transfer.

The DMA controller is usually on a separate board that plugs into the main bus of the computer. The controller board usually contains hardware to calculate the CRC check codes. It also has receive clock circuitry. It frequently has circuitry to drive directly the read/write head and head stepper motors.

General troubleshooting and maintenance of hard disks is similar to that of floppies, with the following exceptions. The circuitry is usually more complicated. The data transfer rate is often faster. Most hard disk systems require periodic cleaning or replacement of their air filters.

15.9 CASSETTE TAPE DRIVES

15.9.1 Discussion

Several standard cassettes are used in cassette tape drives. The most popular size is the Philips cassette. This is the cassette used for audio recording. The Lear cartridge, similar to the eight-track tapes used for some home audio recording, is another popular cassette. The 3M cassette is also used; it is somewhat smaller than the Philips cassette. Audio cassette drives usually connect through the microphone, auxiliary, and earphone jacks common on most audio cassette recorders.

Two different recording techniques are used for cassette tape drives. Higher-performance drives use digital signals on digital tape. Lower-performance drives record FSK (frequency shift keying) on audio tape.

Data are usually recorded on tape in blocks. Several formats are used. Digital formats are similar to the floppy disk formats mentioned earlier. The most popular audio recording standard is the Kansas City standard.

Most digital cassette drives connect to the computer through a common communication interface such as the RS-232C, RS-449, or IEEE-488 interface. Some cassette tape units include the electronics to follow the data transfer protocol. Most do not. The computer usually supplies the protocol bytes that surround the data when writing to the tape. The computer also calculates the check code when writing or reading data.

CRC check codes are usually calculated with hardware. Checksum codes and most of the protocol are usually implemented with software. The transmit and receive circuits for most audio cassette recorders are similar to modem transmitters and receivers.

15.9.2 Troubleshooting

Troubleshooting cassette tape drives involves many of the skills required for troubleshooting communications interfaces, floppy disk drives, and modems. Fortunately, cassette tape drive interfaces are less complicated.

Writing and reading a test pattern (such as all ones, all zeros, or an alternate one–zero pattern) is a good overall test of the interface. It is helpful to make a test tape prior to any servicing call. The test tape can be used as a signal generator for troubleshooting the read circuitry. Some digital recorder manufacturers will provide standard test tapes for aligning tape heads. Cassette repair kits are available from several tape manufacturers for repairing defective cassettes.

A problem unique to cassette tape interfaces does arise with some less experienced users. They will use common audio cassette tape in a digital drive or, less commonly, digital tape in an audio drive. The drives will rarely work well, if at all, with the wrong kind of tape. Also, the tone and volume control settings of the audio unit can affect the digital recording performance.

15.9.3 Maintenance

Periodic cleaning of dust and dirt with a vacuum cleaner is the only maintenance required on most cassette tape drives. Some manufacturers do recommend periodic cleaning of oxide from the read/write heads. Cleaning solvent kits are available from electronics supply companies. Head cleaning tapes are also available from several tape manufacturers.

15.10 PAPER TAPE PUNCHES

The most common paper tape is 1 inch wide. It is used with seven- or eight-level codes such as ASCII. Narrower tape is available. It is usually used with five-level Baudot codes.

The code for each character is punched across the tape. Guide or alignment holes are punched down the tape. The alignment holes are offset from the tape's center line and are smaller than the data holes.

The tape is forced through the punch by a gear or star wheel that follows the punch. The gear is driven by a stepping motor on newer models. Older models use a solenoid and a ratchet mechanism. The gear moves the tape through the punch in discrete steps. Once the tape is in its new position, the code is punched.

Several formats are used to store data on paper tape. Most are similar to those used on cassette tapes. Data are stored in blocks. The space between blocks is usually filled with null (no holes) or delete (all holes punched) characters.

15.10.1 Troubleshooting

Most problems that occur with paper tape punches are mechanical. The punch or tape guide jams with paper or other foreign objects. Cleaning the punch usually corrects the problem.

Common cleaning solvents (such as acetone or alcohol) can be used to clean punches that become fouled with sticky foreign material such as adhesive. Check the manufacturer's recommendations before cleaning the punch. Some manufacturers discourage use of such solvents; some recommend oiling the punch after cleaning with an organic solvent.

Eventually, the punch dies will wear and need to be replaced. The solenoids driving the dies can also fail and need replacement.

The best test of a punch is to punch a string of deletes (all holes punched) or alternate one–zero patterns (ASCII U or *). A visual inspection of the tape will tell if the punch is working properly.

15.10.2 Maintenance

Paper tape punches require frequent cleaning. Most punches have a catch basket. The basket should be emptied before it fills and clogs the punch. Cleaning the paper residue out of the punch dies when the basket is emptied is a good idea. Running the punch for a few characters without paper tape will sometimes clean the paper residue from the dies. A vacuum cleaner or compressed-air jet can also be used.

Most punches require prelubricated paper to operate properly. Unoiled paper is available. Using the wrong paper will cause excessive wear and premature failure in punches requiring lubricated paper.

15.11 PAPER TAPE READERS

Two styles of paper tape reader are in use. One uses a light source and photo diodes to detect the holes in the tape. The other style uses spring-loaded fingers to sense a hole in the paper. In some models, the fingers throw a switch when they slip through the holes. In other models, the fingers complete a low-voltage circuit when they slip through the holes and contact a ground plate on the other side of the tape.

15.11.1 Troubleshooting

The finger-style reader is very prone to having the fingers bent or broken. Scrap paper or lint can also prevent the continuity-type finger reader from making good electrical contact.

The optical-type reader can also fail due to scrap paper or lint blocking the sensor. The light source can fail. Usually, a simple visual inspection will verify if the light source is working. Occasionally, the photo diode, or the circuitry it drives, will fail.

15.11.2 Maintenance

Maintenance usually consists of periodically cleaning the reader. Keeping a reader clean is usually more critical than keeping a punch clean. The best procedure is to open the top of the reader and remove any debris with a vacuum cleaner or compressed air. A soft brush is useful for loosening stubborn dust and lint.

Sometimes a small tool, such as a screwdriver or relay adjustment tool, may be needed to loosen stuck paper. Avoid damaging the reader with the tool. The finger-type readers are especially susceptible to damage of this type.

15.12 SUGGESTED READING

ADM 3A/3A+ Dumb Terminal Video Display Unit Maintenance Manual. Lear Siegler, Inc., Data Products Division, Anaheim, CA, October 1980.

Artwick, B. A., *Microcomputer Interfacing.* Prentice-Hall, Englewood Cliffs, NJ, 1980.

Brilliott, A., *A Low Component Count Video Data Terminal Using the DP8350 CRT Controller and the INS8080 CPU.* National Semiconductor, Santa Clara, CA, 1978. Application Note 199.

Cayton, B., *Low Cost Versatile CRT Terminal Using the CRT 5027 and CRT 8002.* Standard Microsystems Corporation, Hauppauge, NY. Application Note AN2-1.

Interface Specifications. Centronics Data Computer Corporation, Hudson, NH, 1977.

Lesea, A., and Zaks, R., *Microprocessor Interfacing Techniques.* Sybex, Berkeley, CA, 1977.

Mortensen, H. H., *Simplify CRT Terminal Design with the DP8350.* National Semiconductor, Santa Clara, CA, 1978. Application Note 198.

Motorola Semiconductor Products Inc., Microprocessor Applications Manual. McGraw-Hill, New York, 1975.

MX-80 Epson Dot Matrix Printer Operation Manual. Epson Shinshu Seiki Co., Ltd., Nagano, Japan.

MX-80 Epson Dot Matrix Printer Technical Manual. Epson Shinshu Seiki Co., Ltd., Nagano, Japan.

1983 Components Handbook. Western Digital, Irvine, CA, 1983. (Excellent application notes on floppy and hard disk controllers.)

Peripheral Design Handbook. Intel Corporation, Santa Clara, CA, 1979.

SA600 Fixed Disk Drive OEM Manual. Shugart Associates, Sunnyvale, CA, 1981.

SA 800/801 Diskette Storage Drive Service, Maintenance Manual. Shugart Associates, Sunnyvale, CA, 1977.

SA200 Minifloppy Diskette Storage Drive OEM Manual. Shugart Associates, Sunnyvale, CA, 1982.

SA410/460 96TPI Single/Double-Sided Minifloppy Diskette Storage Drives OEM Manual. Shugart Associates, Sunnyvale, CA, 1981.

Standard Microsystems Corporation Data Catalog 1982/83. Standard Microsystems, Hauppauge, NY, 1982.

APPENDIX A

Basic Microprocessor Concepts

Bernard McIntyre
University of Houston

A.1 INTRODUCTION

Several chapters of this handbook are concerned with data that is obtained or processed by a microprocessor. The logic analyzer, for example, is a micro-computer that is programmed to clock data into its input port and store it in memory when a trigger word is recognized. Depending upon the level of so-phistication of the computer and software making up the logic analyzer, there can be many options available for following a program's execution. Also, a microprocessor based control system can be programmed to strobe digital data into the microprocessor, process it, make logical decisions that affect the op-eration of the system, and provide the results to the system through its output port. The purpose of this appendix is to provide the background necessary for the reader to understand many of the terms used in this handbook, such as microprocessors, central processing unit, and timing diagrams. Some of these terms are best understood by choosing a particular system. The one used for the examples here is the Zilog Z-80 central processing unit.

A.2 THE ROLE OF THE CENTRAL PROCESSING UNIT

A basic model of a computer based on the concept of a central processing unit (CPU) is shown in Figure A-1. The CPU is central to the operation of this computer in that it receives, processes, and transmits all data associated with the operation of a program. Data flow to memory devices and input/output (I/O) devices can be bidirectional, but it is always between a device and the

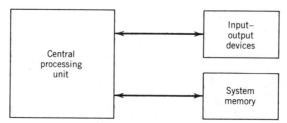

Figure A-1   The central processing unit: all data transfers in the system are through the CPU.

CPU. The CPU must be capable of executing a stored user program or sequence of commands by translating each command into a series of controlled events that is called a microprogram. The simplest example of this is a mechanical music box that utilizes a revolving cylinder having cogs that vibrate a metal strip upon contact. In this case the CPU and memory are the wheel and cogs, whose positions represent a fixed microprogram that can be changed only by using a different cylinder. The relative timing in a sequence of notes is controlled by the position of the cogs on the wheel and several events or notes can be processed at the same time on the cylinder. The I/O port would be called an n bit parallel I/O port where n is the number of metal strips and the music box could be called a dedicated digital computer. This dedicated computer, which has only one function based on its fixed memory, always starts where it left off when it is turned on. This latter property may not be desirable in that the listener may not enjoy starting a tune at some unknown point but may prefer to always have it ready to play at the beginning regardless of its previous history.

The dedicated computer described above, having a CPU and a fixed program, may be suitable for those who wish to listen to music but it would be useless to a composer. The composer would want to use a more general purpose computer with a CPU that can run any of several microprograms, or instructions. The user would have to program the computer by selecting elements of the *instruction set* for the CPU to perform in a sequential manner. The fact that each instruction to the CPU from the user program results in the CPU executing a complete microprogram is usually transparent to the user. As the CPU finishes the execution of one microprogram it must read the next step of the user program to determine the next sequence of events.

The general purpose computer will be defined here as one in which the CPU has stored microprograms which are accessed in a manner determined by the user program. The CPU section of the computer will have the elements shown in Figure A-2. The *arithmetic logic unit* (ALU) operates on data contained in temporary memory registers and returns the result to the temporary register section. The *program counter* (PC) points to the next sequential instruction of the user program in external memory. When the CPU finishes the execution of the current instruction it reads the next code from external memory

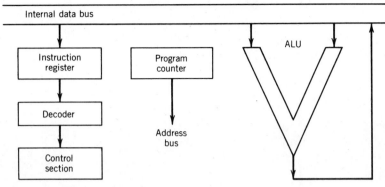

Figure A-2   The CPU section of a computer.

and puts it in the instruction register. From there it is decoded and a stored microprogram is accessed that determines the control functions for the duration of the execution cycle. On a large mainframe computer the ALU may be a large and physically distinct section of the computer and there may be separate address, data, and control wire harnesses to each input/output device and memory section. The advantage of this hard wire approach is that control and data signals can be sent in parallel. Data can be sent to an I/O device while the next operating code is being accessed from memory by the CPU. The disadvantages to this system are that its physical dimensions can present limitation to program speed and the large numbers of wires make it very difficult to trace hardware problems.

The type of CPU of interest here will be one in which the ALU and control sections can be fabricated on a single chip using *very large scale integration* (VLSI) technology. The CPU will also be called a general purpose microprocessor, or simply, a microprocessor. A given computer may have several microprocessors in its system electronics, each designed for a specific purpose, such as a *serial input output* (SIO) IC, or a *direct memory access* (DMA) IC, but the CPU is a more general purpose type of microprocessor and will direct the flow of data to and from the more specialized microprocessors.

A.3 THE CPU CONTROL SECTION

The control section of the CPU affects the order, timing, and direction in which data flows. Its output consists of clock pulses that are used for timing or gating data registers. The design of the control section and its programming is based on the model of Wilkes (1951). His CPU control model, illustrated in Figure A-3, was the first application of a systematic approach to the design of a microprocessor control section. An n bit op code goes into the instruction register and then into a decoder from which there can be at most 2^n output lines. Each output line can be connected to any of m control lines that form a

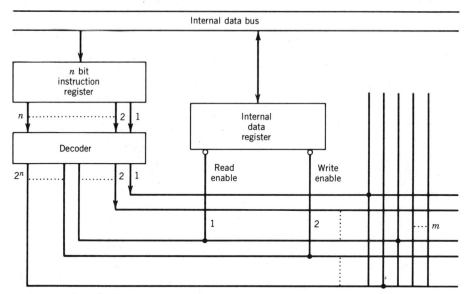

Figure A-3 A CPU control section.

read only memory* (ROM) data bus. The number of control lines in the system can be quite large. The Z-80 CPU has 16 data registers (special purpose memory locations) that can be read or written into. The control lines for these registers would be in the form of tristate (voltage high, voltage low, or high resistance) enable lines in a bidirectional bus as shown in Figure A-3. For a register-to-register data transfer the CPU would enable the output of one register for a read operation and the input of the second register for the write operation. In this way, the register being read remains unchanged. The major change that has been made to Wilkes' design has been to use several decoders in the control section rather than one large one. This is especially useful in a CPU such as the Z-80 or Intel 8085 that have several addressable internal registers. The registers can be encoded with three bits for a group of eight, and the eight control lines coming from that decoder would go to the read or write enable connections for the registers. The decoder with the read lines is called the source decoder and the one with the write control lines is the destination decoder. For a CPU with its control lines organized this way, the op codes can be read in octal form to see the explicit breakdown of register encoding in terms of source and destination registers.

Read only memory (ROM), when addressed, provides information to the data bus (*n* parallel lines). The information is put into the memory at the time of manufacture, or in the case of programmable ROM (PROM) by the user with a specially designed machine. One cannot put information into the memory (write) by programming as in the case of a *random access memory* (RAM), which can accept or provide information when the memory locations in the memory are addressed and it is told which to do, write into a memory location or read it.

A.4. CPU ARCHITECTURE

Figure A-4 is a block diagram of a Z-80 CPU minus some of the features not found on most microprocessors. The microprocessor utilizes the single bus system in which all data transfers take place on an internal or external bus. There is a separate bus for address and control lines also. Data registers in the CPU are grouped into 8 or 16 bit units that are typically constructed as dynamic random access memory (DRAM) and must be automatically refreshed (stored memory information renewed) by the CPU. This is one reason why the clock to the CPU has a lower frequency limit. The small capacitors that hold the bits of information will discharge if they are not recharged (refreshed) periodically. When the CPU is turned on the control section senses a restart and resets each register. The restart signal will also be sent to peripheral ICs that are part of the system. The next function of the control section is to move the contents of the program counter register, which in this case is OOOO$_H$, onto the address bus. Now the CPU looks for a command, or op code, from the memory section, so the control section does a memory read and decode operation, called an op code fetch. This first instruction to the CPU might be one that would set the stack pointer, followed by an interrupt vector jump table, and then a keyboard search routine to see what the user wants to do.

In logic and arithmetic operations the CPU reads memory contents and moves that data into a register or the ALU to execute the instruction. In the

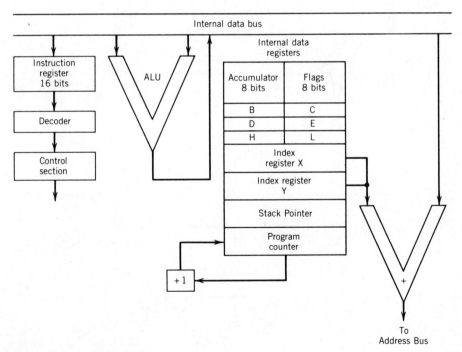

Figure A-4   A Z-80 CPU.

op code-fetch the memory data goes into the instruction register via the internal data bus, to the decoder, and the control section. Note that as the PC contents go to the data bus they also go through an incrementer and back to the PC register. In this way, the PC will always point to the next location in memory that must be accessed in sequential operations. The instruction register of the Z-80 is 16 bits wide so there could be up to 2^n hard-wired microprograms stored in the control section. Actually there are far fewer than that in storage and most of them are very similar in nature. A large number of very different microprograms would be difficult to design and debug so most of the programs are closely related to one another. For example, the Z-80 instruction set includes op codes to test, set, or reset any bit in any register. This subset requires many microprograms, but using the encoding and decoding system described in the control section, the microprograms can be designed in an orderly manner with fewer errors and at less cost.

At this point it is important to understand that the control system of the CPU has not only controlled the flow of data within it in executing an op code

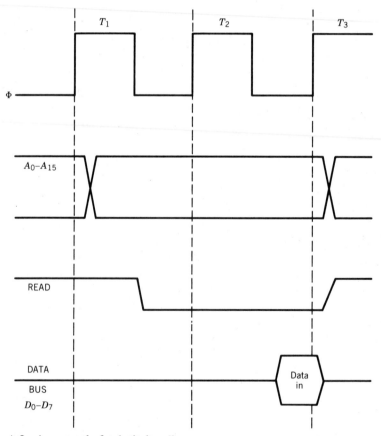

Figure A-5   An op-code fetch timing diagram.

fetch, but it has also synchronized the operation of an external IC with its own timing. In one sense, the design of the control system in the CPU is simplified in that the CPU designer knows what is going to be in the CPU and how the timing might be quickly and efficiently done. But then the CPU must be interfaced with a memory system and other ICs that are unknown to the designer. Still, the larger the variety of ICs that can be interfaced with the CPU, the easier it is to market. This interfacing is accomplished by having the control system establish a rigid protocol, or set of rules, for the data transfer. The protocol can be shown in terms of a timing diagram for the CPU that illustrates the behavior of the address bus, data bus, and control bus during the op code fetch. Figure A-5 shows each bus as time progresses from left to right. An op-code fetch, including decoding, takes 4 clock, or T cycles. It probably could have been designed for 3 cycles, but during cycles T_3 and T_4, while the CPU is decoding the instruction, the address bus is not being used. The CPU allows for the refreshing of external dynamic memory during these two cycles with special control lines not shown in Figure A-5. The CPU control section ensures that every op code fetch can be automatically followed by the refreshing of one row of DRAM with a minimum of external hardware. Only the op code fetch portion, T_1 and T_2 are shown in Figure A-5.

The rising edge of the system clock Φ at T_1 initiates the op code fetch;

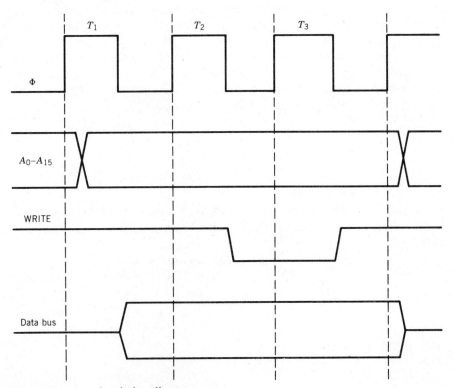

Figure A-6   A write timing diagram.

M_1 goes active low for this operation and the PC goes onto the address bus. In general, some of the address bus lines are at logic level 0 and others are at level 1, so both levels are shown in a timing diagram. Shortly after M_1 goes active the crossover of the address bus lines implies the new address is stable at the CPU output pins. The address bus remains stable until cycle T_3, when the address bus is used for the refresh system. The read line goes active low after the falling edge in T_1 and signifies to the external memory that the input side of the tristate data bus of the CPU is enabled. It remains enabled until just after the start of T_3. The rising read edge signifies that the CPU is latching whatever is on the data bus at that time. The brief transition shown on the data bus near the rising edge of the read line is used to specify the set up and hold time requirements of the CPU. It does not signify that it is the only period of data bus activity. The memory system could have had data on the bus for two full clock cycles but the CPU only latches the data at the rising edge of the read line. Contrast the timing associated with reading an op code with that of writing data to RAM in Figure A-6. This operation is done in three T cycles. The first data bus transition in the timing diagram shows when the CPU has enabled the output side of its data bus, and during T_3 it is disabled. The output data is kept on the bus for the maximum time by the CPU to be useful for slow devices. The two timing diagrams show that the write operation is designed to keep data on the bus for the maximum time while the read operation gives the maximum warning to an external device and latches data at the last instant. This description of timing diagrams is intended only to indicate how a portion of the CPU control system must function in order to synchronize the timing of an external IC to that of the CPU. For a more complete presentation of the CPU timing and architecture the reader should consult the technical manual for the particular CPU; a good discussion of CPU architectures and their control systems can be found in Nesin (1985).

BIBLIOGRAPHY

Nesin, D. J., "Processor Organization and Microprogramming." Science Research Associates Inc. (1985).

Wilkes, M. V., "The Best Way to Design An Automatic Calculating Machine." Report of Manchester University Computer Inaugural Conference, (July 1951), pp. 16–18.

APPENDIX B

BASIC ROBOTIC CONCEPTS

Farrokh Attarzadeh
University of Houston

B.1 INTRODUCTION

The world of robots has many dimensions and robot classification can be based upon a variety of factors, ranging from external form and tasking through control strategies and intelligence (capability of a robot to perform functions that are normally associated with human intelligence, such as reasoning and learning). Intelligent robots may be stationary, operating within a well-defined area and encountering a predictable set of environmental conditions. They may also be mobile, operating in a domestic or hostile environment, and designed such that they can adapt to a changing environment.

The robotic design process involves two major considerations:

Technical Considerations Here we are concerned with specifying the performance requirements, the actual design process, and verification of the design by experiments supported by analysis and modeling.

Economic Considerations The concern here is with the cost effectiveness of the design. This concern is influenced to a great extent by the number of systems to be produced. A second parameter is the amount of time reasonably available for the design process.

Technical and economic considerations are not fully independent of one another. The consequences of the interaction of the two disciplines can lead to conflict as well as opportunity.

In the following pages a brief review of the major robotic components will be presented. The reference portion of this appendix provides additional sources for interested readers.

B.2 BACKGROUND

The industrial robot was invented in the United States about 25 years ago. In the decades since, worldwide confusion developed over a definition of industrial robot. The problem involved the term being used to refer to automated equipment that did not possess the special abilities of robots. Robotic Industries Association (RIA) attempted to solve the problem by creating the following definition that is now recognized by most nations:

> *Industrial robot*: a reprogrammable, multifunctional manipulator designed to move material, parts, tools or specialized devices through variable programmed motions for the performance of a variety of tasks.

The keywords are "reprogrammable, multifunctional manipulator." Unlike other forms of automation, robots can be programmed to do a variety of tasks, making them the most versatile of manufacturing tools. Many advantages

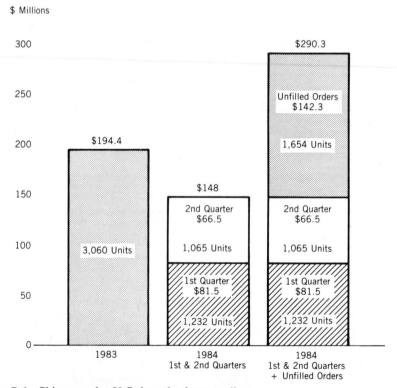

Figure B-1 Shipments by U.S. based robot suppliers.

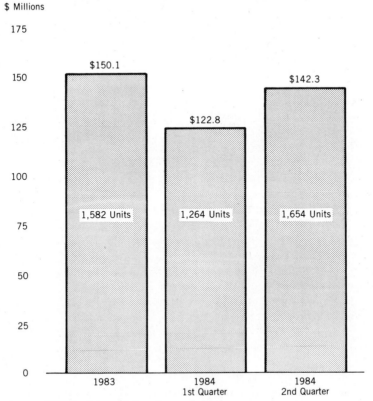

Figure B-2 Unfilled orders—U.S. robot suppliers.

result from the robot's reprogrammability. Since robots can switch tasks with a minimum of start-up and debugging costs, a company is able to maximize its use of a proven design and reduce overall manufacturing costs.

Currently, there are some 13,000 industrial robots in use in the United States. About 35% of the installations are in the auto industry. Other major industries using robots include home appliances, aerospace, consumer goods, electronics, and off-road vehicles. Recent developments that give robots added intelligence such as machine vision, tactile sensing, and mobility make robots suitable for a wider range of industries. The near future will find robots used increasingly in industries such as textiles, food processing, pharmaceuticals, furniture, construction, and health care.

Robots offer substantial gains in manufacturing productivity, particularly when integrated into automated systems. The history of U.S. robot installations indicates that robots increase productivity by 20–30%. Because the majority of robots are applied to existing machinery, companies using robots can accelerate payback on current equipment while reducing the need for new capital investment.

Figures B-1 through B-4 indicate the most recent statistics of the industrial robots, compiled by the RIA.

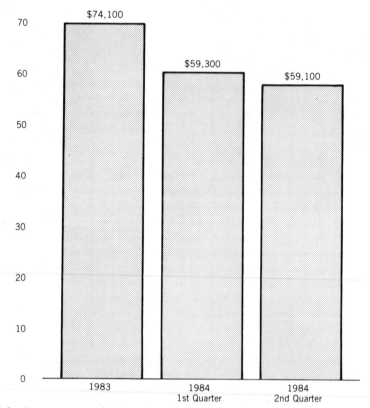

$ Thousands

Figure B-3 Average cost of robots based on new orders.

B.3 ROBOTIC CONFIGURATIONS AND ACTUATORS

There are four types of robotic configurations: cylindrical (rectilinear), polar (spherical), anthropomorphic (revolute or jointed arm), and cartesian (x,y,z), shown in Figures B-5 through B-8 respectively. Unless they are designed for a specific robot, robotic controllers must be able to coordinate the movements of all four robot configurations and any combination of arm and wrist actuators (see Figure B-9). The choice of an actuator for a given application will mainly depend on its suitability for the purpose. Electrical actuators are excellent where high accuracy, repeatability, and quiet operation are needed.

In some plants, air and oil pressure sources are often readily available, making the choice of hydraulic or pneumatic robots economically desirable. Hydraulic actuators are very stiff in action and can produce very high forces, but they are not as clean as electric actuators and are more expensive. A stiff actuator can be very fast acting, so hydraulic actuators tend to be chosen for speed.

Table B-1.   **Advantages and disadvantages of robotic actuators**

ACTUATOR	ADVANTAGES	DISADVANTAGES
Electric	Fast Accurate High availability Relatively inexpensive	Require gear trains Electric arcing might be a problem They have power limitations
Hydraulic	Large lift capacities Moderate speeds Can be accurately controlled	Expensive Pollute the workspace with fluids and noise

Pneumatic	Relatively inexpensive
	High speeds
	Don't pollute the workspace
	  with fluids
		Compressibility of air limits
		  their accuracy
		Positional accuracy and
		  repeatability
		Are sensitive to changes in
		  load
		Noise pollution
		Leakage of air is a major
		  concern
		Air filtering and drying is
		  required

Pneumatic actuators are generally cheap, clean, safe, and can produce moderately high forces. However, the compressibility of air reduces the stiffness of pneumatic systems, which can be a disadvantage. Table B-1 shows the advantages and disadvantages of all three actuators.

B.4 END-OF-ARM TOOLS

There are two types of end-of-arm tools. One works much like a human hand to grasp and hold something. The other is a tool that performs work, such as drilling, routing or painting.

There are many functional ways a robot hand can grasp or hold a part. Suction cups hooked to a vacuum pump or Venturi device are effective. Simple pinch grippers or scoop type hands that slide under a part can also be used. For irregularly shaped or fragile parts, expansion bladders extended from some means of rigid back-up support that, when filled with air (or a fluid), encapsulate the part and apply uniform holding pressure on the part surface (see Figures B-10 through B-12).

B.5 MOTION CONTROL

Motion control can be broken down into point-to-point positioning, velocity control, and both position and velocity control. Many systems, however, are

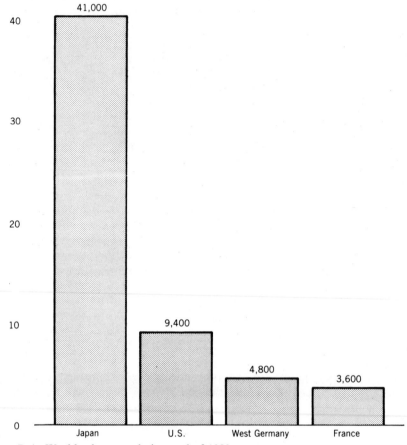

Figure B-4   **World robot population end of 1983.**

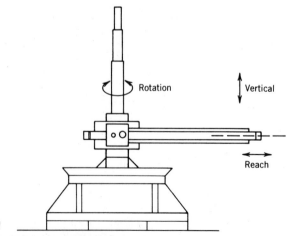

Figure B-5   Cylindrical coordinates.

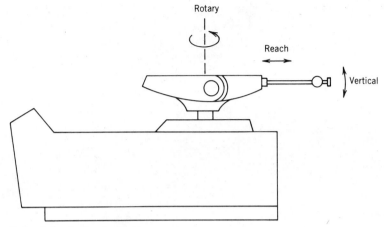

Figure B-6   Spherical coordinates.

a mix of these and may require control of other parameters such as acceleration, temperature, or pressure.

The major requirements of each motion control type are as follows: The critical parameter of point-to-point positioning is usually the positioning of the device relative to a previous position or a fixed reference. A drill press is a good example.

Velocity control requires that the speed be controlled within a predetermined accuracy as in a milling operation.

Position and velocity control systems control both position and velocity.

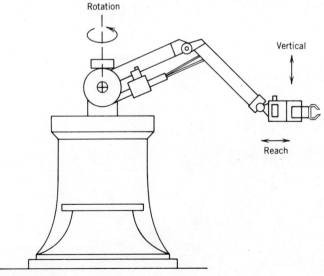

Figure B-7   Jointed arm.

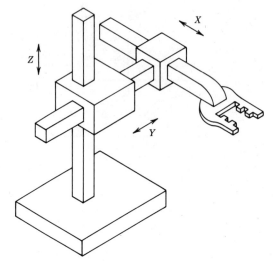

Figure B-8   Cartesian coordinates.

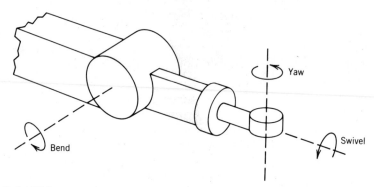

Figure B-9   Wrist axes.

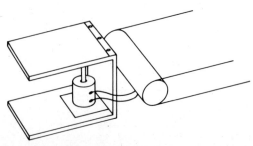

Figure B-10   Two finger mechanism with solenoid.

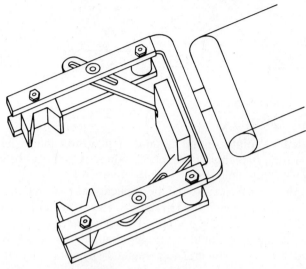

Figure B-11 Two-pincer mechanism.

X-Y plotters are a good example of this type of control setup. The velocity in one axis can be a function of the incremental positioning of the other axis.

All of these forms of motion control can be handled using open- and closed-loop systems. The next section covers the use of servo and stepping motors in these systems.

B.5.1. Open-Loop Control

Figure B-13 shows a generalized open-loop motion control system. This type of setup can be used for both simple and complex applications. For instance,

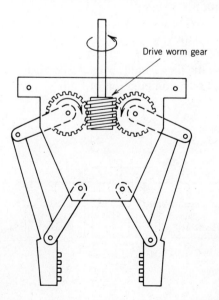

Figure B-12 Gripper with antislip pads.

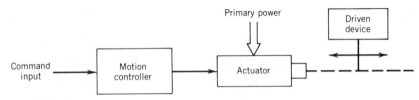

Figure B-13   Open loop control system.

it might be used for point-to-point motion, such as the familiar pawl and ratchet operated by a rotary solenoid design; thus, there is no need for feedback. The command input to the solenoid is simple—on or off.

Open-loop control can also handle applications in which the driven device has precise position or velocity requirements. Here, the motion controller could be a mix of digital and analog circuitry driving an electromechanical actuator. In this case, the actuator would require predictable motion characteristics because no feedback is involved. The command signal may be either on or off, or groups of encoded functional commands. In the first case, the on-off commands would have to be encoded in the controller, as in a typical microprocessor-controlled system. In the more complex version, the commands would normally be fed in by either a computer or an operator-controlled device. These different methods will be examined in more detail later on.

B.5.2 Closed-Loop Control

Figure B-14 is a typical block diagram of a closed-loop system. The complexity of these systems is determined by the motion needs of the driven device. A simple point-to-point reciprocating motion, for example, would only require two limit switches to reverse the motor direction. But in a high performance system, such as a copier or plotter, more feedback is involved due to the addition of position, velocity and acceleration sensors.

Many types of motion controllers are used in these systems. The actuator may be electromechanical, hydraulic, pneumatic, or a combination of these. The command input signals are a function of system needs, but often are related to position. The electronics used to process the input and feedback signals and to drive the actuator may be analog, digital or a combination of both.

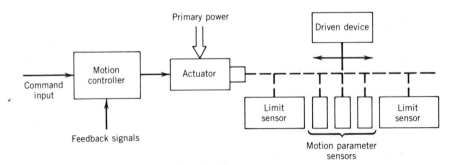

Figure B-14   Closed loop control system.

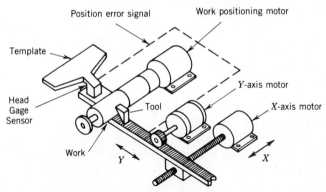

Figure B-15 Two axis machine control system.

An example of a simple dual-axis closed-loop system is a machine tool cutter, as shown in Figure B-15. Here the cross-feed motor follows the contour of a template as detected by a gauge head sensor. The sensor subtracts the desired position (template edge) from the present position of the carriage to produce an error signal. The error signal is amplified and used to drive the motor until the error signal is driven to zero. A cutter produces the work during this process. The longitudinal motor normally operates at a constant speed. Its control may be either open-loop or closed-loop depending upon the system requirements.

B.5.3 Servos in Closed-Loops

Figure B-16 shows a typical closed-loop servo driven system. In a servo system, the forward control portion tends to be the most complex. It normally consists of the actuating element (a servomotor), the motor drive circuitry (controller), and compensation networks for stability.

Electrical servomotors generally fall into one of three categories: dc servomotors, ac servomotors, and brushless dc servomotors.

Because of its linear torque/current relationship, the dc motor is often used where there's a need for high peak power with quick starts and stops. High performance numerically controlled (NC) machinery is a good example. Direct current servomotors cover the power range from about 0 to 10k watts.

In an explosive environment, or where maintenance is a problem, the dc motor's commutating brushes should be avoided. Here, an ac servomotor is a

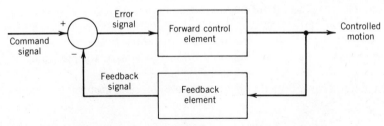

Figure B-16 Major elements of a closed loop, servo driven system.

good alternative. However, to maintain a linear speed/torque relationship, resistance is sometimes added to the armature. This will decrease efficiency and may cause a heat buildup. AC servomotors most often fall into the power range of 3 to 2400 watts.

Brushless ac servomotors overcome the problems of conventional ac and dc servos. The wound rotor is replaced by permanent magnets that provide a speed/torque curve closer to that of a dc motor. The power range of brushless ac motors is about 0 to 10k watts.

Controllers for servomotors are available in many analog and digital types. Linear amplifiers, for example, are used to provide a straightforward, reliable design. These can be a problem, however, when driving higher power motors because their inefficiency creates unwanted heat. To minimize this, nonlinear methods, ranging from silicon controlled rectifiers (SCR's) to pulse width and frequency modulated devices using high power transistors are used.

The number and quality of the servo system's feedback elements are dictated by the motion specification. Point-to-point control requires a feedback device that gives an analog or digital signal for position. Potentiometers or synchros provide an analog signal, and shaft encoders supply a digital output. Speed control requires a velocity pickup device such as that supplied by an analog tachometer or digital incremental shaft encoder.

The overall transfer function (the ratio of the controlled motion over the reference command) is normally a complex mathematical expression. For high accuracy, and to make the motor closely track the input command, the error signal must approach zero. To put it another way, high forward-loop gain is required. Unfortunately, this leads to stability problems such as overshoot and continuous hunting.

Ideally, the transfer function would be independent of variations in controller, motor, and load parameters, but in practice this "ideal" can only be approximated. Care must be taken in the design (particularly of compensation networks) to ensure that a change in the system parameters, or even the operating point, doesn't bring about instability or a runaway condition.

To summarize, closed-loop motion control can provide a very accurate and responsive system. Often, this may be the only feasible way to go. This approach does cost more though, both for hardware and the engineering time spent in the design and stability analysis.

B.5.4 Open-Loop Systems

Using open-loop control is much more appealing because the systems are less complex. However, as mentioned before, it can't always be used. It requires an actuator that can perform precisely over its operating range. For example, if the controlled variable is speed, a synchronous motor or a stepping motor could be used. Which type depends upon the speed, torque, and power requirements.

In general, one should consider using a stepping motor if the requirements are:

Torque—fraction of an ounce-inch to several thousand ounce-inches,
Power—one horsepower or less,
Speed—0 to 3000 rpm,
Positional accuracy—1 to 5% of the motor step angle.

Because of its digital nature, the stepping motor provides many advantages as an actuator in motion control systems. These include direct compatibility with digital control methods, accurate positioning with noncumulative errors, and a position and/or speed control with moderate performance is available in the open-loop mode. With such a control, other transducers and components such as tachometers, gear trains, and encoders can often be eliminated. If a very high performance control is desired, closed-loop control (in a logic sense) with step motors can be used (Figure B-17). This requires an encoder on the motor shaft.

Other stepping motor advantages are:
Construction is simple and rugged with only two bearings in mechanical contact,
Bidirectional rotation and control with no additional control complexity, and
The motor can be stalled repeatedly without damage to the motor or load.

B.5.5 Open-Loop Control of a Stepping Motor

Figure B-18 shows the general arrangement of an open-loop stepping-motor control system. For a given system, motor selection should be based on the load requirements and the manufacturer's data. Torque, speed, acceleration, and inertia are some of the parameters that must be considered (see Table B-2 for performance comparison of stepping motor types).

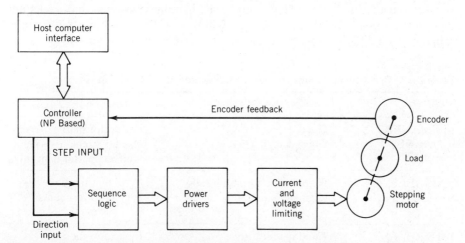

Figure B-17 Closed-loop system using encoder-position feedback to intelligent microprocessor-based controller that interfaces with host computer.

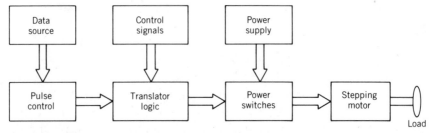

Figure B-18 Open-loop stepping motor control system. Data from the source is converted into appropriate pulses. At the translator, pulses are converted into motor control signals. The drive power for the selected motor windings is handled by the power switches. The effect is to amplify the translator output to a power level high enough to turn the motor.

B.5.6 Application of Stepping Motors

Principal application areas for stepping motors and stepping motor controllers include: machine tools (metal cutting and forming), metalworking operations such as coil winding and cutting to length, electronic component bonding and slicing, food products packaging, electromedical control apparatus, industrial controls, textile manufacture, printing, robotics, engineering and scientific instrumentation, environmental and process control, woodworking, photographic, optical, business machines, and computer peripherals.

B.6 SENSORS

In the ultimate automated factory there will be few, if any, operators, production foremen, or quality control inspectors to check operating status. Instead, sensors attached to the machines, materials handling systems, and packaging equipment spread throughout the plant will look at the conditions within their range and send data through a network of data highways directly to program-

Table B-2. **Performance comparison of PM, VR, and PM-VR hybrid steppers**

TYPE	ADVANTAGES	DISADVANTAGES
PERMANENT MAGNET (PM)	Good detent torque Good damping	High rotor inertia Magnet strength variation affects performance
VARIABLE RELUCTANCE (VR)	High torque to inertia ratio High stepping rate	No detent torque Subject to resonance
PM-VR HYBRID	Good detent torque High stepping rate High resolution	High rotor inertia Resonance at some speeds

mable controllers, computers, or self-contained microprocessors. These units will process the data and send actuation commands back through the system— all without human interaction.

Sensors haven't yet replaced people as the main form of information gathering in the factories, but the technology exists now to handle almost any needed sensing. As the degree of automation increases, sensor applications will move away from individual sensor-control-actuator loops. The trend will then be toward complex, interactive groups of sensors tied together by a hierarchy of computer networks.

In this section, some of the sensors used to monitor position, including electromechanical limit switches, proximity switches, photoelectric sensors, potentiometers, and encoders will be discussed.

These devices can be split into two basic groups: presence sensors, which detect the presence (or absence) of an object, and displacement sensors, which measure the displacement of an object from a reference position. Limit switches, proximity switches, and photoelectric sensors are members of the first group, while potentiometers and encoders are members of the second.

B.6.1 Electromechanical Limit Switches

Limit switches are the oldest of the presence sensing devices. They provide a simple operating mechanism that is well understood and easily applied. When a lever or plunger is moved through a prescribed distance, a snap action switch either opens or closes a contact. Limit switches can be used for all types of motion sensing, parts counting, and position indicating applications.

Operating mechanisms include horizontal push rollers, vertical push rollers, push plungers (top and side), omnidirectional rods, and a myriad of rotary operated styles. However, limit switches are restricted to applications that allow direct contact by a cam or the target itself. Their cycle rate is much lower than that of the proximity switches and photoelectric switches. Also, because they are electromechanical devices with moving parts, they have frictional and electrical contact wear points that limit their useful life to about 10 to 50 million operations.

The simplicity and relatively low cost of electromechanical limit switches will probably keep them in control schemes for quite some time, even in computer controlled, automated factories.

B.6.2 Proximity Switches

Proximity switches perform the same basic task as limit switches, but without contacting the target. While several types of proximity switch are currently in use, including inductive (metal sensing), capacitive (any material), and Hall effect (magnetic sensing) switches, the inductive types are the most widely used. If the object is metallic, an inductive proximity switch will sense its presence and trigger an output signal as soon as the target moves within range.

These switches have no moving parts and no wear points. Their logic circuitry consists of an rf-generating oscillator circuit connected to an amplifier, and signal conditioning components that provide an output.

They are limited to fairly small sensing ranges (usually under 1 in., with long-range units to 2 in.). Generally, published maximum ranges apply to ferrous metals; reduction factors must be applied for nonferrous metals.

Some proximity switches are almost completely unaffected by environmental contaminants (such as dust, moisture, and lubricants). Because they contain no moving parts or maintenance items, they can be completely and permanently sealed. Care must be taken, though, to prevent extraneous metals from getting into the sensor's range and creating false signals.

Proximity switches are made as direct replacements for electromechanical limit switches. Their housings have the same mounting dimensions and configurations, and their output is at the 120 Vac signal level of limit switches. However, the proximity switch's solid state circuitry gives it the potential to become much smaller than the limit switch. In fact, the newest proximity switches are very small, self-contained units that provide an 8 Vdc output signal. They can provide direct inputs into specially conditioned programmable controllers and microcomputers.

These small switches cost much less than older, larger types, can be flush mounted in almost any material, and are encapsulated so they meet NEMA 1, 3, 4, 6, and 13 enclosure requirements.

B.6.3 Photoelectric Sensors

Photoelectric sensors offer an operating range and functions not available with either limit switches or proximity switches. The newest units use a pulsed GaAs (galium arsenide) infrared LED (light emitting diode) and a phototransistor receiver. Because the receiver is set to respond only to the pulse frequency of the emitter, pulsed LED units are practically impervious to ambient light. The infrared beam penetrates contaminants such as coal dust, mist, and oil vapor much better than older incandescent bulb units.

Photoelectric switches consist of a separate emitter and receiver, or an emitter and receiver in one housing. The first type is used for through-scan, the second as a reflective scan or proximity scan device. Unlike limit switches and proximity switches, photoelectric switches generate an analog signal proportional to the amount of light reflected back to the receiver. This signal varies with the object's distance from the receiver.

B.6.4 Presence Sensors

Every one of these devices can be used in the automated factory. Often, more than one type of presence sensor can be used for a particular job. Proper selection depends on many factors. Things to consider include:

Effects of the mechanical interface on machine design,
Environment,
Size and material of the target,
Minimum and maximum possible distance from sensor to object,

Accuracy of position data required,

Speed of the target,

Ultimate destination of the output signal (i.e., the coil circuit of electromechanical relay or the logic circuit of a programmable controller),

Duty cycle of the sensor,

Space available for the sensor and any associated controls,

Type of NEMA enclosure rating required for the environment.

Here's an example of how these sensors might be applied. A conveyor carries rough-machined billets (a small, unfinished bar of iron) to a 5-axis CNC machining center. Periodically, a ram slides across the conveyor to push a billet onto the center's indexing table. The table turns through 270° and stops under a cutting head turret. The turret descends and an end mill begins its cutting pattern. To signal the ram to push a billet off the conveyor onto the indexing table, you could use a limit switch next to the conveyor, with an operating lever extending into the path of the billet. Or, you could install a proximity switch next to the conveyor so that the billet would pass within its range and switch on the ram. You might also consider using a photoelectric switch next to the conveyor, aligned so the passing billet interrupts or reflects its beam.

If only a few billets a minute move along the conveyor, a limit switch would do the job at the lowest cost, and still provide a useful life of 10 to 40 years. However, if the conveyor carries one billet per second past the switch, its life would be limited to a year and a half, or less. Here, a higher priced proximity switch would be more cost effective.

If the billet sizes vary more than an inch or two, or their position on the conveyor cannot be tightly controlled, they could move in and out of a proximity switch's range. In this case, a photoelectric sensor would provide the necessary sensing range, regardless of billet size.

Obviously, choosing a presence sensor for this application requires a careful analysis of more than the few points just mentioned. Once the billet is on the indexing table, though, the precise movements of the machining center can be better sensed by displacement sensors than any of the presence sensors.

B.6.5 Displacement Sensors

In the automated factory, displacement sensors measure the movement of machine tool components and robot arms, and provide feedback to positioning devices such as servomotors. They can measure either linear displacement or angular rotation with great accuracy, and can operate fast enough to provide accurate position signals at fairly high speed.

Three general types of displacement sensors are potentiometers, resolvers, and encoders. The first two provide analog signals proportional to displacement. Encoders generate a digital pulse train that indicates relative position.

Displacement sensors have two interfaces with the system in which they operate. One is the electrical connection with the control device. The second is the mechanical connection to the moving component in the system. The electrical signal must often be very precise, with high resolution, high accuracy, and as little signal noise as possible. On the other hand, the mechanical connection must be strong, with enough stiffness in the direction of travel to prevent deflection-induced errors. The displacement sensor must also withstand high-torque machine component movements, high shaft bearing loads, and in some cases, abrupt direction reversals.

B.6.6 Potentiometers

Potentiometers are the limit switches of displacement sensors. They've been around the longest, and are based on the oldest technology. They measure relative displacement by movement of a slider or wiper along a resistor. The wiper contact splits the resistor into two parts, and the ratio of voltage drop across each segment corresponds to the displacement of the wiper along the resistor.

Displacement potentiometers, typically either wirewound or conductive plastic types, can measure linear displacement or angular displacement. Linear potentiometers usually cover the range of $\frac{1}{4}$ in. to 36 in., though special models can be made for unusual applications. Rotary potentiometers typically cover up to about 340°, but can be specified with a travel of 360° where shaft position through the complete revolution must be known.

The internal resistor structure can be modified to produce a variety of signal forms, but the output is always an analog signal. Inputting the signal directly to a logic device requires an analog to digital converter.

Potentiometers are the least expensive of the displacement sensors, but they suffer from wear caused by the wiper movement. Also, contact between the wiper and the resistor can generate *radiofrequency interference* (RFI).

B.6.7 Encoders

Optical encoders are opto-electronic devices that translate linear or rotational movement into a digital signal. In a typical shaft encoder a transparent disk with small, evenly spaced radial lines printed on its outer arc rotates between an LED emitter and a phototransistor receiver. A similarly stationary disk, called a mask, is installed very close to the rotating disk to create a shutter mechanism. The lines and spaces on the rotating disk alternately interrupt and pass the light beam. This causes the photoelectric circuit to generate a square wave. Output circuitry converts the square wave to a pulse train. By counting the pulses with an up/down counter, relative shaft position can be tracked. If a time-based frequency counter is substituted for the simple up/down counter, rotational speed can be measured. In either case, the encoder output is digital.

Much more complicated absolute position encoders generate a digitally coded signal that is unique for any incremental shaft position within the encoder's resolution capability. Absolute encoders are immune to power failures and they can drive a readout directly, but they're much more expensive than the simpler incremental encoders. Industrial shaft encoders for heavy shaft loading and harsh environments of manufacturing plants are capable of resolutions of $\pm 0.036°$ (2500 cycles/revolution), but with a maximum allowable frequency of 50 kHz, high shaft speeds may require a lower resolution encoder.

Because incremental encoders produce a digital output, they can be used in closed loop control systems to provide direct digital feedback to stepper motors or servo systems. Two motor drive/encoder pairs can be installed with shafts at 90° to each other to form a numerically controlled X-Y positioning system. By using precision linkages, shaft encoders can track linear movement of machine arms, tool cradles, and telescoping robot mechanisms. Again, the direct generation of a digital logic signal makes them especially compatible with computers and microprocessors.

Because LED light sources and integrated circuitry have replaced the original lamp and discrete component circuitry, the reliability of encoders has greatly improved. They're still the most expensive device for measuring shaft position, but their prices are dropping. In addition, when the cost of processing electronics for potentiometers and resolvers is considered, the total cost difference shrinks considerably.

B.6.8 Tactile Sensors

Tactile sensors, sensors responding to actual contact or touch, fall into two broad classifications—those which merely sense contact, and those which not only sense contact but also the degree of contact (e.g., the amount of grip produced by a robot hand). The latter are the more important type, and the more complicated.

Simple touch sensors include capacitors as a touch sensitive switch, and mechanical *feelers*, operating a microswitch on contact. Simplest of all is the mechanical stop that brings movement to a halt by a part of the movement running up against the stop.

Tactile sensors, which incorporate a sensor of grip or feel, are usually force sensors. The degree of contact or grip is measured as a force that generates a control signal proportional to that force. The control circuit can then be preset so that once the force has risen to the required maximum safe level, movement stops and there is no further increase in grip.

Force sensors employ a transducer transforming force or pressure into an electrical signal. Various types of transducers may be used, according to the force levels involved. *Piezoelectric* transducers are suitable for a wide range of forces with excellent sensitivity and signal response. Carbon graphite force sensors, working on the same principle as a carbon graphite microphone, are less sensitive. Strain gauges have a wide range of possibilities, especially where high force levels are involved. Two or more strain gauges arranged in a bridge circuit can also be used to measure resultant force where more than one force is involved in the tactile function. For example, the gripping force when holding something and the force resulting from turning it at the same time could be determined separately, or as a resultant force on the object involved.

B.6.9 Robot Vision

The robot vision (also sometimes referred to as artificial, intelligent, or synthetic vision) is defined as "the automatic acquisition and analysis of images to obtain desired data for interpreting a scene or controlling an activity."

What sets intelligent robot vision systems apart from other visual sensors—such as electronic bar code scanners and photoelectric technology—is their ability to interpret, identify, and distinguish objects or entities within a visual scene, or to perform measurements and assessments on the objects that are recognized in a scene (Figure B-19). For example, a vision system application may require that the dimensions of a part be determined rather than part identity or orientation. This *dimensionality* information must be extracted from the visual scene. Thus, quite often, the information coming from a vision system may have no visual content whatsoever—only a message which describes the quality or nature of objects observed by the system.

The most common robot vision systems use one or more (for three-dimensional viewing) solid-state electronic cameras to look at the object in conventional or laser light, and produce a digitized image for computer processing. The frequency with which the camera captures the image is a function of its application: standard television cameras capture 60 images per second, but industrial cameras may provide more or fewer images, as the application dictates.

The information from a vision system can be fed to a variety of peripherals or other automated equipment. For human feedback, a TV monitor and/or keyboard are necessary. For machine feedback, information can be electronically communicated to robot control for arm manipulation, or a host computer as part of a factory-wide data collection system.

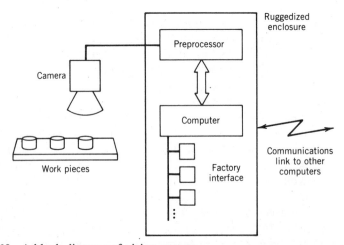

Figure B-19   A block diagram of vision system.

B.6.10 Voice I/O

One of the most practical applications of voice I/O is for quality and inventory control on the factory floor. In many cases, workers on production lines have to manually record or key-in quality and defect data. Typically, the worker's hands are occupied with production tasks, so data entry is difficult. Voice data entry via a headset/microphone permits uninterrupted hand operation (see Figure B-20).

In general, voice I/O provides the following benefits: reduced hand/eye occupation, improved operator mobility, capture of data at point of origin, no keyboard skills needed, real-time data collection, and data verification prior to computer input.

Some of the most important developments in voice I/O have been improvements in speech recognition. Speech recognition involves the on-line, real-time recognition of spoken words by the computer. Two techniques are used: speaker dependent devices, which require the user to "train" the system in advance with his or her own vocabulary; and speaker independent devices, which can recognize inputs from any speaker, without prior training.

Important specifications for both methods are accuracy and robustness. Accuracy is a measure of how often the device correctly interprets an input, and robustness reflects the ability of a system to operate under adverse conditions. For example, a speech input system has to contend with high background noise, changing inflections, and various rates of speaking.

Most speech-recognition devices have several functional modules (Figure B-21). The heart of the device is an acoustic-pattern classifier that transforms user utterances into digital code. The classifier itself consists of a spectrum analyzer, an analog multiplexer and A/D converter, a programmed digital pro-

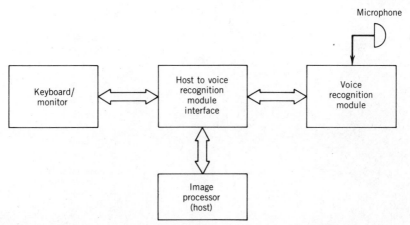

Figure B-20 Functional block diagram of voice input control system. Notice that host to voice recognition module interface subsystem acts as node between user's voice or keyboard input and host system.

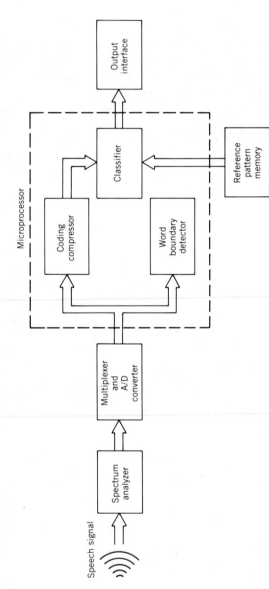

Figure B-21   Basic functional components of speech-recognition device.

cessor, a reference-pattern recognition facility, and an output interface. The spectrum analyzer divides speech signals from input microphones into 16 frequency bands. The coding compressor compensates for changes in rate of articulation (time alignment) and reduces the spectral data generated by each utterance to a fixed-length bit string. Because of technological limitations, most available systems are isolated, or discrete, word recognizers, which detect word boundaries by "listening" for a minimum period of silence between utterances. After establishing this word-boundary gap, the classifier compares the digitally represented utterances to previously captured reference patterns from a speaker or group of speakers. The resulting best match is taken as recognition of the utterances.

Further information on robots and robotics may be obtained through the following sources:

International Institute for Robotics, Box 210708,
Dallas, Texas 75211

Robot Institute of America, P.O. Box 930,
Dearborn, Michigan 48128

Robotics International (RI/SME), P.O. Box 930,
One SME Drive, Dearborn, Michigan 48128

Robotics Age (monthly magazine), Robotics Age, Inc.,
174 Concord Street, Peterborough, New Hampshire 03458

Robotics Today (monthly magazine), Robotics International
of Society of Manufacturing Engineers, One SME Drive,
P.O. Box 930, Dearborn, Michigan 48128

Robot Experimenter (monthly magazine, first publication
was due in September 1985), P.O. Box 458, Peterborough,
New Hampshire 03458

BIBLIOGRAPHY

Attarzadeh, F., and R. M. Judy, "Drive and Steering Considerations for a Microprocessor-Based Land Mine Sweeper," Sixth IASTED International Symposium on Robotics and Automation '85, Paper 078-941-1, June 1985.

Bucchere, T., "Servos and Steppers: Good Choices for Motion Control," *Instrument and Control Systems*, pp. 63-65, May 1982.

Critchlow, A. J., "An Overview of Robotics Technology," *IEEE Potential*. pp. 12-16, Fall 1983.

Dorf, R. C., *Robotics and Automated Manufacturing*, Reston, VA: Reston Publishing Co., Inc., 1983.

Fisher, E. L., Editor, *Robotics and Industrial Engineering Selected Readings*, Atlanta, GA: Industrial Engineering and Management Press, IIE, 1983.

Hardcastle, F. L., "How to Apply Position Sensors," *Instruments and Control Systems*, pp. 52-57, May 1982.

Heath, L., *Fundamentals of Robotics, Theory and Applications*, Reston, VA: Reston Publishing Co., Inc., 1985.

Holland, J. M., *Basic Robotic Concepts*, Indianapolis, IN: Howard W. Sams and Co., Inc., 1983.

Kafrissen, E., and M. Stephans, *Industrial Robots and Robotics*, Reston, VA: Reston Publishing Co., Inc., 1984.

Makin, E., "Using Voice I/O for Industrial Applications," *The Industrial and Process Control Magazine*, pp. 47–49, June 1984.

Masterson, J. W., E. C. Poe, and S. W. Fardo, *Robotics*, Reston, VA: Reston Publishing Co., Inc., 1985.

Morris, H. M., "Industry Begins to Apply Vision Systems Widely," *Control Engineering*, pp. 68-70, January 1985.

—— "Robot Vision: Bringing Eyesight to the Blind," *Control Engineering*, pp. 64-67, January 1985.

—— "Powering the Robotic Arm," *Control Engineering*, pp. 92-95, March 1985.

Robillard, M. J., *Microprocessor Based Robotics*, Indianapolis, IN: Howard W. Sams and Co., Inc., 1983.

—— *Advanced Robot Systems*, Indianapolis, IN: Howard W. Sams and Co., Inc., 1984.

Rony, P. R., K. E. Rony, and P. A. Rony, *Introduction to Robot Programming in Basic*, Reston, VA: Reston Publishing Co., Inc., 1985.

Rothberg, M., "Applying Speech-Recognition to Data Entry," *Mini-Micro Systems*, pp. 153-162, November 1980.

Warring, R. H., *Robots and Robotology*, Blue Ridge Summit, PA: Tab Books, Inc., 1983.

Weiss, M., "Guidelines for Controlling Robots," *The Industrial and Process Control Magazine*, pp. 37-40, April 1984.

INDEX